Ring Theory 2007

Proceedings of the Fifth China–Japan–Korea Conference

Ring Theory 2007

Proceedings of the Fifth China–Japan–Korea Conference

Tokyo, Japan 10 – 15 September 2007

editors

H Marubayashi
Tokushima Bunri University, Japan

K Masaike
Tokyo Gakugei University, Japan

K Oshiro
Yamaguchi University, Japan

M Sato
Yamanashi University, Japan

CHINA-JAPAN-KOREA INTERNATIONAL SYMPOSIUM ON RING THEORY

World Scientific

NEW JERSEY • LONDON • SINGAPORE • BEIJING • SHANGHAI • HONG KONG • TAIPEI • CHENNAI

Published by

World Scientific Publishing Co. Pte. Ltd.
5 Toh Tuck Link, Singapore 596224
USA office: 27 Warren Street, Suite 401-402, Hackensack, NJ 07601
UK office: 57 Shelton Street, Covent Garden, London WC2H 9HE

British Library Cataloguing-in-Publication Data
A catalogue record for this book is available from the British Library.

RING THEORY 2007
Proceedings of the Fifth China–Japan–Korea Conference

ISBN-13 978-981-281-832-4
ISBN-10 981-281-832-4

Printed in Singapore by World Scientific Printers

PREFACE

The Fifth China-Japan-Korea International Symposium on Ring Theory was held at National Olympic Memorial Youth Center, Tokyo, Japan during the period 10-15 September, 2007. This was the joint Symposium with the 40th Symposium on Ring Theory and Representation Theory, Japan. The present volume contains the texts of selected talks delivered at the conference. We would like to express our hearty thanks to all of the referees for their helpful suggestions and advices.

Although the place of meeting is located at the downtown of Tokyo, the circumstance is calm enough for participants to develop the active discussion and investigation. We would like to express our appreciation to all the participants who have taken time to join us in this symposium. We were quite happy to realize the symposium which devotes extremely important topics.

The past conferences were held at Guilin (China), Okayama (Japan), Kyongju (Korea) and Nanjing (China). The main aim of this conference is to foster the exchange of investigation among ring theorists in this area. The number of participants for this conference has increased a lot. It is very delightful for us that more than 120 persons from 13 countries and regions came to Tokyo to attend this symposium. We believe the presentations and discussions in these conferences contribute to create collaborative works in this field of mathematics.

In 1991, this symposium has started by the efforts of Prof. Shae Xue Liu, Prof. Hiroyuki Tachikawa and Prof. Manabu Harada. It was a great pleasure that a special session for these three professors were prepared at the symposium in Tokyo to introduce their achievements in mathematics, so that we were able to study the history of ring theory for the past half century.

The fifth symposium is financially supported by JSPS Grant-in Aid for Scientific Research (B) (Principal Researcher: Kiyoichi Oshiro), and JSPS Grant-in Aid for International Scientific Meetings in Japan (Principal Researcher: Masahisa Sato).

Also, this conference is supported by the co-operation of Tokyo Gakugei University, Yamaguchi University and Yamanashi University. We would like to extend our deepest appreciation to staff of these Universities for their most kind efforts and supports in organizing the symposium.

The next China-Japan-Korea International Symposium on Ring Theory will be held in Korea. For more detailed information on this Symposium, you should refer to the following homepage:

http://fuji.cec.yamanashi.ac.jp/~ring/cjk2011/

The Main Organizer
Kanzo Masaike

July, 2008

ORGANIZING COMMITTEES

DEDICATED TO
THE THREE GREAT PROFESSORS

Liu Shaoxue
Hiroyuki Tachhikawa
Manabu Harada

Professor Liu Shaoxue

Professor Hiroyuki Tachhikawa

Professor Manabu Harada

THE FIFTH CHINA-JAPAN-KOREA INTERNATIONAL SYMPOSIUM ON RING THEORY (2007)

September 10 (Mon) - 15 (Sat), 2007

National Olympic Memorial Youth Center
3-1. Yoyogi Kamizono-cho, Shibuya-ku, Tokyo
151-0052, JAPAN

Program

September 10 (Monday)

Registration

Room 107(9:00–22:00)

September 11 (Tuesday)

Opening Ceremony

Room 101(8:50–9:10)

Invited Lectures: Room 101

9:20–10:05 Jae Keol Park
Quasi-Baer Ring Hulls and Applications to C^*-Algebras

10:15–11:00 Jianlong Chen
Some Progress on Clean Rings

11:15–12:00 Charudatta R. Hajarnavis
Symmetry in Noncommutative Noetherian Rings

Branch Sessions: Rooms 403 & 405

13:20–13:40

Room 403: C.M.Ringel
The Relevance and the Ubiquity of Prüfer Modules

Room 405: Jin Yong Kim
On Quasi Continuous Rings

13:50–14:10

Room 403: Kunio Yamagata

On Lifting of Idempotents in the Endomorphism Ring of a Module

Room 405: Nanqing Ding

On Divisible and Torsionfree Modules

14:20–14:40

Room 403: Ji-Wei He, Quanshui Wu

Koszul Differential Graded Algebras and BGG Correspondence

Room 405: Zhixi Wang

The Fundamental Theorems of Entwined Modules

14:50–15:10

Room 403: Hongbo Shi

A Note on The Finitistic Dimension Conjecture

Room 405: A. Mehdi

Totally Projective Modules And The Extentions Of Bounded ***QTAG***-Modules

Coffee Break

15:40–16:00

Room 403: Hailou Yao

On Simple Connectedness of Minimal Representation-Infinite Algebras

Room 405: Derya Keskin Tütüncü, Nil Orhan Ertaş

On the Relative (Quasi-) Discreteness of Modules

16:10–16:30

Room 403: Ryo Takahashi

On Contravariantly Finite Subcategories of Finitely Generated Modules

Room 405: Derya Keskin Tütüncü, Fatma Kaynarca, M. Tamer Ksşsn

On Non-δ-M-cosingulab Completely $\oplus$-δ_M-supplemented Modules

16:40–17:00

Room 403: Jun Zhang

Pushout Artin-Schelter Regular algebras of Global Dimension Four

Room 405: Akira UEDA

Finitely Generated Modules Over Non-commutative Valuation Rings

September 12 (Wednesday)

Invited Lectures: Room 101

9:00–9:45 Izuru Mori
An Introduction to Noncommutative Algebraic Geometry

10:00–10:45 Hong Kee Kim
The Generalized Finite Intersection Property

11:00–11:45 Quanshui Wu
Derived Categories over Auslander-Gorenstein Rings

Memorial Photography

Branch Sessions: Rooms 403 & 405

13:20–13:40
Room 403: Masahiko Uhara, Yoshihisa Nagatomi, Kiyoichi Oshiro
Skew-matrix Ring and Applications to QF-ring

Room 405: Lia Vaš
Extending Ring Derivations to Rings and Modules of Quotients

13:50–14:10
Room 403: Hiroshi Nagase
Hochschild Cohomology of Algebras with Stratifying Ideals

Room 405: Mohammad Shadab Khan
On Derivations in Near Rings

14:20–14:40
Room 403: Liang Shen
Characterizations of QF Rings

Room 405: Yoshitomo Baba
On Colocal Pairs

14:50–15:10
Room 403: Yanhua Wang
Construct Bi-Frobenius Algebras via Quivers

Room 405: Cosmin S. Roman
On Direct Sums of Extending and Baer Modules

Coffee Break

15:40–16:00

Room 403: M. Tamer KOŞAN, A. Çiğdem ÖZCAN
δ-M-small and δ-harada Modules

Room 405: Chen-Lian Chuang, Tsiu-Kwen Lee
Ore Extensions which are GPI-rings

16:10–16:30

Room 403: Ken-ichi Iwase
Applications of Harada Rings and Kupisch Series for Harada Rings

Room 405: Mohammad Ashraf
On Lie Ideals and Generalized (θ, ϕ)-dervations in Rings

16:40–17:00

Room 403: Takao Hayami
On Hochschild Cohomology Ring of an Order of a Quaternion Algebra

Room 405: Hisaya Tsutsui and Yasuyuki Hirano
Fully k-primary Rings

Honors Sessions

Room 309

18:50–19:30 Award for Proferssors Tachikawa, Liu and Harada

19:30–20:10 Yingbo Zhang
Professor Liu - Live and Work

20:20–21:20 C.M.Ringel
The Work of Tachikawa on Finite-Dimensional Algebras and Their Representations

21:30–22:00 Kiyoichi Oshiro
Professor Harada - Person and Work

September 13 (Thursday)

Invited Lectures: Room 101

9:00–9:45 Patrick F. Smith
Ascending Chain Conditions in Groups, Rings and Modules

10:00–10:45 Naoko Kunugi
On Broué's Abelian Defect Group Conjecture in Representation Theory of Finite Groups

11:00–11:45 Pere Ara
The Realization Problem for von Neumann Regular Rings

Branch Sessions: Rooms 403 & 405

13:20–13:40

Room 403: S. Tariq Rizvi
On Structure of Rings of Quotients

Room 405: Nadeem-ur-Rehman
Generalized Derivations in Prime Rings

13:50–14:10

Room 403: M. Siles Molina
Algebras of Quotients of Leavitt Path Algebras

Room 405: Shakir Ali
On Generalized Jordan Left Derivations in Certain Classes of Rings

14:20–14:40

Room403: M. Brešar, F. Perera, J. Sánchez Ortega*, M. Siles Molina
Computing the Maximal Algebra of Quotients of a Lie Algebra

Room 405: Chan Huh
Characterizations of Elements in Prime Radicals of Skew Polynomial Rings and Skew Laurent Polynomial Rings

14:50–15:10

Room 403: Fujio Kubo
Compartible Algebra Structures of Lie Algebras

Room 405: P.-H. Lee and E. R. Puczyłowski
On the Behrens Radical of Matrix Rings and Polynomial Rings

Coffee Break

15:40–16:00

Room 403: Guilong Liu
Duals of Modules and Quasitriangular Hopf Algebras

Room 405: John S. Kauta
Valuation Rings in the Quotient Ring of a Skew Polynomial Ring

16:10–16:30

Room 403: Yong Uk Cho
A Class of Near-Rings from Rings

Room 405: Wagner de Oliveira Cortes
Partial Skew Polynomial Rings and Jacobson Rings

16:40–17:00

Room 403: Chang Ik Lee, Yang Lee and Sung Ju Ryu
On Strongly NI Rings

Room 405: Sudesh K. Khanduja
On Irreducible Factors of the Polynomial $\mathbf{f}(\mathbf{x}) - \mathbf{g}(\mathbf{y})$

18:30–21:00 **Banquet (Restaurant Toki)**

September 14 (Friday)

Invited Lectures: Room 101

9:00–9:45 Zhaoyong Huang
Auslander-type Conditions

10:00–10:45 Kazutoshi Koike
Morita Duality and Recent Development

11:00–11:45 Juncheol Han
Group Actions in a Unit-Regular Ring, II

Branch Sessions: Rooms 403 & 405

13:20–13:40

Room 403: Hiroki Abe and Mitsuo Hoshino
Derived Equivalences and Serre Duality for Gorenstein Algebras

Room 405: Kaoru Motose
Some Congruences Concerning Finite Groups

13:50–14:10

Room 403: Hisaaki Fujita and Akira Oshima
A Tiled Order of Finite Global Dimension with No Neat Primitive Idempotent

Room 405: Tsunekazu Nishinaka
Primitivity of Group Rings of Extensions of Free Groups

14:20–14:40

Room 403: Tai Keun Kwak
Extended Armendariz Rings and Rigid Rings

Room 405: Gene Abrams & Gonzalo Aranda Pino
Leavitt Path Algebras for Countable Graphs

14:50–15:10

Room 403: Kanzo Masaike
On τ-coherent Rings

Room 405: Tang Gaohua
Zero-Divisor Semigroups of Some Simple Graphs

Coffee Break

15:40–16:00

Room 403: Lixin Mao
Weak Global Dimension of Coherent Rings

Room 405: Chan Yong Hong, Yang Lee
On Property (A)

16:10–16:30

Room 403: Hiroshi Yoshimura
Local QF Rings with Radical Cubed Zero II

Room 405: Angelina Yan Mui Chin
Clean Rings and Related Classes of Rings

16:40–17:00

Room 403: John Clark
A Study of Uniform One-Sided Ideals in Simple Rings

Room 405: Tokuji Araya
Auslander-Reiten Conjecture on Gorenstein Rings

17:00–17:10

Room 403: Hisaya Tsutsui, Yasuyuki Hirano
Fully k-Primary Rings

Memorial Lecture and Closing Ceremony Interenational Meeting Room No.1

19:00–19:30 Vlastimil Dlab
Approximations of Algebras

19:50–20:50 Shigeo Koshitani
Representation Theory from a Certain Point of View

Closing Ceremony(21:00–21:30)

September 15 (Saturday)

Excursion to Hakone by Sightseeing Bus (8:00–15:00)

PARTICIPANTS LIST

Shakirs	Ali	Aligarh Muslim University shakir50@rediffmail.com
Hiroki	Abe	University of Tsukuba abeh@math.tsukuba.ac.jp
Takuma	Aihara	Chiba University 04um0101@g.math.s.chiba-u.ac.jp
Pere	Ara	Universitat Autonoma de Barcelona para@mat.uab.cat
Gonzalo	Aranda Pino	Vila Universitaria garanda@crm.es
Tokuji	Araya	 araya@math.okayama-u.ac.jp
Hideto	Asashiba	Shizuoka University shasash@ipc.shizuoka.ac.jp
Mohammad	Ashraf	Aligarh Muslim University mashraf80@hotmail.com
Yoshitomo	Baba	Osaka Kyoiku University ybaba@cc.osaka-kyoiku.ac.jp
Jianlong	Chen	Southeast University jlchen@seu.edu.cn
Yong Uk	Cho	Silla University yucho@silla.ac.kr
John	Clark	University of Otago jclark@maths.otago.ac.nz
Nanqing	Ding	Nanjing University nqding@nju.edu.cn
Vlastimil	Dlab	Carleton University vdlab@math.carleton.ca
Hisaaki	Fujita	University of Tsukuba fujita@math.tsukuba.ac.jp
Tang	Gaohua	Guangxi Teachers' College tanggaohua@163.com
Liu	Guilong	Beijing Language and Culture University liuguilong@blcu.edu.cn
Charudatta	Hajanavis	University of Warwick crh@maths.warwick.ac.uk

Juncheol	Han	Pusan National University jchan@pusan.ac.kr
Akihide	Hanaki	Shinshu University hanaki@math.shinshu-u.ac.jp
Takao	Hayami	Tokyo University of Science hayami@rs.kagu.tus.ac.jp
Akihiko	Hida	Saitama University ahida@math.edu.saitama-u.ac.jp
Yoshimasa	Hieda	Osaka Prefectural College of Technology hieda@las.osaka-pct.ac.jp
Yasuyuki	Hirano	Naruto University of Education yahirano@naruto-u.ac.jp
Zhaoyong	Huang	Nanjing University huangzy@nju.edu.cn
Chan	Huh	Pusan National University chuh@pusan.ac.kr
Atsushi	Ishii	Nagoya university atsushi-ishii@ca2.so-net.ne.jp
Yasuo	Iwanaga	Shinshu University iwanaga@shinshu-u.ac.jp
Ken-ichi	Iwase	Osaka-kyoiku university iwase@cc.osaka-kyoiku.ac.jp
Osamu	Iyama	Nagoya University iyama@math.nagoya-u.ac.jp
He	Ji-Wei	Shaoxing College of Arts and Science hejw@fudan.edu.cn
Kiriko	Kato	Osaka Prefecture University kiriko@mi.s.osakafu-u.ac.jp
John S.	Kauta	Universiti Brunei Darussalam Darussalam kauta@fos.ubd.edu.bn
Shigeto	Kawata	Osaka City University kawata@sci.osaka-cu.ac.jp
Fatma	Kaynarca	Afyonkarahisar Kocatepe University fatmakaynarca@hotmail.com
Derya	Keskin	Hacettepe University keskin@hacettepe.edu.tr
Sudesh Kaur	Khanduja	Panjab University skhand@pu.ac.in
Isao	Kikumasa	Yamaguchi University kikumasa@yamaguchi-u.ac.jp
Hong Kee	Kim	Gyeongsang National University hkkim@gsnu.ac.kr

Jin Yong	Kim	Kyung Hee University jykim@khu.ac.kr
Yoshimi	Kitamura	Tokyo Gakugei University kitamura@u-gakugei.ac.jp
Daisuke	Kobayashi	Nagoya university master-koba@syd.odn.ne.jp
Hisatoshi	Koike	Okinawa National College of Technology koike@okinawa-ct.ac.jp
Hiroaki	Komatsu	Okayama Prefectural University komatsu@cse.oka-pu.ac.jp
Shigeo	Koshitani	Chiba University koshitan@math.s.chiba-u.ac.jp
Fujio	Kubo	Hiroshima University remakubo@amath.hiroshima-u.ac.jp
Naoko	Kunugi	okyo University of Science kunugi@rs.kagu.tus.ac.jp
Yoshiki	Kurata	 ykurata@mwc.biglobe.ne.jp
Yosuke	Kuratomi	Kitakyushu National College of Technology kuratomi@kct.ac.jp
Mamoru	Kutami	Yamaguchi University kutami@yamaguchi-u.ac.jp
Tai Keun	Kwak	Daejin University tkkwak@daejin.ac.kr
Chang Ik	Lee	Pusan National University cilee@pusan.ac.kr
Pjek-Hwee	Lee	National Taiwan University phlee@math.ntu.edu.tw
Tsiu-Kwen	Lee	National Taiwan University tklee@math.ntu.edu.tw
Yang	Lee	Pusan National University ylee@pusan.ac.kr
Shao Xue	Liu	Beijing Normal University liusx@bnu.edu.cn
Di-Ming	Lu	Zhejiang University dmlu@zju.edu.cn
Lixin	Mao	Nanjing Institute of Technology maolx2@hotmail.com
Hidetoshi	Marubayashi	Tokushima bunri University marubaya@naruto-u.ac.jp
Uhara	Masahiko	Yamaguchi University yu_uhara@yahoo.co.jp
Kanzo	Masaike	Tokyo Gakugei University masaike@u-gakugei.ac.jp

Alveera	Mehdi	Aligarh Muslim University alveera_mehdi@rediffmail.com
Junichi	Miyachi	Tokyo Gakugei University miyachi@u-gakugei.ac.jp
Hiroki	Miyahara	Shinshu University miyahara_shinshu@yahoo.co.jp
Izuru	Mori	Shizuoka University simouri@ipc.shizuoka.ac.jp
Mari	Morimoto	Akita National College of Technology marim@akita-nct.jp
Kaoru	Motose	 moka.mocha_no_kaori@snow.ocn.ne.jp
Yao	Musheng	Fudan University yaomusheng@gmail.com
Hiroshi	Nagase	Nara National College of Technology nagase@libe.nara-k.ac.jp
Haruhisa	Nakajima	Josai University nakaziejp@yahoo.co.jp
Kazunori	Nakmoto	Yamanashi University nakamoto@yamanashi.ac.jp
Sangbok	Nam	Kyungdong University sbnam@k1.ac.kr
Yasushi	Ninomiya	Shinshu University ysninom@shinshu-u.ac.jp
Kenji	Nishida	Shinshu University kenisida@math.shinshu-u.ac.jp
Tsunekazu	Nishikawa	Osaka City University d06sa007@ex.media.osaka-cu.ac.jp
Tsunekazu	Nishinaka	Okayama Shoka University nishinak@po.osu.ac.jp
Hiroshi	Okuno	Juntendo University School of Medicine okuno@sakura.juntendo.ac.jp
Go	Okuyama	Hokkaido Institute of Technology gokuyama@hit.ac.jp
Wagner de	Oliveira Cortes	Federal University of Rio Grande do Sul wocortes@hotmail.com
Juana Sanchez	Ortega	Universidad de Malaga (Campus de Teatinos) jsanchez@agt.cie.uma.es
Kiyoichi	Oshiro	Yamaguchi University oshiro@yamaguchi-u.ac.jp
Ayse Cigdem	Ozcan	Hacettepe University ozcan@hacettepe.edu.tr
Jao Keol	Park	Pusan National University jkpark@pusan.ac.kr

Nadeem-ur	Rehman	Aligarh Muslim University rehman100@gmail.com
Claus	Ringel	Bielefeld University ringel@mathematik.uni-bielefeld.de
Tariq	Rizvi	The Ohio State University rizvi.1@osu.edu
Cosmin	Roman	The Ohio State University cosmin@math.ohio-state.edu
Narasaki	Ryo	Osaka University narasaki@cr.math.sci.osaka-u.ac.jp
Yosuke	Sakai	University of Tsukuba ysksakai@math.tsukuba.ac.jp
Katsunori	Sanada	Tokyo University of Science sanada@rs.kagu.tus.ac.jp
Hiroki	Sasaki	Shinshu University sasaki@math.shinshu-u.ac.jp
Masahisa	Sato	Yamanashi University msato@yamanashi.ac.jp
Katsusuke	Sekiguchi	Yamaguchi University sekiguch@kokushikan.ac.jp
Mohammad	Shadab Khan	Aligarh Muslim University km_shadab@rediffmail.com
Liang	Shen	Southeast University lshen@seu.edu.cn
Hongbo	Shi	Nanjing university of finance and economics dshi782000@yahoo.com
Mercedes	S. Molina	Malaga Universitaria mercedes@agt.cie.uma.es
Patrick	Smith	University of Glasgow pfs@maths.gla.ac.uk
Manabu	Suda	Tokyo University of Science suda@ma.kagu.tus.ac.jp
Takao	Sumiyama	Aichi Institute of Technology sumiyama@aitech.ac.jp
Hiroyuki	Tachikawa	
Ryo	Takahashi	Shinshu University takahasi@math.shinshu-u.ac.jp
Shinsuke	Takahashi	 perimetros@infoseek.jp
Yasuhiko	Takehana	Hakodate National College of Tecnology takehana@hakodate-ct.ac.jp
Fuminori	Tasaka	Chiba University ftasaka@g.math.s.chiba-u.ac.jp

Yusuke	Tsujioka	Nagoya university esupe-0415@docomo.ne.jp
Hisaya	Tsutsui	 tsuts157@erau.edu
Akira	Ueda	Shimane University ueda@riko.shimane-u.ac.jp
Morio	Uematsu	Jobu University uematsu@jobu.ac.jp
Katsuhiro	Uno	Osaka Kyoiku University uno@cc.osaka-kyoiku.ac.jp
Lia	Vas	University of the Sciences in Philadelphia l.vas@usip.edu
Tomoyuki	Wada	Tokyo University of Agriculture and Technology wada@cc.tuat.ac.jp
Michihisa	Wakui	Kansai University, wakui@ipcku.kansai-u.ac.jp
Yanhua	Wang	Shanghai University of Finance and Economics yhw@mail.shufe.edu.cn
Zhixi	Wang	Capital Normal University wangzhx@mail.cnu.edu.cn
Takayoshi	Wkamatsu	Saitama University wakataka@post.saitama-u.ac.jp
Min	Wu	Tsinghua University wumin20022003@yahoo.com.cn
Quanshui	Wu	Fudan University qswu@fudan.edu.cn
Kunio	Yamagata	Tokyo University of Agriculture and Technology yamagata@cc.tuat.ac.jp
Kota	Yamaura	Nagoya university shiroinu@pure.ocn.ne.jp
Angelina	C.Y. Mui	University of Malaya acym@um.edu.my
Kouji	Yanagawa	Kansai University yanagawa@ipcku.kansai-u.ac.jp
Hailou	Yao	Beijing University of Technology Yaohl@bjut.edu.cn
Dong	Yeo	Pusan National University dongeunyeo@pusan.ac.kr
Usami	Yoko	Ochanomizu University usami@cc.ocha.ac.jp
Kenji	Yokogawa	Okayama University of Science yokogawa@mdas.ous.ac.jp
Hiroshi	Yoshimura	Yamaguchi University yoshi@yamaguchi-u.ac.jp

Yuji	Yoshino	Okayama University yoshino@math.okayama-u.ac.jp
Ohnuki	Yosuke	Suzuka National College of Technology ohnuki@genl.suzuka-ct.ac.jp
Doi	Yukio	Okayama university ydoi@cc.okayama-u.ac.jp
Yoshii	Yutaka	Chiba University 04um0109@g.math.s.chiba-u.ac.jp
Jun	Zhang	University of Washington zhangjun19@gmail.com
Yingbo	Zhang	Beijing Normal University zhangyb@bnu.edu.cn

CONTENTS

PART A

Special Lectures

MANABU HARADA - THE MAN AND HIS WORK

Kiyoichi Oshiro

Department of Mathematics, Yamaguchi University, Yoshida, Yamaguchi

At the Fifth China-Japan-Korea International Symposium on Ring Theory in Tokyo, we celebrated Professor Shaoxue Liu, Professor Hiroyuki Tachikawa and Professor Manabu Harada who had planned and originated the First Symposium at Guangxi in China in 1991. It was a great pleasure for all participants to celebrate these three eminent ring theorists.

In this paper, as one of Harada's students, I would like to briefly describe Professor Harada's career and work.

He was born in 1931 and raised in Osaka City. He spent his undergraduate and graduate years at Osaka City University from 1949 through to 1955. After he received his master's degree in 1955, he joined the staff of the Department of Mathematics of Osaka City University as an assistant professor. In those days, Keizo Asano was a professor and Hiroshi Nagao was an associate professor at the University and Harada was influenced by them. Professor Asano is well-known for pioneering classical quotient rings. He studied orders in simple rings in the 1950s, nowadays called "Asano orders" ([1]). On the other hand, Professor Nagao studied global dimensions of residue rings of hereditary semiprimary rings (with Eilenberg and Nakayama ([9])), and is known as the coauthor of the book [17] on the modular representation of finite groups.

Under such excellent circumstances, Harada began studying algebras and soon began writing a sequence of fundamental articles on ring theory. His first paper is:

> M. Harada: *Note on the dimension of modules and algebras*, J. Inst. Polytecnics, Osaka City Univ., 1956.

In this paper, he showed that von Neumann regular rings can be characterized as those rings whose weak global dimensions are zero. This is a beautiful theory on von Neumann regular rings. After writing nine papers from 1956 through to 1960, he studied Asano orders and maximal orders using homological algebra. In particular, he wrote four papers on orders in 1963, including the famous paper:

M. Harada: *Hereditary orders*, Trans. Amer. Math. Soc., 1963.

Hereditary orders are hereditary noetherian algebras. Consequently, these algebras encouraged the study of hereditary noetherian rings and so Harada is recognized as a pioneer in the study of these rings.

In the early 1960s, Goldie's article [13] on semi-prime rings with maximal conditions and Mitchell's book [15] on category theory were published.

By the Goldie Theorem, the study of hereditary orders was developed in a more general setting, that is, hereditary noetherian rings were studied as a branch of non-commutative noetherian rings. Chatters [5] showed that hereditary noetherian rings can be represented as direct sums of prime rings and artinian rings. Moreover, Harada completely determined hereditary artinian rings in the article:

M. Harada: *Hereditary semiprimary rings and triangular matrix rings*, Nagoya J. Math., 1966.

Eisenbud-Griffith-Robson ([10], [11]) showed that if R is a hereditary noetherian prime ring, then R/I is a Nakayama ring for every non-zero ideal I. Consequently, this theorem turned the spotlight on Nakayama rings.

During 1961-1963, Harada visited Brandeis University and Northwestern University in the U.S.. In those days, M. Auslander was a professor at Brandeis University and he and Goldman had studied maximal orders. In 1962, Harada received his Ph.D. from Brandeis under Auslander's supervision. After Harada returned to Japan in 1963, he also obtained his D.Sc. at Osaka City University in that year.

In the middle of 1960s, Harada began studying category theory. For this study, in 1967, he concentrated on reading Mitchell's book [15] with his student Sai. I can well remember those days, since Sai was one of my graduate classmates and I attended their seminar. Harada was strongly interested in the so-called factor category which is quite different from the quotient category introduced by Gabriel [12] and Serre [22]. For an additive category $\mathbf{A}$, an ideal of $\mathbf{A}$ is defined by a subclass $\mathbf{I}$ with certain conditions on the class of all morphisms of $\mathbf{A}$ and the factor category $\mathbf{A}/\mathbf{I}$ is introduced ([8] and [14]). Auslander had also used the factor category for the representation of finite dimensional algebras ([2]).

Soon, Harada began writing papers on factor categories. For a family $\mathbf{A}$ of R-modules, Harada considered the full sub-additive category $\mathbf{A}$ (of the category of all R-modules) whose objects are modules in $\mathbf{A}$, and defined the "Jacobson radical $J(\mathbf{A})$" of $\mathbf{A}$, and introduced the factor category $\mathbf{A}/J(\mathbf{A})$. Taking good families $\mathbf{A}$ (injective modules, projective modules, completely indecomposable modules, etc.), he wrote about 30 papers on the factor category and its applications from 1967 through to 1978. His work on this topic is remarkable in both the quality and quantity. Well-known in particular are the following papers written jointly with his students Y. Sai and H. Kanbara:

M. Harada and Y. Sai: *On categories of indecomposable modules, I*, Osaka J. Math., 1970.

M. Harada and H. Kanbara: *On categories of projective modules*, Osaka J. Math.,1971.

In these papers, the famous Harada-Sai Lemma appeared, and the Krull-Remak-Schmidt-Azumaya Theorem was studied by making use of his factor category, and the final version of Krull-Remak-Schmidt-Azumaya's Theorem was completed by introducing the very useful $LsTn$ (local semi-T-nilpotency) condition.

In 1970, Harada became a professor at Osaka City University, and after he wrote various papers on the factor category, his concern turned to other topics, and from the end of 1970s, he produced numerous new important concepts on modules, such as small modules, cosmall modules, simple injective modules, mini-injective modules, extending modules and lifting modules, etc.. In 1978, he introduced two new important artinian rings concerning non-small modules and non-cosmall modules in the paper:

M. Harada:*Non-small modules and non-cosmall modules*, in "Ring Theory. Proceedings of 1978 Antwerp Conference" Dekker, New York, 1979.

This paper considered artinian rings with the condition

($\star$): *Every non-small modules contains a non-zero injective summand,*

and also artinian rings with the condition

$(\star)^*$: *Every non-cosmall module contains a non-zero projective summand.*

These artinian rings are generalizations of QF-rings and Nakayama rings. Several years later, however, it was shown that these two classes of artinian rings R coincide, since ($\star$) holds for left R-modules if and only if $(\star)^*$ holds for right R-modules (Oshiro [21]). These new artinian rings are now named "Harada rings" in his honour, and the structure theory of Harada rings was established in connection with QF-rings and Nakayama rings (Oshiro [19]-[20]). Actually, Harada rings R have the frame QF-subrings $F(R)$ from which R can be constructed as an upper staircase factor ring of a suitable block extension of $F(R)$. Furthermore, the structure of Harada rings gives much information about the structure of Nakayama rings. Therefore, Harada rings lead us to study QF-rings and Nakayama rings from a new point of view. Through Harada rings, there is keen interest in Nakayama permutations and Nakayama automorphisms on QF-rings, constructions of local QF-rings, skew-matrix rings, structures of Nakayama rings, Fuller's theorem on i-pairs, artinian rings with self-duality, lifting modules, extending modules, and Nakayama QF-group algebras, etc.. By the study on Harada rings, many new theorems and facts were produced; for example, a complete classification of Nakayama rings, the existence of QF-rings without Nakayama automorphisms, and new artinian rings with self-duality, etc. (cf. Baba-Oshiro [4]).

Furthermore, it is worthwhile to note that lifting modules, extending modules and simple injectivity have been extensively studied by many people. These fields were witnessed in the following books: Mohamed-Müller [16]:*Continuous Modules and Discrete Modules* (1990), Dung-Huynh-Smith-Wisbauer [7]: *Extending Modules* (1994), Yousif-Nicholson [18]: *Quasi-Frobenius Rings* (2003) and Clark-Lomp-Vanaja-Wisbauer [6]: *Lifting modules* (2007).

Harada retired from Osaka City University in 1994. During 1956-1994, he had written approximately 90 articles and published three books, including the book:

> M. Harada: *Factor Categories with Applications to Direct Decomposition of Modules*, Lecture Notes in Pure and Applied Mathematics, 88, Marcel Dekker, 1983.

As witnessed above, Harada's works are outstanding and important in the progress of the ring theory. After retiring, he also completely retired from mathematics and enjoys his happy retirement as an Emeritus Professor at Osaka City University.

References

1. K. Asano: Über die Quotientenbilding von Shiefringen, J.Math.Soc.Japan, 1949.
2. M. Auslander and O. Goldman: Maximal orders, Trans. Amer. Math. Soc., 1960.

3. M. Auslander: Functors and Morphisms Determined by Objects, Lecture notes in pure and applied mathematics 37, 1978.
4. Y. Baba and K. Oshiro: Classical Artinian Rings and Related Topics, preprint.
5. A. W. Chatters: A decomposition theorem for Noetherian hereditary rings, J. London Math. Soc, 1971.
6. J. Clark, C. Lomp, N. Vanaja and R. Wisbauer: Lifting modules, Birkhauser Boston, Boston, 2007.
7. N. V. Dung, D.V. Huynh, P. F. Smith, and R. Wisbauer: Extending Modules, Pitman Research Notes in Mathematics Series, 313, London, 1994.
8. C. Ehresmann: Catégories et structures: Paris: Dunod, 1965.
9. S. Eilenberg, H. Nagao and T. Nakayama: On the dimension of modules and algebras IV, Dimension of residue rings of hereditary rings, Nagoya Math. J., 1956.
10. G. Eisenbud and P. Griffith: Serial rings: J. Algebra, 1971.
11. G. Eisenbud and J. C. Robson: Hereditary noetherian prime rings, J. Algebra, 1970.
12. P. Gabriel: Des catégories abeliennes, Bull. Soc. Math. France, 1962
13. A. W. Goldie: Semi-prime rings with maximal condition, Proc. London Math., 1960.
14. M. Kelly: On the radical of a category: J. Austral. Math. Soc., 1964.
15. B. Mitchell: Theory of Categories, Academic Press, New York, 1965
16. S. H. Mohamed and B. H. Müller: Continuous Modules and Discrete Modules, London Mathematical Society, Lecture Notes 147, Cambridge Univ. Press, 1990.
17. H. Nagao and Y. Tsushima: Representations of Finite Groups, Academic Press, Boston, 1989.
18. W. K. Nicholson and M. F. Yousif: Quasi-Frobenius Rings, Cambridge Tracts in Mathematics 158, Cambridge University Press, Cambridge, 2003.
19. K. Oshiro: Structure of Nakayama rings, Proceedings of 20th Symposium on Ring Theory, 1987.
20. K. Oshiro: Theories of Harada in Artinian rings, Proceedings of the third International Symposium on Ring Theory, Birkhauser Boston, Boston, 2001.
21. K. Oshiro: On Harada rings I, II, III, Math. J. Okayama Univ., 1989, 1990.
22. J. P. Serre: Faiseaux algebriques, coherents, Ann of Math., 1955.

The work of Tachikawa on Ring Theory in the world

Masahisa Sato
Interdisciplinary Graduate School of Medicine and Engineering,
University of Yamanashi,
Kofu, Yamanashi 400-8511 JAPAN
E-mail: msato@yamanashi.ac.jp

Professor Hiroyuki Tachikawa is considered by many mathematicians as an authority on the representation theory of algebras. However, tracing all the ways in which Professor Tachikawa approached work and mathematicians, we can realize the history of general ring theory.

Professor Tachikawa was born in 1930 and entered Tokyo University of Education in 1950. His supervisor of mathematics was Kiichi Morita and he started to study the representation theory of finite groups. Professor Tachikawa has a good successor in this field Shigeo Koshitani and he used to say that Koshitani has studied what Tachikawa had hoped very much to study. This shows Tachikawa was concerned with the representation theory of finite groups.

In his first paper [1], Tachikawa gave a nice and considerably short proof of "Brauer's Characterization of Generalized Characters" by making use of Peter Roquette's result. This Brauer's theorem is closely related to another well-known result "Brauer's Induction Theorem", which has many applications such as Artin's L-functions and so on. Actually, Tachikawa was delighted to receive a letter from R.Brauer dated June 30, 1954, in which Brauer showed his interest in Tachikawa's result. Tachikawa is very proud of this in his mathematical career.

In the era around 1940, the ring theoretical interest to the group rings made Tadashi Nakayama study Quasi-Frobenius rings (for short QF-ring). Gradually Tachikawa became interested in Nakayama conjecture, dualities between module categories and etc,.

On the other hand, Tachikawa obtained a doctorate under the supervisor Morita on the characterization of local-colocal algebras. In that time many results were obtained by Victor Camillo, Spencer E. Dickson and Kent R.

Fuller on balanced rings and uniserial rings according to Jans conjecture. However Vlastimil Dlab and Claus Michael Ringel got marvelous result on exceptional rings (i.e., non-uniserial rings over which every module has the double centralizer property) that exceptional rings were local-colocal rings. Later T. Sumioka simplify the proofs of Tachikawa's paper [5].

In the late 1970's, Professor Tachikawa studied a trivial extension of hereditary algebra and its standard duality module. The success of this study led to the study of Frobenius trivial extensions. It is to be noted that Tachikawa's concern was always related to Frobenius algebras and this had a big influence on the members of the Tachikawa school. Among his concerns, Nakayama conjecture is a special one. In fact, at his last lecture, Professor Tachikawa talked about this conjecture with the memory of Nakayama. Professor Tachikawa replaced this conjecture by the following two statements and made this conjecture public to all the ring theorists.
(1) If $\mathrm{Ext}^i_A(DA, A) = 0$ for any $i > 0$, then A is self-injective.
(2) Assume that A is self-injective and M is a finitely generated A-module. If $\mathrm{Ext}^i_A(M, M) = 0$ for any $i > 0$, then M is projective.

Relating to QF-rings, QF-1, QF-2 and QF-3 rings in the sense of Robert M. Thrall are important. There is a summary on these rings in the Springer Lecture Note 357. Almost all results until this period were given for algebras or bimodules. The most significant point of this lecture note is that Tachikawa discussed theories on rings and modules but not algebras and bimodules. Later Tachikawa proved relating problem in [22] that commutative QF-1 rings are self-injective and cogenerator. There are several conjectures for these rings, but Tachikawa never lost his concern to study structures of these rings.

This lecture note involves a lot of important notations and tools in ring theory. For example, classical quotient, double centralizer and minimal faithful module and etc,. These are well behaved notions to know their rolls not only in ring theory but also in representation theory. This fact impresses upon us that ring theoretical importance is basically important in any stage and we must admire his spirit going ahead of the time.

One of the important contents in this lecture note is the famous result due to his first student Kanzo Masaike, who is the main organizer of CJK5, to give beautiful characterization of right classical quotient to be left classical quotient. By glancing at the traditional non-commutative ring theory due to Goldie, torsion theory due to Gabriel etc., we can realize that ring theory includes its diversity and beautiful inner structure.

In this lecture note, it was first pointed out that the Auslander theorem of artin algebras of finite representation type can be generalized to artin rings which are not necessarily algebras. Further it was proved that in this case any module is a direct sum of indecomposable modules. These results came of a joint work with Ringel [21].

Morita equivalences, Morita duality and theory of category equivalences are basic methods in modern ring theory and its frontier was his supervisor Morita. Professor Tachikawa had been naturally concerned with category equivalences, but had the foresight to study general category equivalences not restricting himself to module categories.

In 1970's, Goro Azumaya visited Tokyo University of Education and motivated Tachikawa, his colleague Toyonori Kato and graduate students to study category equivalences, localization and torsion theory. Azumaya had been the professor of Indiana University after studying with Nakayama at Nagoya University. Azumaya completed Morita duality theory in the almost same time as Morita himself did. Tachikawa's paper [4] preceded them and proved the duality theorem of finitely generated modules over artinian rings by use of injective cogenerator bimodules. With this background, we can say that Tachikawa passes through the modernized ages of ring theory starting from the creation of Homological Algebra due to Cartan and Eilenberg.

Among ring theorists at this time, there were Hisao Tominaga and Manabu Harada contributing to development of ring theory in Japan by cooperating with each other. By the great effort of Tominaga, Harada and Tachikawa, the first Japan ring theory symposium was held in 1968 and since then this symposium has been held every year without interruption. The symposium attained to 40th anniversary in 2007.

It is a remarkable contribution that Tachikawa supported the Japan ring theory symposium for a long time and motivated young ring theorists and trained many leading ring theorists. It was very impressive that Tachikawa felt the deep sadness over Tominaga's early passing away in his 60's. Also Tachikawa's classmate Yutaka Kawada was famous for the characterization of Köthe algebras as a mathematician in this time.

Concerning the Representation theory of algebras, Roiter and Gabriel published their papers and Dlab visited Tachikawa to leave his draft of the book with coauthor Ringel published by AMS. They motivated the Tachikawa school to study representation theory of algebras. At the early stage of representation theory of algebras around 1970, the Tachikawa school resumed to study representation theory of algebras and finite groups. The first achievement was brought by Kunio Yamagata for the study of Auslander-Reiten quivers and algebras of finite representation type around

1978. In the latter half of 1970's, Professor Tachikwa proposed the study of trivial extensions of hereditary algebras of finite representation type by the standard duality modules and this was completed by Yamagata and Tachikawa.

Professor Tachikawa and his student Takayoshi Wakamatsu proved the important theorem [27] which suggests the existence of stable equivalences between categories of finitely generated modules over trivial extension QF-algebras of any algebras and their tilting algebras. The equivalences extend tilting functors between given algebras and tilting algebras. Further the equivalences teach us concretely not only the correspondence between connected components of Auslander-Reiten quivers of trivial extension QF-algebras but also the correspondence of (stable) homomorphisms between indecomposable modules which belong not necessarily to the same component of Auslander-Reiten quivers. Later the equivalences are known to be closely connected with the equivalences of derived categories of modules.

Yasuo Iwanaga was also Tachikawa's student who studied noncommutative Gorenstein rings. He stayed long in Canada to research under V. Dlab at Carleton University. This visit made the Tachikawa school a close contact with V. Dlab and C.M. Ringel who had a big influence on the promotion of research for representation theory of algebras. Professor Tachikawa has continued to study tilting theory for many years with this background. The Tachikawa school in 1970's was active for representation theory of algebras, but it was no more than one topic of ring theory. As Tachikawa's colleague, Toyonori Kato from Tohoku University and Youichi Miyashita from Hokkaido University were gathered at Tokyo University of Education and major topics of ring theory had been studied as one of the key stations of the study of ring theory in the world. It was concerned with the wide area of ring theory including the theory of category equivalences, the fundamental theory of localizations, the Artin problem with respect to division rings and Frobenius extension theory which is due to Tachikawa's student Yoshimi Kitamura. The author also wrote the paper to give the concrete description of colocalization under the supervisor Tachikawa. It is Tachikawa's advantage in addition to his academic career that he made superior specialists in each field of ring theory by planning to study category theory and fundamental papers with the members of the Tachikawa school for keeping up with the progress of representation theory which progressed rapidly later. It is a remarkable fact that Tachikawa gained the honor to be chosen as the unique ICRA organizer member through Asia in the first ICRA (International Conference on Representation theory of Algebras) held by Dlab in Carleton University in Canada in 1974. ICRA has been an important conference in the sense that it has always led the progress of representation theory of algebras in the world. The author attended for the first time and felt exited to talk with famous mathematicians like Walter Feit, Maurice Auslander, Michael Artin and etc., face to face. It is remarkable that many young mathematicians attended ICRA at that time and became active in the world as the leaders of representation theory nowadays.

In 1990 at a satellite conference of IMU congress at Kyoto, Professor Tachikawa organized the fifth ICRA at Tsukuba. Both were held in Japan for the first time. It was hard work to hold such a conference at that time because it was rare to obtain academic funds for international conferences. When Professor Tachikawa visited Bielefeld University, he met Professor Lie Shao Xue who was one of leaders of ring theory in China. Since then Tachikawa had a deep friendship with Liu Shao Xue. It is a wonder that there were no contact between the Chinese and Japanese ring theorists. The main themas in each country were completely different due to this phenomenon. In some sense, it is nice that ring theory develops in wide ranges since ring theory includes diversity in itself.

In 1991 just one year after Liu and Tachikawa's proposal of mathematical exchange among ring theorists in both countries, this was realized as the first China-Japan International Symposium on Ring Theory at Guiling in China, which was organized by Liu and Tachikawa. It was expected to hold this symposium every four years. The second symposium was held in Japan with Korea as observer and it has been held as China-Japan-Korea International Symposium on Ring Theory since the third symposium held in Korea. We are indebted to Hidetoshi Marubayashi for these development. It is delightful that the symposium will be held regularly.

Professor Tachikawa has still been active in his research and giving strong motivation to ring theorists. We have learnt sincere and earnest attitude from his life and personality. We strongly hope Professor Tachikawa supervises worldwide ring theory with his good health on the occasion of his seventy-seventh birthday which is considered a happy age in Asia.

Professor Tachikawa's Achievement and his paper lists

The details of Tachikawa's work are introduced in the following web page by C.M.Ringel who gave the lecture about Tachikawa's work on finite-dimensional algebras and their representations in the lecture CJK5 (2007).

http://www.math.uni-bielefeld.de/~ringel/lectures/tachi/Tachikawa/

References

1. H. Tachikawa, A remark on generalized characters of groups, Science Reports of Tokyo Kyouiku Daigaku Section A. Vol. 4 No. 109, 332-334 (1954).
2. K. Morita and H.Tachikawa, Character modules, submodules of a free module and quasi-Frobenius rings, Math. Zeitschr. 65, 414-428 (1956).
3. K. Morita, Y. Kawada and H.Tachikawa, On injective modules, Math. Zeitschr. 68, 217-226 (1957).
4. H.Tachikawa, Duality theorem of character modules for rings with minimum condition. Math. Zeitschr. 68, 479-487 (1958).
5. H.Tachikawa, On rings for which every indecomposable right module has a unique maximal submodule, Math. Zeitschr. 71, 200-222 (1959).
6. H.Tachikawa, A note on algebras of unbounded representation type, Proc. Japan Acad. 36, 59-61 (1960).
7. H.Tachikawa, On algebra of which every indecomposable representation has an irreducible one as the top or the bottom Loewy constituent, Math. Zeitschr. 75, 215-227 (1961).
8. H.Tachikawa, A characterization of QF-3 algebras, Proc. Amer. Math. Soc.13, 701-703 (1962), Correction, 14, 995 (1963).
9. H. Tachikawa, On dominant dimension of QF-3 algebras, Trans. Amer. Math. Soc. 112, 249-266 (1964).
10. H.Tachikawa, A generalization of quasi-Frobenius rings, Proc. Amer. Math. Soc. 20, 471-476 (1969).
11. H.Tachikawa, On splitting of module categories, Math. Zeitschr. 111, 145-150 (1969).
12. H.Tachikawa, Double centralizers and dominant dimensions, Math. Zeitschr. 116, 79-88 (1970).
13. H.Tachikawa, Localization and artinian quotient rings, Math. Zeitschr. 119, 239-253 (1970).
14. H. Tachikawa, On left QF-3 rings, Pacific Journal of Math. 32, 255-268 (1970).
15. H. Tachikawa and Y. Iwanaga, Morita's F_h-condition and double centralizers I, Journal of Algebra 26, 167-171 (1973).
16. H. Tachikawa, Quasi-Frobenius rings and generalizations, QF-3 and QF-1 rings, Lecture Notes in Math. 351 (1973).
17. Tachikawa, QF-3 rings and categories of projective modules, Journal of Algebra 28, 408-413 (1974).
18. H. Tachikawa and Y. Iwanaga, Morita's F_h-condition and double centralizers II, Journal of Algebra 31, 73-90 (1974).
19. H. Tachikawa, Blancedness and left serial algebras of finite type, Lecture Notes in Math. 488, 351-378 (1975).
20. H. Tachikawa, Balanced modules and constructible submodules, Science Reports of Tokyo Kyoiku Daigaku Section A, Vol. 4 No. 351, 37-45 (1975).
21. C. M. Ringel and H. Tachikawa, QF-3 rings, J. reine und angew. Math. 272, 49-72 (1975).
22. H.Tachikawa, Commutative perfect QF-1 rings, Proc. Amer. Math. Soc. 68, 471-476 (1978).

23. H. Tachikawa and K. Ohtake, Colocalization and localization in Abelian categories, Journal of Algebra 56, 1-23 (1979).
24. H. Tachikawa, Representations of trivial extensions of hereditary algebras, Lecture Notes in Mathematics No. 832, 572-599 (1980), Springer Verlag.
25. H. Tachikawa, Reflection functors and translations for trivial extensions of hereditary algebras, Journal of Algebra 90, 98-118 (1984).
26. H. Tachikawa and T. Wakamatsu, Extensions of tilting functors and QF-3 algebras, Journal of Algebra 103, 662-676 (1986).
27. H. Tachikawa and Y. Wakamatsu, Tilting functors and stable equivalences for selfinjective algebras, Journal of Algebra 109, 138-165 (1987).
28. H. Tachikawa, Reflexive Auslander-Reiten sequences, F. van Oystaeyen and L. Le Bruyn (eds), Perspectives in Ring Theory, 311-320 (1988), Kluwer Academic Publishers.
29. H. Tachikawa and Y. Wakamatsu, Cartan matrices and Grothendieck groups of stable categories, Journal of Algebra 144, 390-398 (1991).

Professor Liu Shaoxue – his Live and Work

Zhang Yingbo
School of Mathematics, Beijing Normal University,
Tokyo, Japan, September 2007

Professor Liu Shaoxue was born on November 6 , 1929 in Liaoyang, a city of Liaoning Province in North China. And his family moved to Beijing in 1937. Upon graduation from high school in 1946, he enrolled in Mathematics Department of National Peping Normal University (the predecessor of Beijing Normal University). In September 1953, he was sent to study in Mechanics and Mathematics Department of Moscow University of Former Soviet Union.

His supervisor was A.G.Kurosh, a prestigious expert in Algebra. Under the guidance of Professor Kurosh, he completed his Ph.D thesis 'On Decomposition of Infinite Algebras', and received Ph.D degree in 1956. He was the first student who received Ph.D degree out of the Chinese students, who studied Mathematics in Former Soviet Union. In his Ph.D thesis he proves the following theorems:

(1) An extension Jordan algebra of a locally finite Jordan algebra by a locally finite Jordan algebra is still locally finite.

(2) An extension Lie algebra of a locally finite Lie algebra by a locally finite Lie algebra is still locally finite in case it is an algebraic Lie algebra. (A Lie algebra is called algebraic, if for any two elements $x, y \in A$ there exists some positive integer $n = n(x, y)$, such that $x, xy, xy^2 = (xy)y, \cdots, xy^n = (xy^{n-1}y)$ are linearly dependent over k).

(3) Based on the result of (1), he proves the existence of the Livitzki radical of Jordan algebras independently of K.A.Zhevlakov.

(4) Based on the result of (2), it is obtained immediately that an algebraic solvable Lie algebra is locally finite.

Professor Liu's Ph.D thesis gives some nice and important results in ring theory, and many experts on ring theory quote his results in their papers. For example in the paper 'On local finiteness in the sense of Shirshov' published in Algebra and Logical in 1973, the authors K.A.Zhevlakov and I.P.Shstakov quoted Professor Liu's Ph.D thesis 5 times. These results are the main contents of chapter 4 in the book "Rings that are nearly associative" published in 1978 in Russian and 1982 in English written by K.A. Zhevlakov, A.M.Slinko, I.P.Shestakov, the former students of Kurosh.

Professor Liu Shaoxue went back to Department of Mathematics of Beijing Normal University in 1956, and was promoted to be an associate Professor in 1961. From 1956 to 1966, because of some political policy, professors in China were asked to solve so called the problems of practice in farmland or factories, which was named the association of theory and practice. Professor Liu had no more chance to concentrate on his research work. But he still wrote more than ten papers, one of them was 'On algebras in which every sub-algebra is an ideal' published in 1964, Chinese Mathematics Acta. An algebra which has the property that every sub-algebra is an ideal is called Hamilton algebra, Professor Liu gives a complete description of Hamilton algebra. The result is simple and beautiful, which is extended to the case of exponential associative algebras by Outcalt, and is quoted in chapter 9 of the book "Nilpotent Rings" by R.L.Kruse and D.T. Price in 1969.

During that time Professor Liu was teaching abstract algebra, calculus, partial differential equation and so on for undergraduate students. Besides his research experience, his lectures were made very clear, vivid and impressive. Many years have passed, still his former students remembered his lectures, and said that Professor Liu's lectures impressed them greatly, and guided them to do research on Algebras.

From 1966 to 1976, our country experienced a disaster called Culture Revolution. All universities were closed and professors were sent to countryside or factors. Research on basic sciences was stopped for more than ten years. Universities returned to normal in 1977, and graduate students enrolled into universities in 1978. Liu Shaoxue was promoted to be a professor in 1979. Since the year of 1978, he restarted his research work and his training of graduate students. He wrote a book "Rings and Algebras" in 1983, which was used widely as a textbook for graduate students on Algebra in China.

In 1988, Professor Liu visited Tsukuba University of Japan. One day Professor H.Tachikawa drove him to the Tsukuba Mountain. During the trip, Professor Liu suggested to hold a symposium on ring theory hosted by China and Japan. Professor Tachikawa accepted the suggestion happily and said that "We are able to compete with European experts on ring theory when we have good cooperation between Japan and China."

In autumn 1991, the First China-Japan international Symposium on ring theory was held in Guangxi Normal University in Guilin, the most beautiful city in South China.The main organizers were Liu and Tachikawa. Some algebraists from North America and Europe also attended the symposium, for example C.M.Ringel, B.J.Muller, and M.Beattie. Professor Li Baifei from Taiwan and Academician Wan Zhexian from Beijing also attended the symposium. The Proceedings were published in 1992 in Japan.

Around the year of 1985, after several visits to the United states and Europe, and having a lot of discussions with foreign algebraists, Professor Liu decided to find a new research area for his students pursing Ph.D degree. It should be new but related to ring theory. In May 1985, Professor Liu went to Belgium to visit Professor F.Van Oystaeyen, and went to Germany to visit Professor Claus Ringel. Then he decided that his students should start the study on representation theory of algebras. Professor Liu is a kind-hearted, honest and humorous person. Because of his personality Professor Liu made a lot of good friends.

Professor Liu was very lucky that God introduced a helpful and faithful friend, Claus Ringel, to him at the difficult time to change the research project. Claus visited China almost every year from 1987. He gave a series of lectures on representation theory of algebras for graduate students. There have been 3 students, who obtained Ph.D. degree under a program of joint training between China and Germany, and 4 students received Humboldt Research Award respectively to visit Germany and studied under his guidance for 2 years. Besides Claus, M.Auslande, I. Reiten, V.Dlab, P.Gabriel and many foreign algebraists also helped the algebra group in China. In the year of 2000, the Ninth International Conference on representation theory of algebras was held in Beijing Normal University. The Chinese group was merging more and more into the international family of representation theory of algebras.

There were altogether 17 Master degree students and 19 Ph.D degree students of Professor Liu. Now most of them are working in some of the best universities as Professors, some of them being outstanding mathematicians.

Professor Liu has established a strong group on representation theory of algebras in China. Because of his efforts in educating graduate students, Professor Liu was awarded a first prize for excellent teaching in the City of Beijing in 1991. A person, who was already an expert on ring theory, but still determined to change with his students to representation theory of algebras at the age of 56, deserves to be respected.

Professor Liu has published more then 50 papers in mathematics and 40 of them after 1978. In the papers 'Isomorphism problem of path algebras' and 'Isomorphism problem for tensor algebras over valid quivers', Professor Liu and other authors proved that two path algebras are isomorphic if and only if their corresponding quivers are isomorphic. Based on this, P.A.Grillet started a systematical study of isomorphism problem of semi-group algebras. In the paper 'Group graded rings, Smash Products and Additive categories', Professor Liu and F.Van.Oystaeyen have extended the concept of the smash product of group G and G-graded ring A to the case of G being an infinite group, and obtained the corresponding dual and co-dual theorems. In the paper 'Comparing graded version of the prime radical' by Beattie, Liu and Stewart, the structure theorem of the graded rings is proved.

Professor Liu has written, translated and edited several books. He wrote the book 'Rings and algebras' and the textbook 'Basic Modern Algebra'; translated the books 'Group Theory and 'General Algebra' by Kurosh, 'Finite dimensional algebras' by Drozd and Gilchinko from Russian into Chinese. He edited 'Ring and radical' with foreign algebraists, and edited 'Rings, Groups and algebras in China' as a chief editor. He won a second prize of National Committee's prize for Scientific and Technological Development with the work of title 'The structure and representation theory of rings' in 1988.

Professor Liu is a good professor and nice person.

PART B

Invited Lectures

THE REALIZATION PROBLEM FOR VON NEUMANN REGULAR RINGS

PERE ARA

Departament de Matemàtiques, Universitat Autònoma de Barcelona, 08193, Bellaterra (Barcelona), Spain
E-mail: para@mat.uab.cat

We survey recent progress on the realization problem for von Neumann regular rings, which asks whether every countable conical refinement monoid can be realized as the monoid of isoclasses of finitely generated projective right R-modules over a von Neumann regular ring R.

Keywords: von Neumann regular ring; Leavitt path algebra; refinement monoid.

This survey consists of four sections. Section 1 introduces the realization problem for von Neumann regular rings, and points out its relationship with the separativity problem of [7]. Section 2 surveys positive realization results for countable conical refinement monoids, including the recent constructions in [5] and [4]. We analyze in Section 3 the relationship with the realization problem of algebraic distributive lattices as lattices of two-sided ideals over von Neumann regular rings. Finally we observe in Section 4 that there are countable conical monoids which can be realized by a von Neumann regular K-algebra for some countable field K, but they cannot be realized by a von Neumann regular F-algebra for any uncountable field F.

1. The problem

All rings considered in this paper will be associative, and all the monoids will be commutative.

For a unital ring R, let $\mathcal{V}(R)$ denote the monoid of isomorphism classes of finitely generated projective right R-modules, where the operation is defined by

$$[P] + [Q] = [P \oplus Q].$$

This monoid describes faithfully the decomposition structure of finitely generated projective modules. The monoid $\mathcal{V}(R)$ is always a *conical monoid*, that is, whenever $x + y = 0$, we have $x = y = 0$. Recall that an *order-unit* in a monoid M is an element u in M such that for every $x \in M$ there is $y \in M$ and $n \geq 1$ such that $x + y = nu$. Observe that $[R]$ is a canonical order-unit in $\mathcal{V}(R)$. By results of Bergman [11 Theorems 6.2 and 6.4] and Bergman and Dicks [12, page 315] , any conical monoid with an order-unit appears as $\mathcal{V}(R)$ for some unital hereditary ring R.

A monoid M is said to be a *refinement monoid* in case any equality $x_1 + x_2 = y_1 + y_2$ admits a refinement, that is, there are z_{ij}, $1 \leq i, j \leq 2$ such that $x_i = z_{i1} + z_{i2}$ and $y_j = z_{1j} + z_{2j}$ for all i, j, see e.g. [8]. If R is a von Neumann regular ring, then the monoid $\mathcal{V}(R)$ is a refinement monoid by [20, Theorem 2.8] .

The following is still an open problem:

R1. Realization Problem for von Neumann Regular Rings Is every countable conical refinement monoid realizable by a von Neumann regular ring?

A related problem was posed by K.R. Goodearl in [22]:

FUNDAMENTAL OPEN PROBLEM Which monoids arise as $\mathcal{V}(R)$'s for a von Neumann regular ring R?

It was shown by Wehrung in [35] that there are conical refinement monoids of size $\aleph_2$ which cannot be realized. If the size of the monoid is $\aleph_1$ the question is open. Wehrung's approach is related to Dilworth's Congruence Lattice Problem (CLP), see Section 3. A solution to the latter problem has recently appeared in [42].

Problem R1 is related to the separativity problem. A class $\mathcal{C}$ of modules is called *separative* if for all $A, B \in \mathcal{C}$ we have

$$A \oplus A \cong A \oplus B \cong B \oplus B \implies A \cong B.$$

A ring R is *separative* if the class $FP(R)$ of all finitely generated projective right R-modules is a separative class. Separativity is an old concept in semigroup theory, see [16]. A commutative semigroup S is called *separative* if for all $a, b \in S$ we have $a + a = a + b = b + b \implies a = b$. An alternative characterization is that a commutative semigroup is separative if and only if it can be embedded in a product of monoids of the form $G \sqcup \{\infty\}$, where G is an abelian group. Clearly a ring R is separative if and only if $\mathcal{V}(R)$ is a separative semigroup. Separativity provides a key to a number of outstanding cancellation problems for finitely generated projective modules over exchange rings, see [7].

Outside the class of exchange rings, separativity can easily fail. In fact it is easy to see that a commutative ring R is separative if and only if $\mathcal{V}(R)$ is cancellative. Among exchange rings, however, separativity seems to be the norm. It is not known whether there are non-separative exchange rings. This is one of the major open problems in this area. See [3] for some classes of exchange rings which are known to be separative. We single out the problem for von Neumann regular rings. (Recall that every von Neumann regular ring is an exchange ring.)

SP. Is every von Neumann regular ring separative?

We have ($R1$ has positive answer) $\Longrightarrow$ (SP has a negative answer). To explain why we have to recall results of Bergman and Wehrung concerning *existence* of countable non-separative conical refinement monoids.

Recall that every monoid M is endowed with a natural pre-order, the so-called *algebraic pre-order*, by $x \leq y$ iff there is $z \in M$ such that $y = x+z$. This is the only order on monoids that we will consider in this paper. A monoid homomorphism $f\colon M \to M'$ is an *order-embedding* in case f is one-to-one and, for $x, y \in M$, we have $x \leq y$ if and only if $f(x) \leq f(y)$.

Proposition 1.1. *(cf. [36]) Let M be a countable conical monoid. Then there is an order-embedding of M into a countable conical refinement monoid.*

Let us apply the above Proposition to the conical monoid M generated by a with the only relation $2a = 3a$. Then

$$a + a = a + (2a) = (2a) + (2a)$$

but $a \neq 2a$ in M. By Proposition 1.1 there exists an order-embedding $M \to M'$, where M' is a countable conical refinement monoid, and M' cannot be separative.

Thus if R1 is true we can represent M' as $\mathcal{V}(R)$ for some von Neumann regular ring and R will be non-separative.

2. Known cases

It turns out that only a few cases of R1 are known. In this section I will describe the positive realization results of which I am aware.

The first realization result is by now a classical one. Recall that a monoid M is said to be *unperforated* if, for $x, y \in M$ and $n \geq 1$, the relation $nx \leq ny$ implies that $x \leq y$. A *dimension monoid* is a cancellative, refinement, unperforated conical monoid. These are the positive cones of the dimension groups [20, Chapter 15]. Recall that, by definition, an *ultramatricial* K-algebra R is a direct limit of a sequence of finite direct products of matrix algebras over K. Clearly every ultramatricial algebra is von Neumann regular.

Theorem 2.1. *(18, [19, Theorem 3.17], [20, Theorem 15.24(b)]) If M is a countable dimension monoid and K is any field, then there exists an ultramatricial K-algebra R such that $\mathcal{V}(R) \cong M$.*

A K-algebra is said to be *locally matricial* in case it is a direct limit of a directed system of finite direct products of matrix algebras over K, see [23, Section 1]. It was proved in [23, Theorem 1.5] that if M is a dimension monoid of size $\leq \aleph_1$, then it can be realized as $\mathcal{V}(R)$ for a locally matricial K-algebra R. Wehrung constructed in [35] dimension monoids of size $\aleph_2$ which cannot be realized by regular rings. Indeed the monoids constructed in [35] are the positive cones of dimension groups which are vector spaces over $\mathbb{Q}$. A refinement of the method used in [35] gave a dimension monoid counterexample of size $\aleph_2$ with an order unit of index two [41], thus answering a question posed by Goodearl in [21].

Another realization result was obtained by Goodearl, Pardo and the author in [6].

Theorem 2.2. *[6, Theorem 8.4]. Let G be a countable abelian group and K any field. Then there is a purely infinite simple regular K-algebra R such that $K_0(R) \cong G$.*

Recall that a simple ring R is *purely infinite* in case it is not a division ring and, for every nonzero element $a \in R$ there are $x, y \in R$ such that $xay = 1$ (see [6, Section 1], especially Theorem 1.6). Since $\mathcal{V}(R) = K_0(R) \sqcup \{0\}$ for a purely infinite simple regular ring [6, Corollary 2.2], we get that all monoids of the form $G \sqcup \{0\}$, where G is a countable abelian group, can be realized.

As Fred Wehrung has kindly pointed out to me, another class of conical refinement monoids which can be realized by von Neumann regular rings is the class of *continuous dimension scales*, see [28, Chapter 3]. All the monoids in this class satisfy the property that every bounded subset has a supremum, as well as some additional axioms, see [28, Definition 3-1.1].

Indeed, if M is a commutative monoid, then $M \cong \mathcal{V}(R)$ for some regular, right self-injective ring R if and only if M is a continuous dimension scale with order-unit [28, Corollary 5-3.15]. These monoids have unrestricted cardinality, indeed they are usually quite large.

A recent realization result has been obtained by Brustenga and the author in [5]. As we will note later, the two results mentioned above (Theorem 2.1 and Theorem 2.2) can be seen as particular cases of the main result in [5]. Before we proceed with the statement of this result, and in order to put it in the right setting, we need to recall a few monoid theoretic concepts.

Definition 2.1. Let M be a monoid. An element $p \in M$ is *prime* if for all $a_1, a_2 \in M$, $p \leq a_1 + a_2$ implies $p \leq a_1$ or $p \leq a_2$. A monoid is *primely generated* if each of its elements is a sum of primes.

Proposition 2.1. *[14, Corollary 6.8] Any finitely generated refinement monoid is primely generated.*

We have the following particular case of question R1.

R2. Realization Problem for finitely generated refinement monoids: Is every finitely generated conical refinement monoid realizable by a von Neumann regular ring?

We conjecture that R2 has a positive answer. Our main tool to realize a large class of finitely generated refinement monoids is the consideration of some regular algebras associated with quivers.

Recall that a *quiver* (= *directed graph*) consists of a 'vertex set' E^0, an 'edge set' E^1, together with maps r and s from E^1 to E^0 describing, respectively, the range and source of edges. A quiver $E = (E^0, E^1, r, s)$ is said to be *row-finite* in case, for each vertex v, the set $s^{-1}(v)$ of arrows with source v is finite. For a row-finite quiver E, the graph monoid $M(E)$ of E is defined as the quotient monoid of $F = F_E$, the free abelian monoid with basis E^0 modulo the congruence generated by the relations

$$v = \sum_{\{e \in E^1 \mid s(e) = v\}} r(e)$$

for every vertex $v \in E^0$ which emits arrows (that is $s^{-1}(v) \neq \emptyset$).

It follows from Proposition 2.1 and [8, Proposition 4.4] that, for a finite quiver E, the monoid $M(E)$ is primely generated. Note that this is not always the case for a general row-finite graph E. An example is provided by the graph:

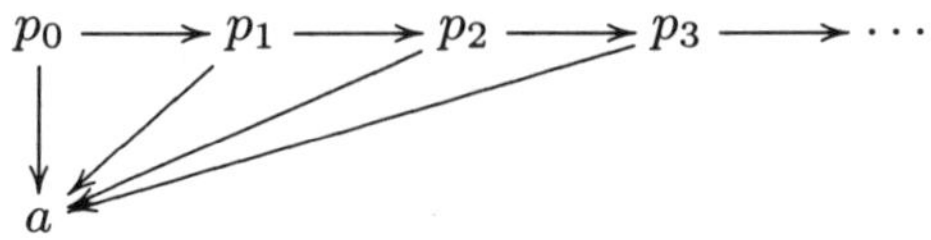

The corresponding monoid M has generators $a, p_0, p_1, \ldots$ and relations given by $p_i = p_{i+1} + a$ for all $i \geq 0$. One can easily see that the only prime element in M is a, so that M is not primely generated.

Now we have the following result of Brookfield:

Theorem 2.3. *[14, Theorem 4.5 and Corollary 5.11(5)] Let M be a primely generated refinement monoid. Then M is separative.*

In fact, primely generated refinement monoids enjoy many other nice properties, see [14] and also [37].

It follows from Proposition 2.1 and Theorem 2.3 that a finitely generated refinement monoid is separative. In particular all the monoids associated to finite quivers are separative. For a row-finite quiver E the result follows by using the fact that the monoid $M(E)$ is the direct limit of monoids associated to certain finite subgraphs of E, see [8, Lemma 2.4].

Theorem 2.4. *([5, Theorem 4.2, Theorem 4.4]) Let $M(E)$ be the monoid corresponding to a finite quiver E and let K be any field. Then there exists a unital von Neumann regular hereditary K-algebra $Q_K(E)$ such that $\mathcal{V}(Q_K(E)) \cong M(E)$. Furthermore, if E is a row-finite quiver, then there exists a (not necessarily unital) von Neumann regular K-algebra $Q_K(E)$ such that $\mathcal{V}(Q_K(E)) \cong M_E$.*

Note that, due to unfortunate lack of convention in this area, the Cuntz-Krieger relations used in [5] are the opposite to the ones used in [8], which are the ones we are following in this survey, so that the result in [5, Theorem 4.4] is stated for *column-finite* quivers instead of row-finite ones. The regular algebras $Q_K(E)$ are related to the *Leavitt path algebras* $L_K(E)$ of [1], [2], [8].

We now observe that Theorem 2.1 and Theorem 2.2 are particular cases of Theorem 2.4. This follows from the fact that the monoids considered in these theorems are known to be graph monoids $M(E)$ for suitable quivers E. Indeed, taking into account [8, Theorem 3.5 and Theorem 7.1], we see that the case of dimension monoids follows from [31, Proposition 2.12] and the case of monoids of the form $\{0\} \sqcup G$, with G a countable abelian group, follows from [34, Theorem 1.2]. As Pardo pointed out to me, the quiver E can be chosen to be finite in case $M = \{0\} \sqcup G$ for a finitely generated abelian group G. To see this, note that such a monoid admits a presentation given by a finite number of generators $a_1, \dots, a_n$, and relations of the form $a_i = \sum_{j=1}^{n} \gamma_{ji} a_j$, where all γ_{ji} are strictly positive integers and $\gamma_{ii} \geq 2$ for all i. The corresponding finite quiver will have γ_{ji} arrows from the vertex i to the vertex j.

Now we would like to describe how this construction sheds light on problem R2. The answer is completely known for the class of antisymmetric finitely generated refinement monoids. The monoid M is said to be *antisymmetric* in case the algebraic pre-order is a partial order, that is, in case $x \leq y$ and $y \leq x$ imply that $x = y$. Note that every antisymmetric monoid is conical.

We say that a monoid M is *primitive* if it is an antisymmetric primely generated refinement monoid [30, Section 3.4]. A primitive monoid M is completely determined by its set of primes $\mathbb{P}(M)$ together with a transitive and antisymmetric relation $\lhd$ on it, given by $q \lhd p$ iff $p + q = p$. Indeed given such a pair $(\mathbb{P}, \lhd)$, the abelian monoid $M(\mathbb{P}, \lhd)$ defined by taking as a set of generators $\mathbb{P}$ and with relations given by $p = p + q$ whenever $q \lhd p$, is a primitive monoid, and the correspondences $M \mapsto (\mathbb{P}(M), \lhd)$ and $(\mathbb{P}, \lhd) \mapsto M(\mathbb{P}, \lhd)$ give a bijection between isomorphism types of primitive monoids and isomorphism types of pairs $(\mathbb{P}, \lhd)$, where $\mathbb{P}$ is a set and $\lhd$ a transitive antisymmetric relation on $\mathbb{P}$, see [30, Proposition 3.5.2].

Let M be a primitive monoid and $p \in \mathbb{P}(M)$. Then p is said to be *free* in case $p \not\lhd p$. Otherwise p is *regular*, see [10, Section 2]. So giving a primitive monoid is equivalent to giving a poset $(\mathbb{P}, \leq)$ which is a disjoint union of two subsets: $\mathbb{P} = \mathbb{P}_{\mathrm{free}} \sqcup \mathbb{P}_{\mathrm{reg}}$. If M is a finitely generated primitive monoid then $\mathbb{P}(M)$ is a finite set, indeed $\mathbb{P}(M)$ is the minimal generating set of M.

We can now describe the finitely generated primitive monoids which are graph monoids. Recall that a *lower cover* of an element p of a poset $\mathbb{P}$ is an element q in $\mathbb{P}$ such that $q < p$ and $[q, p] = \{q, p\}$. The set of lower covers of p in $\mathbb{P}(M)$ is denoted by $\mathrm{L}(M, p)$, and $\mathrm{L}_{\mathrm{free}}(M, p)$ and $\mathrm{L}_{\mathrm{reg}}(M, p)$ denote the sets of free and regular elements in $\mathrm{L}(M, p)$ respectively.

Theorem 2.5. *(cf. [10, Theorem 5.1]) Let M be a finitely generated primitive monoid. Then the following statements are equivalent:*

(1) M is a graph monoid.

(2) M *is a direct limit of graph monoids.*
(3) $|\mathrm{L}_{\mathrm{free}}(M,p)| \leq 1$ *for each* p *in*$\mathbb{P}_{\mathrm{free}}(M)$.

For the monoids as in the statement there is a hereditary, von Neumann regular ring $Q(E)$ such that $\mathcal{V}(Q(E)) = M_E = M$ (Theorem 2.4). It is worth mentioning that, in some cases, an infinite quiver is required in Theorem 2.5. A slightly more general result is indeed presented in [10, Theorem 5.1]. Namely the same characterization holds when M is a primitive monoid such that $\mathrm{L}(M,p)$ is finite for every p in $\mathbb{P}(M)$.

In view of Theorem 2.5, the simplest primitive monoid which is not a graph monoid is the monoid

$$M = \langle p, a, b \mid p = p + a = p + b \rangle.$$

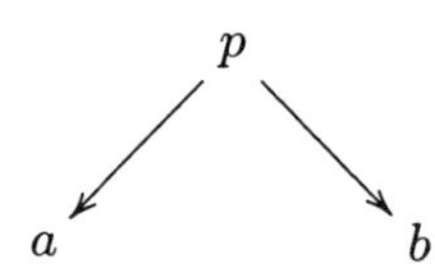

Fig. 1. The poset $\mathbb{P}(M)$ for the monoid $M = \langle p, a, b \mid p = p + a = p + b \rangle$.

In this example $\mathbb{P}(M) = \mathbb{P}_{\mathrm{free}}(M) = \{p, a, b\}$, and p has two free lower covers a, b. Thus, by Theorem 2.5, the monoid M is not even a direct limit of graph monoids (with monoid homomorphisms as connecting maps). However M can be realized as the monoid of a suitable von Neumann regular ring, as follows. Fix a field K and consider two indeterminates t_1, t_2 over K. We consider the regular algebra $Q_{K(t_2)}(S_1)$ over the quiver S_1 with two vertices $v_{0,1}, v_{1,1}$ and two arrows e_1, f_1 such that $r(e_1) = s(e_1) = v_{1,1} = s(f_1)$ and $r(f_1) = v_{0,1}$. The picture of S_1 is as shown in Figure 2.

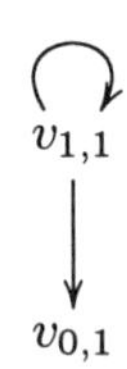

Fig. 2. The graph S_1.

The algebra $Q_1 := Q_{K(t_2)}(S_1)$ has a unique non-trivial (two-sided) ideal

M_1, which coincides with its socle, so that we get an extension of rings:

$$0 \longrightarrow M_1 \longrightarrow Q_1 \xrightarrow{\pi_1} K(t_2)(t_1) = K(t_1,t_2) \longrightarrow 0$$

However the element t_1 does not lift to a unit in Q_1, rather there are $z_1, \overline{z}_1$ in Q_1 such that $\overline{z}_1 z_1 = 1$, but $z_1\overline{z}_1 \neq 1$, and $\pi_1(z_1) = t_1$. Some additional information on the algebra $Q_F(S_1)$, where F denotes an arbitrary field, can be found in [5, Examples 4.3].

Let S_2 be a copy of S_1, now with vertices labelled as $v_{0,2}, v_{1,2}$ and arrows labelled as e_2, f_2, and set $Q_2 = Q_{K(t_1)}(S_2)$. There is a corresponding diagram

$$0 \longrightarrow M_2 \longrightarrow Q_2 \xrightarrow{\pi_2} K(t_1)(t_2) = K(t_1,t_2) \longrightarrow 0$$

Let P be the pullback of the maps π_1 and π_2, so that P fits in the following commutative square:

$$\begin{array}{ccc} P & \xrightarrow{\rho_1} & Q_1 \\ {\scriptstyle \rho_2}\big\downarrow & & \big\downarrow{\scriptstyle \pi_1} \\ Q_2 & \xrightarrow{\pi_2} & K(t_1,t_2) \end{array} \qquad (1)$$

Then P is a von Neumann regular ring and $\mathcal{V}(P) = M$, see [4]. Indeed a wide generalization of this method gives the following realization result:

Theorem 2.6. *([4, Theorem 2.2]) Let M be a finitely generated primitive monoid such that all primes of M are free and let K be a field. Then there is a unital regular K-algebra $Q_K(M)$ such that $\mathcal{V}(Q_K(M)) \cong M$.*

Moreover both the regular algebras associated with quivers [5] and the regular algebras constructed in [4] are given explicitly in terms of generators and relations (including universal localization [33]).

Recall that a monoid M is *strongly separative* in case $a+a = a+b$ implies $a = b$ for $a, b \in M$. A ring R is said to be *strongly separative* in case $\mathcal{V}(R)$ is a strongly separative monoid, see [7] for background and various equivalent conditions. As we mentioned above, every primely generated refinement monoid is separative [14, Theorem 4.5]. In particular every primitive monoid is separative. Moreover, a primitive monoid M is strongly separative if and only if all the primes in M are free, see [14, Theorem 4.5, Corollary 5.9]. Thus, the class of monoids covered by Theorem 1.4 coincides exactly with the strongly separative finitely generated primitive monoids. The case of a general finitely generated primitive monoid remains open, although it seems amenable to analysis in the light of [5] and [4].

3. Realizing distributive lattices

Let R be a regular ring. Then the lattice $\mathrm{Id}(R)$ of all (two-sided) ideals of R is an algebraic distributive lattice. Here an *algebraic lattice* means a complete lattice such that each element is the supremum of all the compact elements below it. The set of compact elements in $\mathrm{Id}(R)$ is the set $\mathrm{Id}_c(R)$ of all finitely generated ideals of R, and it is a distributive semilattice, see for example [27]. Here a *semilattice* means a $\vee$-semilattice with least element 0. A semilattice is the same as a monoid M such that $x = x + x$ for every x in M, and a distributive semilattice is just a semilattice satisfying the refinement axiom [27, Lemma 2.3]. Observe that, if M is a semilattice then $x \vee y = x + y$ gives the supremum of x, y in M.

The famous *Congruence Lattice Problem* (CLP) asks whether an algebraic distributive lattice is the congruence lattice of some lattice; equivalently, whether every distributive semilattice is the semilattice of all the compact congruences of a lattice. This problem has been recently solved in the negative by Fred Wehrung [42], who constructed for each $\aleph \geq \aleph_{\omega+1}$ an algebraic distributive lattice with $\aleph$ compact elements such that it cannot be represented as the congruence lattice of any lattice, see also [25]. His methods have been refined by Růžička [32] to cover the case $\aleph \geq \aleph_2$. It is worth to remark that Wehrung had previously shown in [39] that every algebraic distributive lattice with $\leq \aleph_1$ compact elements can be realized as the ideal lattice for some von Neumann regular ring R, and thus is isomorphic to the congruence lattice of the lattice $L(R_R)$ of principal right ideals of R [38, Corollary 4.4]. So the formulation of Růžička is best possible (concerning the number of compact elements). The "hard core" of Wehrung's proof in [39] is a ring-theoretic amalgamation result proved by P. M. Cohn in [17, Theorem 4.7].

One can ask: What is the relationship between the CLP, or more concretely, the representation problem of algebraic distributive lattices as lattices of ideals of regular rings, and our problem R1? The answer is that, for a regular ring R, the lattice $\mathrm{Id}(R)$ is only a small piece of information compared with the information contained in the monoid $\mathcal{V}(R)$, in the sense that $\mathcal{V}(R)$ determines $\mathrm{Id}(R)$, but generally the structure of $\mathcal{V}(R)$ can be much more complicated than the structure of $\mathrm{Id}(R)$, e.g. for simple rings. Indeed we have a lattice isomorphism $\mathrm{Id}(R) \cong \mathrm{Id}(\mathcal{V}(R))$, where for a conical monoid M, $\mathrm{Id}(M)$ is the lattice of all order-ideals of M, cf. [27, Proposition 7.3]. Recall that an *order-ideal* of M is a submonoid I of M with the property that whenever $x \leq y$ in M and $y \in I$, we have $x \in I$. If L is an algebraic distributive lattice which is not the congruence lattice of any lattice and M is any conical refinement monoid such that $\mathrm{Id}(M) \cong L$, then M cannot be realized as $\mathcal{V}(R)$ for a regular ring R. For every algebraic distributive lattice L there is at least one such conical refinement monoid, namely the semilattice L_c of compact elements of L, but we should expect a myriad of such monoids to exist. Wehrung proved in [39] that if $|L_c| \leq \aleph_1$ then the semilattice L_c can be realized as $\mathcal{V}(R)$ where R is a von Neumann regular ring, and he showed in [40] that there is a distributive semilattice S_{ω_1} of size $\aleph_1$ which is not the semilattice of finitely generated, idempotent-generated ideals of any exchange ring of finite stable rank. In particular there is no locally matricial K-algebra A over a field K such that $\mathrm{Id}_c(A) \cong S_{\omega_1}$; see [40] for details. This contrasts with Bergman's result [13] stating that every distributive semilattice of size $\leq \aleph_0$ is the semilattice of finitely generated ideals of an ultramatricial K-algebra, for every field K.

Say that a subset A of a poset $\mathbb{P}$ is a *lower subset* in case $q \leq p$ and $p \in A$ imply $q \in A$. The set $\mathcal{L}(\mathbb{P})$ of all lower subsets of $\mathbb{P}$ forms an algebraic distributive lattice, which is a sublattice of the Boolean lattice $\mathbf{2}^{\mathbb{P}}$. Now if L is a *finite* distributive lattice, then by a result of Birkhoff ([24, Theorem II.1.9]) there is a finite poset $\mathbb{P}$ such that L is the lattice of all lower subsets of $\mathbb{P}$. In the case of a finite Boolean algebra $\mathbf{2}^n$ with n atoms, the poset $\mathbb{P}$ is just an antichain with n points, and $\mathbf{2}^n = \mathcal{P}(\mathbb{P}) = \mathcal{L}(\mathbb{P})$. Our construction in [4] gives a realization of L as the ideal lattice of a regular K-algebra $Q_K(\mathbb{P})$, where K is an arbitrary fixed field, such that the monoid $\mathcal{V}(Q_K(\mathbb{P}))$ is the monoid $M(\mathbb{P})$ associated with $(\mathbb{P}, <)$, with all elements in $M(\mathbb{P})$ being free, that is, $M(\mathbb{P})$ is the abelian monoid with generators $\mathbb{P}$ and relations given by $p = q + p$ whenever $q < p$ in $\mathbb{P}$. Moreover $Q_K(\mathbb{P})$ satisfies the following properties ([4, Proposition 2.12, Remark 2.13, Theorem 2.2]):

(a) There is a canonical family of commuting idempotents $\{e(A) : A \in L\}$ such that
   (i) $e(A)e(B) = e(A \cap B)$
   (ii) $e(A) + e(B) - e(A)e(B) = e(A \cup B)$
   (iii) $e(\emptyset) = 0$ and $e(\mathbb{P}) = 1$.
   (iv) $e(A)Q_K(\mathbb{P})e(A) \cong Q_K(A)$.

(b) Let $I(A)$ be the ideal of $Q_K(\mathbb{P})$ generated by $e(A)$. Then the assignment

$$A \mapsto I(A)$$

defines a lattice isomorphism from $L = \mathcal{L}(\mathbb{P})$ onto $\mathrm{Id}(Q_K(M))$.

(c) The map $M(\mathbb{P}) \to \mathcal{V}(Q_K(\mathbb{P}))$ given by $p \mapsto [e(\mathbb{P} \downarrow p)]$, for $p \in \mathbb{P}$, is a monoid isomorphism. Here $\mathbb{P}\downarrow p=\{q\in\mathbb{P}: q\leq p\}$ is the lower subset of $\mathbb{P}$ generated by p.

The set $\mathrm{Idem}(R)$ of idempotents of a ring is a poset in a natural way, by using the order $e\leq f$ iff $e=fe=ef$. This poset is a partial lattice, in the following sense: every two commuting idempotents e and f have an infimum ef and a supremum $e+f-ef$ in $\mathrm{Idem}(R)$. Say that a map $\phi\colon L\to\mathrm{Idem}(R)$ from a lattice L to $\mathrm{Idem}(R)$ is a *lattice homomorphism* in case $\phi(x)$ and $\phi(y)$ commute and $\phi(x\vee y)=\phi(x)\vee\phi(y)$ and $\phi(x\wedge y)=\phi(x)\wedge\phi(y)$, for every $x,y\in L$.

The above results can be paraphrased as follows: The canonical mapping $\mathrm{Idem}(Q_K(\mathbb{P}))\to\mathrm{Id}(Q_K(\mathbb{P}))=\mathcal{L}(\mathbb{P})$ sending each element e in $\mathrm{Idem}(Q_K(\mathbb{P}))$ to the ideal generated by e has a distinguished section $e\colon\mathrm{Id}(Q_K(\mathbb{P}))\to\mathrm{Idem}(Q_K(\mathbb{P}))$, $A\mapsto e(A)$, which is a lattice homomorphism.

Write $Q=Q_K(\mathbb{P})$. Observe that we have lattice isomorphisms

$$L=\mathcal{L}(\mathbb{P})\cong\mathrm{Id}(Q)\cong\mathrm{Id}\mathcal{V}(Q)\cong\mathcal{V}(Q)/{\asymp}.$$

Here $\mathcal{V}(Q)/{\asymp}$ is the maximal semilattice quotient of the monoid $\mathcal{V}(Q)$, see [27, Section 2]. The finite distributive lattice L can be represented in many other ways as an ideal lattice of a regular ring, for instance using ultramatricial algebras [13], but the monoids corresponding to these ultramatricial algebras have little to do with $M(\mathbb{P})$. Indeed as soon as $\mathbb{P}$ is not an antichain we will have that $\mathcal{V}(R)$ is non-finitely generated for every ultramatricial algebra R such that $\mathrm{Id}(R)\cong\mathcal{L}(\mathbb{P})$.

4. The dependence on the field

We like to work with von Neumann regular rings which are algebras over a field K. A natural question is whether the field K plays any role concerning the realization problem. So we ask the following variant of R1.

R(K). Realization Problem for von Neumann Regular K-algebras Let K be a fixed field. Is every countable conical refinement monoid realizable by a von Neumann regular K-algebra?

The answer to this question is known for uncountable fields, thanks to an observation due to Wehrung. Indeed the basic counter-example comes from a construction due to Chuang and Lee [15]. Their remarkable example gave a negative answer to five open questions in Goodearl's book [20] on von Neumann regular rings.

We take this opportunity to present the complete argument, including a result of Goodearl, generalizing Wehrung's observation, and a version of the Chuang-Lee construction.

We will need the notion of (pseudo-)rank function, as given in [20, Chapter 16]. Recall that a *pseudo-rank function* N on a unital regular ring R is a function $N\colon R \to [0,1]$ such that

(a) $N(1) = 1$.
(b) $N(xy) \le N(x)$ and $N(xy) \le N(y)$ for all $x, y \in R$.
(c) $N(e+f) = N(e) + N(f)$ for all orthogonal idempotents $e, f \in R$.

A *rank function* is a pseudo-rank function N such that $N(x) > 0$ for all nonzero $x \in R$.

Proposition 4.1. *(Goodearl) Let (M,u) be a conical refinement monoid with order-unit admitting a faithful state s, i.e. a monoid homomorphism $s\colon M \to \mathbb{R}^+$ such that $s(u) = 1$ and $s(x) > 0$ for every nonzero x in M. Assume that M is not cancelative. Then there is no regular algebra R over an uncountable field F such that $\mathcal{V}(R) \cong M$.*

Proof. Assume that R is a regular F-algebra over an uncountable field F with $\mathcal{V}(R) \cong M$. Clearly we can assume that R is unital and that $[1]$ corresponds to u under the isomorphism $\mathcal{V}(R) \cong M$.

By [26, Theorem 2.2], it suffices to prove that there is no uncountable independent family of nonzero right or left ideals of R. Since $\mathcal{V}(R) \cong \mathcal{V}(R^{\mathrm{op}})$, where R^{op} is the opposite ring of R, we see that it suffices to show this fact for right ideals. Indeed, once this is established, we get from [26, Theorem 2.2] and [7, Proposition 4.12] that R is unit-regular, and thus $\mathcal{V}(R)$ must be cancellative by [7, Theorem 4.5] , a contradiction with our hypothesis.

By [20, Proposition 17.12] there exists a pseudo-rank function N on R such that $N(x) = s([xR])$ for every $x \in R$. Since s is faithful, we see that N is indeed a rank function. Now by [20, Proposition 16.11] we get that R contains no uncountable direct sums of nonzero right ideals, as desired. □

Now we are going to recall the example of Chuang and Lee [15]. We will give a presentation which is a little bit more general. Let R be a σ-unital regular ring, that is, a regular ring having an increasing sequence (e_n) of idempotents in R such that $R = \bigcup_{n=1}^{\infty} e_n R e_n$. Put $R_n = e_n R e_n$. Recall that the multiplier ring $\mathcal{M}(R)$ is the completion of R with respect to the strict topology; see [9]. Write

$$\mathcal{R} = \{(x_n) \mid (x_n) \text{ is a Cauchy sequence in the strict topology }\} \subseteq \prod_{n=1}^{\infty} R_n.$$

By the continuity of operations, $\mathcal{R}$ is a unital subring of $\prod_{n=1}^{\infty} R_n$. There is an obvious canonical surjective homomorphism $\Phi\colon \mathcal{R} \to \mathcal{M}(R)$ whose kernel is $I = \{(x_n) \mid x_n \to 0\}$, where the convergence is with respect to the strict topology.

Lemma 4.1. *I is always a (non-unital) regular ring. If each e_nRe_n is unit-regular, then I is unit-regular, meaning that eIe is unit-regular for every idempotent e in I.*

Proof. Let $x = (x_n) \in I$. Choose a sequence of integers $m_1 < m_2 < \cdots$ such that for all $m \geq m_i$ we have $x_m e_i = e_i x_m = 0$. Now for $m_i \leq m < m_{i+1}$, choose a quasi-inverse y_m of x_m in $(e_m - e_i)R(e_m - e_i)$. (Note that $x_m \in (e_m - e_i)R(e_m - e_i)$ for $m_i \leq m < m_{i+1}$.) We get a quasi-inverse $y = (y_n)$ of x such that $y_n \to 0$ strictly, so $y \in I$ and I is regular. The last part is easy, and is left to the reader. □

Observe that if Q is any regular ring such that $Q \subseteq \mathcal{M}(R)$, then $\Phi^{-1}(Q)$ is a regular ring ([20, Lemma 1.3]) which is a subdirect product of the regular rings (R_n). In particular $\Phi^{-1}(Q)$ is stably finite if each R_n is so.

Now we see that when K is a *countable field* the regular algebra $Q_K(E)$ of the quiver E with $E^0 = \{v_0, v_1\}$ and $E^1 = \{e, f\}$, with $r(e) = s(e) = s(f) = v_1$ and $r(f) = v_0$ gives an example that fits in the above picture. (Note that the quiver E is the same as the quiver S_1 of Figure 2.) Since the field K is countable, the algebra $Q_K(E)$ is also countable.

Write $Q = Q_K(E)$, and let I be the ideal of Q generated by v_0. Then $I = \mathrm{Soc}(Q)$ is a simple (non-unital) ring, and I is countable, so we have $I \cong M_\infty(K)$ (because v_0Qv_0 is isomorphic to K), see [9, Remark 2.9] . Here $M_\infty(K)$ denotes the K-algebra of countably infinite matrices with only a finite number of nonzero entries. Since this is a crucial argument here, let us recall the details. The ring I is countable and simple with a minimal idempotent v_0, so by general theory there is a dual pair V, W of K-vector spaces such that $I \cong \mathcal{F}_W(V)$, the algebra of all adjointable operators on V of finite rank. Since I is countable, both V and W are countably dimensional K-vector spaces. By an old result of G. W. Mackey [29, Lemma 2] , there are dual bases (v_i) and (w_j) for V and W respectively, that is, we have $\langle v_i, w_j\rangle = \delta_{ij}$ for all i, j, which shows that $\mathcal{F}_W(V) \cong M_\infty(K)$. Since I is essential in Q we get an embedding of Q into the multiplier algebra $\mathcal{M}(I)$. Observe that $\mathcal{M}(I) \cong RCFM(K)$, the algebra of row-and-column-finite matrices with coefficients in K, and that $Q/I \cong K(t)$. So the algebra Q has the same essential properties as the Chuang and Lee algebra, see [15]. Now $M_\infty(K)$ is clearly σ-unital and unit-regular. Indeed there is a σ-unit (e_n) for I consisting of idempotents such that $e_nIe_n \cong M_n(K)$. Now Lemma 4.1 together with [20, Lemma 1.3] gives that $S := \Phi^{-1}(Q)$ is regular, and it is residually artinian. The ring S is not countable but it can be easily modified to get a countable algebra with similar properties. Indeed consider the K-subalgebra S_0 of S generated by $\oplus_{n=1}^{\infty} M_n(K)$ and a, b where a, b are elements in S such that $\Phi(a)\Phi(b) = 1$ and $\Phi(b)\Phi(a) \neq 1$. Observe that S_0 is countable. We can build a sequence of countable K-subalgebras of S:

$$S_0 \subseteq S_1 \subseteq S_2 \subseteq \cdots \subseteq S$$

such that each element in S_i is regular in S_{i+1} for all i. It follows that $S_\infty = \bigcup S_i$ is a countable, regular K-algebra, which is embedded in $\prod_{n=1}^{\infty} M_n(K)$. Moreover S_∞ cannot be unit-regular because it has a quotient ring which is not directly finite. It follows that $M = \mathcal{V}(S_\infty)$ is not cancellative and it is a countable monoid satisfying the hypothesis of Proposition 4.1, because there is a rank function on $\prod_{n=1}^{\infty} M_n(K)$, e.g. the function $N((x_n)) = \sum_{n=1}^{\infty} 2^{-n} N_n(x_n)$, where N_n is the unique rank function on $M_n(K)$. Therefore M gives a counterexample to R(F) for uncountable fields F, although by definition it can be realized over *some* countable field K.

Acknowledgments

This work has been partially supported by the DGI and European Regional Development Fund, jointly, through Project MTM2005-00934, and by the Comissionat per Universitats i Recerca de la Generalitat de Catalunya.

It is a pleasure to thank Gene Abrams, Ken Goodearl, Kevin O'Meara and Enrique Pardo for their helpful comments. I am specially grateful to Fred Wehrung for his many valuable comments and suggestions.

References

1. G. Abrams, G. Aranda Pino, The Leavitt path algebra of a graph, *J. Algebra*, **293** (2005), 319–334.
2. G. Abrams, G. Aranda Pino, Purely infinite simple Leavitt path algebras, *J. Pure Appl. Algebra*, **207** (2006), 553–563.

3. P. ARA, Stability properties of exchange rings, International Symposium on Ring Theory (Kyongju, 1999), 23–42, Trends Math., Birkhäuser Boston, Boston, MA, 2001
4. P. ARA, The regular algebra of a poset, *Preprint.*
5. P. ARA, M. BRUSTENGA, The regular algebra of a quiver, *J. Algebra*, **309** (2007), 207–235.
6. P. ARA, K. R. GOODEARL, E. PARDO, K_0 of purely infinite simple regular rings, *K-Theory*, **26** (2002), 69–100.
7. P. ARA, K. R. GOODEARL, K. C. O'MEARA, E. PARDO, Separative cancellation for projective modules over exchange rings, *Israel J. Math.*, **105** (1998), 105–137.
8. P. ARA, M. A. MORENO, and E. PARDO, Nonstable K-theory for graph algebras, *Algebr. Represent. Theory*, **10** (2007), 157–178.
9. P. ARA, F. PERERA, Multipliers of von Neumann regular rings, *Comm. Algebra*, **28** (2000), 3359–3385.
10. P. ARA, F. PERERA, F. WEHRUNG, Finitely generated antisymmetric graph monoids, *Preprint*, HAL ccsd-00156906.
11. G. M. BERGMAN, Coproducts and some universal ring constructions, *Trans. Amer. Math. Soc.*, **200** (1974), 33–88.
12. G. M. BERGMAN, W. DICKS, Universal derivations and universal ring constructions, *Pacific J. Math.*, **79** (1978), 293–337.
13. G. M. BERGMAN, Von Neumann regular rings with tailor-made ideal lattices. *Unpublished note*, 1986.
14. G. BROOKFIELD, Cancellation in primely generated refinement monoids, *Algebra Universalis*, **46** (2001), 342–371.
15. C. L. CHUANG, P. H. LEE, On regular subdirect products of simple Artinian rings, *Pacific J. Math.*, **142** (1990), 17–21.
16. A. H. CLIFFORD, G. B. PRESTON, "The algebraic theory of semigroups. Vol. I", Mathematical Surveys, No. 7 American Mathematical Society, Providence, R.I. 1961.
17. P. M. COHN, On the free product of associative rings, *Math. Z.*, **71** (1959), 380–398.
18. G. A. ELLIOTT, On the classification of inductive limits of sequences of semisimple finite-dimensional algebras, *J. Algebra*, **38** (1976), 29–44.
19. K. R. GOODEARL, "Partially ordered abelian groups with interpolation", Mathematical Surveys and Monographs, 20. American Mathematical Society, Providence, RI, 1986
20. K. R. GOODEARL, "Von Neumann Regular Rings", Pitman, London 1979; Second Ed., Krieger, Malabar, Fl., 1991.
21. K. R. GOODEARL, K_0 of regular rings with bounded index of nilpotence, Abelian group theory and related topics (Oberwolfach, 1993), 173–199, Contemp. Math., 171, Amer. Math. Soc., Providence, RI, 1994.
22. K. R. GOODEARL, Von Neumann regular rings and direct sum decomposition problems, Abelian groups and modules (Padova, 1994), Math. Appl. 343, 249–255, Kluwer Acad. Publ., Dordrecht, 1995.
23. K. R. GOODEARL, D. E. HANDELMAN, Tensor products of dimension groups

and K_0 of unit-regular rings, *Canad. J. Math.* **38** (1986), 633–658.

24. G. GRÄTZER, "General Lattice Theory, Second Edition", Birkhäuser Verlag, Basel, 1998.
25. G. GRÄTZER, Two problems that shaped a century of lattice theory. *Notices Amer. Math. Soc.* **54** (2007), no. 6, 696–707.
26. K. R. GOODEARL, P. MENAL, Stable range one for rings with many units, *J. Pure Applied Algebra* **54** (1988), 261–287.
27. K. R. GOODEARL, F. WEHRUNG, Representations of distributive semilattices in ideal lattices of various algebraic structures, *Algebra Universalis*, **45** (2001), 71–102.
28. K. R. GOODEARL, F. WEHRUNG, The complete dimension theory of partially ordered systems with equivalence and orthogonality. *Mem. Amer. Math. Soc.*, **176** (2005), no. 831.
29. L. G. W. MACKEY, Note on a theorem of Murray, *Bull. Amer. Math. Soc.*, **52** (1946), 322–325.
30. R. S. PIERCE, Countable Boolean Algebras, in *Handbook of Boolean Algebras*, edited by J. D. Monk with R. Bonnet, Elsevier, 1989, 775–876.
31. I. RAEBURN, Graph algebras, CBMS Reg. Conf. Ser. Math., vol. 103, Amer. Math. Soc., Providence, RI, 2005.
32. P. RŮŽIČKA, Free trees and the optimal bound in Wehrung's theorem, to appear in *Fund. Math.*
33. A. H. SCHOFIELD, "Representations of Rings over Skew Fields", LMS Lecture Notes Series 92, Cambridge Univ. Press, Cambridge, UK, 1985.
34. W. SZYMAŃSKI, The range of K-invariants for C^*-algebras of infinite graphs, *Indiana Univ. Math. J.*, **51** (2002), 239–249.
35. F. WEHRUNG, Non-measurability properties of interpolation vector spaces, *Israel J. Math.*, **103** (1998), 177–206.
36. F. WEHRUNG, Embedding simple commutative monoids into simple refinement monoids, *Semigroup Forum*, **56** (1998), 104–129.
37. F. WEHRUNG, The dimension monoid of a lattice, *Algebra Universalis*, **40** (1998), 247–411.
38. F. WEHRUNG, A uniform refinement property for congruence lattices, *Proc. Amer. Math. Soc.*, **127** (1999), 363–370.
39. F. WEHRUNG, Representation of algebraic distributive lattices with $\aleph_1$ compact elements as ideal lattices of regular rings, *Publicacions Matemàtiques*, **44** (2000), 419-435.
40. F. WEHRUNG, Semilattices of finitely generated ideals of exchange rings with finite stable rank, *Trans. Amer. Math. Soc.*, **356** (2004), 1957-1970.
41. F. WEHRUNG, A K_0-avoiding dimension group with an order-unit of index two, *J. Algebra*, **301** (2006), 728–747.
42. F. WEHRUNG, A solution to Dilworth's congruence lattice problem, *Adv. Math.*, **216** (2007), 610- 625.

SOME PROGRESS ON CLEAN RINGS

JIANLONG CHEN† and ZHOU WANG‡

Department of Mathematics
Southeast University
Nanjing 210096, P. R. China
E-mail: †*jlchen@seu.edu.cn*
‡*zhouwang@seu.edu.cn*

A ring is called clean if each element is the sum of a unit and an idempotent, which was introduced by Nicholson in 1977. Since then, it has attracted many experts in ring theory to do further researches. During the study of clean rings, some related clean ring classes and many challengeable open questions are arisen. In this paper, we give a survey of recent development on clean rings.

Keywords: Clean rings; Strongly clean rings; Uniquely clean rings; Uniquely strongly clean rings.

1. Introduction

If $A = \bigoplus_{i\in I} P_i = \bigoplus_{j\in J} Q_j$, then do these two direct sum decompositions of the module A have isomorphic refinements? A module M_R is said to have the (finite) exchange property if, for any (finite) index set I, whenever M is a direct summand of a direct sum $A = \bigoplus_{i\in I} A_i$, then $A = M \bigoplus(\bigoplus_{i\in I} B_i)$ for some submodule B_i of A_i $(i \in I)$. This notion was introduced by Crawley and Jonsson [18] in 1964, and they showed that modules with the exchange property often have isomorphic refinements for direct sum decompositions. Warfield [41] called a ring R an exchange ring if R_R has the finite exchange property and showed that this definition is left-right symmetric. A module M_R has the finite exchange property iff $End(M_R)$ is exchange. Furthermore, Goodearl [20] and Nicholson [29] obtained independently very useful characterizations that R is exchange iff for any $a \in R$, there exists $e^2 = e \in R$ such that $e \in aR$ and $1 - e \in (1 - a)R$ iff for any $a \in R$, there exists $e^2 = e \in R$ such that $e - a \in (a - a^2)R$.

An element of a ring is called clean if it can be written as the sum of a unit and an idempotent. A ring is clean if each of its elements is clean. This notion was introduced by Nicholson [29] in 1977 as a sufficient condition for a ring to have the exchange property. Since then, various results on this notion have been obtained, and some related clean ring classes are introduced and are researched. In 2004, Nicholson and Zhou [34] gave a survey on clean rings in the Proceeding of the 4th China-Japan-Korea International Conference on Ring Theory. In this paper, we will bring out some new progress about clean rings.

Throughout this paper all rings are associative with identity and all modules are unitary. We denote the group of units of the ring R by $U(R)$, and the Jacobson radical is denoted by $J(R)$.

2. Clean rings

Nicholson [29] showed that every clean ring R is an exchange ring, and the converse holds if all idempotents of R are central. As observed by Camillo and Yu [14], the ring constructed by Bergman (see Example 1 [23]) is an exchange ring which is not clean. Camillo and Yu also proved that a ring R is semiperfect if and only if R is a clean ring containing no infinite set of orthogonal idempotents, and that every unit regular ring is clean. Furthermore, Camillo and Khurana gave a new characterization of unit regular rings.

Theorem 2.1 [8] A ring R is unit regular if and only if every element of R can be written as $a = e + u$ such that $aR \cap eR = 0$, where $e^2 = e \in R$, $u \in U(R)$.

Han and Nicholson [22] showed that the matrix ring $M_n(R)$ is clean for any clean ring R. On the other hand, it was observed in [31] that if V_D is a vector space of countably infinite dimension over a division ring D, then $End(V_D)$ is clean. Later, Nicholson, Varadarajan and Zhou [32] proved that if P_R is a projective module over a right perfect ring R, then $End(P_R)$ is clean. Recently, this result has been extended as the following.

Theorem 2.2 [9] If M_R is discrete or continuous, then $End(M_R)$ is clean.

Let $C(X)$ be the ring of continuous real-valued functions on a complete regular and Hausdorff space X. The cleanness of $C(X)$ is characterized in terms of the topological properties of X (see [4],[27],[28]). Recently, Hager and Kimber have obtained some new results:

Theorem 2.3 [21] Let X be completely regular and Hausdorff and let

$F \leq R$ be a subfield. $C(X,\ F)$ is clean if and only if βX is zero-dimensional.

Recall that a pm-ring is one in which every prime ideal is contained in a unique maximal ideal.

Theorem 2.4 [21] Let $1 \in A$ be a dense clean subring of $\mathbb{R}$ that is not a field and let X be zero-dimensional. The following are equivalent:

(1) $C(X,\ A)$ is clean.
(2) $C(X,\ A)$ is a pm-ring.
(3) X is a P-space.

If G is a group, we denote the group ring of G over R by RG. C_n stands for the cyclic group of order n and C_∞ denotes the infinite cyclic group. It was proved in [22] that if R is a boolean ring and G is a locally finite group, then RG is clean. This is a consequence of the next result, the proof of which used an idea in [17]. Recall that a ring R is strongly regular if, for any $a \in R$, $a \in a^2R$. A ring R is called strongly π-regular if the chain $aR \supseteq a^2R \supseteq \cdots$ terminates for every $a \in R$ (or equivalently, the chain $Ra \supseteq Ra^2 \supseteq \cdots$ terminates for every $a \in R$ by Dischinger [19])

Theorem 2.5 [16] If R is a strongly regular or commutative strongly π-regular ring and G is a locally finite group, then RG is strongly π-regular.

The above result gives an affirmative answer to the question in [22] whether the group ring RG of a locally finite group G over a commutative regular ring R is clean.

Proposition 2.1 [16] Let R be a commutative ring and G be an abelian group. If RG is clean then G is locally finite.

It was proved in [14] that a ring R is semiperfect if and only if R is a clean ring containing no infinite set of orthogonal idempotents. This result can be used to give many examples of clean and non-clean rings. For example, for a finite group G and a prime p, $Z_{(p)}G$ is Noetherian where $\mathbb{Z}_{(p)}$ is the localization of the ring $\mathbb{Z}$ of integers at the prime p, so $Z_{(p)}G$ is clean if and only if it is semiperfect. It is known [22] that if R is semiperfect, then RC_2 is clean. Below we will see that C_2 is the only non-trivial cyclic group having this property.

Example 2.1 [16]

(1) If R is a commutative ring, then RC_∞ is not clean.
(2) If $k \geq 2$, then $\mathbb{Z}_{(5)}C_{2^k}$ is not clean.
(3) If $p \neq 2$ is a prime, then there exists a prime q such that $\mathbb{Z}_{(q)}C_p$ is not clean.

Proposition 2.2 [16] Let $n \geq 2$. The following are equivalent:

(1) RC_n is clean for every semiperfect ring R.
(2) RC_n is clean for every local ring R.
(3) $n = 2$.

If $C = C(R)$ denotes the center of a ring R and $g(x)$ is a polynomial in $C(R)[x]$, Camillo and Simón [10] called a ring $g(x)$-clean if every element is the sum of a unit and a root of $g(x)$, and showed that if V is a vector space of countable infinite dimension over a division ring D, then $End(V_D)$ is $g(x)$-clean provided that $g(x)$ has two distinct roots in $C(D)$. Later, Nicholson and Zhou extended the result to the case when V is any semisimple module (see Theorem 1 [35]).

Theorem 2.6 [39] Let $g(x) = (x-a)(x-b)$ with $a,\ b \in C$. Then R is $g(x)$-clean if and only if R is clean and $b - a \in U(R)$.

The above theorem implies that in some sense the notion of $g(x)$-clean rings in the Nicholson-Zhou Theorem [35] and in the Camillo-Simón Theorem [10] is indeed equivalent to the notion of clean rings. However, for a general $g(x) \in C[x]$, $g(x)$-clean rings need not be clean.

Example 2.2 [39] Let $\mathbb{Z}_{(7)} = \{m/n \in Q \mid m, n \in \mathbb{Z} \text{ and } gcd(7,n) = 1\}$ be a commutative local ring and let C_3 be a cyclic group of order 3. Then $\mathbb{Z}_{(7)}C_3$ is $g(x)$-clean where $g(x) = x^6 - 1 = (x-1)(x+1)(x^4+x^2+1)$ and $2 \in U(\mathbb{Z}_{(7)})$. However, Han and Nicholson [22] showed that $\mathbb{Z}_{(7)}C_3$ is not clean. Thus we obtain an example which is $g(x)$-clean but not clean.

A ring is called n-good if every element is a sum of n units. Wolfson [42] and Zelinsky [45] showed, independently, that every element of the ring of all linear transformations of a vector space over a division ring of characteristic not 2 is 2-good. Later, Henriksen [24] found that the matrix ring $M_n(R)$ is 3-good for any ring R and $n > 1$. Lastly, Vámos [37] proved further that for any ring R, the endomorphism ring of a free R-module of rank at least 2 is 3-good.

On the other hand, Xiao and Tong [43] called a ring n-clean if every element is the sum of an idempotent and n units. The class of these rings contains clean rings and n-good rings. Motivated by the result of Henriksen and Vámos, we obtain the following theorem.

Theorem 2.7 [40] Let R be a ring and let the free R-module F be (isomorphic to) the direct sum of $\alpha \geq 2$ copies of R where α is a cardinal number. Then the ring of endomorphisms of F is 2-clean.

Theorem 2.8 [40] Let R be a ring. Then the $\omega \times \omega$ row and column-finite matrix ring $B(R)$ is 2-clean.

A related question in [31] appeared to be still open:

Question 2.1 [31] For a field F, is the $\omega \times \omega$ row and column-finite matrix ring $B(R)$ clean?

3. Strongly clean rings

In [30], Nicholson defined the notion of strong cleanness. An element of a ring is strongly clean if it is the sum of a unit and an idempotent which commute. A ring is strongly clean if each of its elements is strongly clean. Note that clean and strongly clean rings are the "additive analogs" of unit-regular and strongly regular rings, respectively, because a ring R is unit regular iff every element of R is the product of an idempotent and a unit (in either order) and R is strongly regular iff every element of R is the product of an idempotent and a unit that commute. Local rings are obviously strongly clean. By Burgess and Menal [7], every strongly π-regular ring is strongly clean. In particular, all one-sided perfect rings are strongly clean.

Responding to two questions in [30], it was proved in [38] that $M_2(\mathbb{Z}_{(2)})$ is not strongly clean. This is also proved in [36] where it is shown that if R is strongly clean so also is eRe for any idempotent $e \in R$. Thus, strong cleanness is not a Morita invariant and a semiperfect ring need not be strongly clean. Later, variant results were obtained to show when $M_n(R)$ is strongly clean by many authors in several different contexts.

Theorem 3.1 [13] Let R be a commutative ring. If $M_2(R)$ is strongly clean, then for any $\omega \in J(R)$, $x^2 - x + \omega = 0$ is solvable in R.

In [13], Chen, Yang and Zhou obtained a criterion in terms of solvability of a simple quadratic equation in R for $M_2(R)$ to be strongly clean. Recently, Li [26] has given several more equivalent conditions for the 2×2 matrix ring over a commutative local ring to be strongly clean.

Theorem 3.2 [26] Let R be a commutative local ring. Then the following statements are equivalent:

(1) $M_2(R)$ is strongly clean.
(2) For every $A \in M_2(R)$ with $detA \in J(R)$ and $det(A - I) \in J(R)$, the characteristic equation of A, $x^2 - (trA)x + detA = 0$ is solvable in R.
(3) For every $A \in M_2(R)$ with $detA \in J(R)$ and $det(A - I) \in J(R)$, the equation $\frac{x^2 - x - detA}{(trA)^2 - 4detA} = 0$ is solvable in R.
(4) Every $A \in M_2(R)$ with $detA \in J(R)$ and $det(A - I) \in J(R)$ is diagonalizable in $M_2(R)$.
(5) For every $\omega \in J(R)$, the matrix $B = \begin{pmatrix} 1 & 1 \\ -\omega & 0 \end{pmatrix}$ is strongly clean.
(6) For every $\omega \in J(R)$, the equation $x^2 - x + \omega = 0$ is solvable in R.
(7) For every $\omega \in J(R)$ and $u \in U(R)$, the equation $x^2 - ux + \omega = 0$ is solvable in R.

Theorem 3.3 [13] Let R be a commutative local ring and $n \geq 1$. The following are equivalent:

(1) $M_2(R)$ is strongly clean.
(2) $M_2(R[[x]])$ is strongly clean.
(3) $M_2(R[x]/(x^n))$ is strongly clean.
(4) $M_2(RC_2)$ is strongly clean.

Theorem 3.4 [14] Let $R = \hat{\mathbb{Z}}_p$ be the ring of p-adic integers where p is a prime. Then $M_2(R)$ is strongly clean.

ω denotes a complex number such that $\omega^2 \in \mathbb{Z}$ and $\omega \in \mathbb{Q}$, and let $\mathbb{Z}(\omega) = \{n + m\omega : n, m \in \mathbb{Z}\}$. Then $\mathbb{Z}(\omega)$ is a domain and the representation $n + m\omega$ of elements of $\mathbb{Z}(\omega)$ is unique. The quotient field of a commutative domain R is denoted $\mathbb{Q}_c(R)$.

Theorem 3.5 [14] Suppose that $\mathbb{Z}(\omega)$ is a UFD. Let R be a subring of $\mathbb{Q}_c(\mathbb{Z}(\omega))$ such that $S \subseteq R \subseteq \mathbb{Q}_c(S)$ for some subring S of $\mathbb{Z}(\omega)$. Then the following are equivalent:

(1) $R = \mathbb{Q}_c(R)(= \mathbb{Q}_c(S))$.
(2) $M_n(R)$ is strongly clean for all $n \geq 1$.
(3) $M_n(R)$ is strongly clean for some $n > 1$.
(4) $M_2(R)$ is strongly clean.

Theorem 3.6 [14] Let S be a commutative domain of characteristic $\neq 2$ and R be a subring of $\mathbb{Q}_c(S[x])$ with $S[x] \subseteq R$. Then the following are equivalent:

(1) $R = \mathbb{Q}_c(R)(= \mathbb{Q}_c(S[x]))$.
(2) $M_n(R)$ is strongly clean for all $n \geq 1$.
(3) $M_n(R)$ is strongly clean for some $n > 1$.
(4) $M_2(R)$ is strongly clean.

More general, several necessary and sufficient conditions are obtained for $M_n(R)$ over a commutative local ring to be strongly clean by Borooah, Diesl and Dorsey [5]. Let R be a commutative local ring. A factorization $h(t) = h_0(t)h_1(t)$ in $R[t]$ of a monic polynomial $h(t)$ is said to be an SRC factorization if $h_0(0), h_1(1) \in U(R)$ and $h_0(t), h_1(t)$ are co-prime in the PID $\overline{R}[t](= R/J(R)[t])$. The ring R is an n-SRC ring if every monic polynomial of degree n in $R[t]$ has an SRC factorization.

Theorem 3.7 [5] For a commutative local ring R and $n \geq 1$, the following are equivalent:

(1) $M_n(R)$ is strongly clean.
(2) Every companion matrix in $M_n(R)$ is strongly clean.
(3) R is an n-SRC ring.
(4) For every $x_0, x_1 \in R$ with $x_0 - x_1 \in U(R)$, every monic polynomial $h \in R[t]$ of degree n has an SRC factorization relative to (x_0, x_1).
(5) For every $x_0, \cdots, x_k \in R$, with $x_i - x_j \in U(R)$ whenever $i \neq j$, every monic polynomial $h \in R[t]$ of degree n has an SRC factorization relative to $(x_0, x_1, \cdots, x_k)$.

Theorem 3.8 [44] Let R be a commutative local ring and let $n,\ k \in N$. Then the following are equivalent:

(1) $M_n(R)$ is strongly clean.
(2) $M_n(R[[x]])$ is strongly clean.
(3) $M_n(\frac{R[x]}{(x^k)})$ is strongly clean.
(4) $M_n(R[[x_1, x_2, \cdots, x_k]])$ is strongly clean.
(5) $M_n(\frac{R[x_1, x_2, \cdots, x_k]}{(x_1^{n_1}, x_2^{n_2}, \cdots, x_k^{n_k})})$ is strongly clean.

Next, Chen, Yang and Zhou identify new families of strongly clean rings through upper triangular matrix rings.

Theorem 3.9 [14] If R is a commutative local ring, then $T_n(R)$ is a strongly clean ring for every $n \geq 1$.

The above theorem is also proved by Borooah, Diesl and Dorsey [6] independently.

Corollary 3.1 [14] Let R be a commutative semilocal ring. Then the following are equivalent:

(1) R is a semiperfect.
(2) $T_n(R)$ is strongly clean for every $n \geq 1$.
(3) $T_n(R)$ is strongly clean for some $n \geq 1$.

Recall skew triangular matrix ring over R: For a ring R and an endomorphism σ of R, let $T_n(R, \sigma) = \{(a_{ij})_{n\times n} : a_{ij} \in R \text{ and } a_{ij} = 0 \text{ if } i > j\}$. For $(a_{ij}), (b_{ij}) \in T_n(R, \sigma)$, define $(a_{ij})+(b_{ij}) = (a_{ij}+b_{ij})$ and $(a_{ij})*(b_{ij}) = (c_{ij})$, where $c_{ij} = 0$ for $i > j$, and $c_{ij} = \Sigma_{k=i}^{j} a_{ik}\sigma^{k-i}(b_{kj})$ for $i \leq j$.

Theorem 3.10 [14] If R is a commutative local ring and σ is an endomorphism of R with $\sigma(J(R)) \subseteq J(R)$ then $T_n(R, \sigma)$ is a strongly clean ring for every $n \geq 1$.

Recall that an element $a \in R$ is called strongly π-regular if both chains $aR \supseteq a^2R \supseteq \cdots$ and $Ra \supseteq Ra^2 \supseteq \cdots$ terminate, or equivalently, there exist $e^2 = e \in R$ and $u \in U(R)$ and $n \geq 1$ such that $a^n = eu = ue$. By Nicholson [30], every strongly π-regular element is strongly clean.

If R is a local ring, then $R[[x]]$ is local and hence is strongly clean. But for any ring R, $R[[x]]$ is not strongly π-regular because x is never a strongly π-regular element in $R[[x]]$. If R is a ring and σ is a ring endomorphism of R, let $R[[x, \sigma]]$ denote the ring of skew formal power series over R (with multiplication defined by $xr = \sigma(r)x$ for all $r \in R$).

Theorem 3.11 [16] Let R be a ring and $r = \Sigma r_i x^i \in R[[x, \sigma]]$. If either r_0 or $1 - r_0$ is a strongly π-regular element of R, then r is a strongly clean element of $R[[x, \sigma]]$.

Corollary 3.2 [16] If R is a strongly π-regular ring and σ is an endomorphism of R, then $R[[x,\sigma]]$ is a strongly clean ring.

Theorem 3.12 [16] Let σ be an endomorphism of R, $n \geq 1$ and $r = \Sigma_{i=0}^{n-1} r_i x^i \in R[x,\sigma]/(x^n)$. If r_0 or $1 - r_0$ is strongly π-regular in R, then r is strongly clean in $R[x,\sigma]/(x^n)$.

Corollary 3.3 [16] If R is a strongly π-regular ring and σ is an endomorphism of R, then $R[x,\sigma]/(x^n)$ is strongly clean for all $n \geq 1$.

A ring R is said to have stable range 1 if, whenever $aR + bR = R$ where $a, b \in R$, $a + by$ is a unit for some $y \in R$. A ring R is called directly finite if $ab = 1$ in R always implies $ba = 1$. Every unit regular ring is clean by [14], and every strongly $\pi-$regular ring has stable range one [2] and is directly finite. But the following questions, all raised in [30], still remain open.

Question 3.1 Does every strongly clean ring have stable range one?

Question 3.2 Is every strongly clean ring directly finite?

Question 3.3 Is every unit regular ring strongly clean?

4. Uniquely clean rings

A ring R is called uniquely clean if every element of R has a unique representation as the sum of a unit and an idempotent. Uniquely clean rings were first considered by Anderson and Camillo [1] in the commutative case, where it was showed that any commutative clean ring R with $R/M \cong \mathbb{Z}_2$ for each maximal ideal M of R is uniquely clean, so a commutative local ring is uniquely clean if and only if $R/J(R) \cong \mathbb{Z}_2$. A study of noncommutative uniquely clean rings was carried out by Nicholson and Zhou [33] and the following were proved.

Theorem 4.1 [33] A local ring R is uniquely clean if and only if $R/J(R) \cong \mathbb{Z}_2$.

Theorem 4.2 [33] A ring R is uniquely clean iff for all $a \in R$ there exists a unique idempotent $e \in R$ such that $e - a \in J(R)$ if and only if $R/J(R)$ is boolean and idempotents lift uniquely modulo $J(R)$.

Theorem 4.3 [33] Every image of a uniquely clean ring is again uniquely clean.

We denote the augmentation ideal of RG by $\Delta(RG)$ and it is well known that $RG/\Delta(RG) \cong R$. The group G is called a torsion group if every element has finite order. If p is a prime, g is called a p-torsion element if $o(g) = p^k$ for some $k \geq 0$ and G is called a p-group if every element is p-torsion. A ring R (or a group G) is called locally finite if every finitely generated subring (subgroup) is finite.

Example 4.1 [12] Let R be a ring and let $G \neq \{1\}$ be a locally finite group in which every finite subgroup has odd order. Then RG is not uniquely clean.

Lemma 4.1 [12] If R is any ring and G is a locally finite group then $J(R)G \subseteq J(RG)$.

Theorem 4.4 [12] Let R be a ring and let G be a group. If RG is uniquely clean then R is uniquely clean and G is a 2-group.

If we assume that G is locally finite, we get a converse to Theorem 4.4. The proof requires a number of preliminary lemmas. A family of subsets $\{X_i | i \in I\}$ of a set X is called a direct cover of X if $X = \cup_{i\in I} X_i$ and, for any i, $j \in I$, there exists $k \in I$ such that $X_i \cup X_j \subseteq X_k$. Clearly a locally finite ring (group) is directly covered by its finite subrings (finite subgroups).

Lemma 4.2 [12] Let R be a ring and let G be a group. If $R = \cup R_i$ and $G = \cup G_j$ are direct covers of subrings and subgroups, respectively, and if $R_i G_j$ is clean (uniquely clean) for all i and j, then RG is clean (uniquely clean).

Lemma 4.3 [12] If G is a finite 2-group then $\mathbb{Z}_2 G$ is a local, uniquely clean ring and $J(\mathbb{Z}_2 G) = \Delta(\mathbb{Z}_2 G)$.

Lemma 4.4 [12] If R is a boolean ring and G is a locally finite group, then RG is uniquely clean if and only if G is a 2-group.

Lemma 4.5 [12] If $R/J(R)$ is boolean, G is a locally finite 2-group, and $\omega : RG \to R$ is the augmentation map, then $J(RG) = \{x \in RG | \omega(x) \in$

$J(R)\}$.

If G is a group, let G' denote the derived subgroup of G. Write $G^{(0)} = G$, $G^{(1)} = G'$, and, for each $i \geq 0$, define $G^{(i+1)} = (G^{(i)})'$. Then the series $G = G^{(0)} \supseteq G^{(1)} \supseteq G^{(2)} \supseteq \cdots$ is called the derived series of G, and G is called solvable if $G^{(n)} = \{1\}$ for some n. With this the following theorem is obtained.

Theorem 4.5 [12] If R is a ring and G is a locally finite group, then RG is uniquely clean if and only if R is uniquely clean and G is a 2-group.

Theorem 4.6 [12] Let G be a solvable group, and let R be a ring. Then RG is uniquely clean if and only if R is uniquely clean and G is a 2-group.

Question 4.1 Is the converse of Theorem 4.4 true?

5. Uniquely strongly clean rings

An element a of a ring R is called uniquely strongly clean (or USC for short) if $a = e + u$ where $e^2 = e$, $u \in U(R)$ and $eu = ue$, and the representation is unique. The ring R is called uniquely strongly clean if every element of R is uniquely strongly clean. A ring R is called abelian if all idempotents of R are central. In [33], Nicholson and Zhou showed that every uniquely clean ring is abelian. Thus the following fact is observed: a ring R is uniquely clean if and only if R is an abelian USC ring.

Let R be a ring and $e^2 = e \in R$. If R is USC, then so is eRe. However, for any $n > 1$, no corner ring of a USC ring is isomorphic to an $n \times n$ matrix ring. In particular, any $n \times n$ matrix ring is never USC for $n > 1$. Thus, a USC ring is not a Morita invariant. By [14], $\mathbb{M}_2(\hat{\mathbb{Z}}_p)$ is strongly clean where p is a prime, but it is not USC by the above statement.

Example 5.1 [15] Let $R = \mathbb{Z}_2 \times \mathbb{Z}_2$ and let $\sigma : R \to R$ be given by $(a, b) \mapsto (b, a)$. Then σ is an endomorphism of R. Let $R[x; \sigma]$ be the ring of skew polynomials over R with multiplication defined by $xr = \sigma(r)x$. Then $R[x; \sigma]/(x^2)$ is a USC ring that is not uniquely clean.

The next result provides a family of USC rings through triangular matrix rings.

Theorem 5.1 [15] Let R be a commutative ring. The following are equivalent:

(1) R is USC.
(2) $\mathbb{T}_n(R)$ is USC for all $n \geq 1$.
(3) $\mathbb{T}_n(R)$ is USC for some $n \geq 1$.

By [33], $\mathbb{T}_n(R)$ is never uniquely clean for $n \geq 2$. Thus, a family of USC rings that are not uniquely clean are obtained. Let $R = \mathbb{T}_2(\mathbb{Z}_2)$. Then $\mathbb{T}_2(R)$ is USC. Noting that the ring R is USC but not abelian, one raises the following:

Question 5.1 [15] In Theorem 5.1, does condition (1) imply condition (2) without the assumption that R is commutative? If not, can the word "commutative" be replaced by "abelian"?

A set $\{e_{ij} : 1 \leq i, j \leq n\}$ of nonzero elements of R is said to be a system of n^2 matrix units if $e_{ij}e_{st} = \delta_{js}e_{it}$, where $\delta_{jj} = 1$ and $\delta_{js} = 0$ for $j \neq s$. In this case, $e := \Sigma_{i=1}^{n} e_{ii}$ is an idempotent of R and $eRe \cong \mathbb{M}_n(S)$ where $S = \{r \in eRe : re_{ij} = e_{ij}r \text{ for all } i, j = 1, 2, \cdots, n\}$.

A ring R is called a semipotent ring if every right ideal not contained in $J(R)$ contains a nonzero idempotent, or equivalently if every left ideal not contained in $J(R)$ contains a nonzero idempotent.

Lemma 5.1 [25] Let R be a semipotent ring with $J(R) = 0$ and $n > 1$. If $a^n = 0$ but $a^{n-1} \neq 0$, then the ideal RaR of R contains a system of n^2 matrix units.

Theorem 5.2 [15] A ring R is USC with $J(R) = 0$ if and only if R is boolean.

Theorem 5.3 [15] The following are equivalent for a ring R:

(1) R is USC.
(2) For any $a \in R$, there exists a unique $e^2 = e \in R$ such that $ea = ae$ and $a - e \in J(R)$.

It is an unsolved question whether a strongly clean ring has stable range one (see Question 3.1). But if R is USC, then $\overline{R}$ has stable range one (being boolean), and so R has stable range one.

Question 5.2 [15] Is the image of a USC ring again USC?

Theorem 5.4 [15] Let R be a ring and $r = \Sigma r_i x^i \in R[[x, \sigma]]$. Suppose that r_0 or $1 - r_0$ is a strongly π-regular element of R. If r_0 is a USC element of R, then r is a USC element of $R[[x, \sigma]]$.

If $r_0 \in R$ is idempotent or nilpotent, then r_0 is both USC and strongly π-regular. So the next corollary follows immediately.

Corollary 5.1 [15] If R is a boolean ring, then $\mathbb{T}_n(R)[[x, \sigma]]$ is USC for all $n \geq 1$.

Corollary 5.2 [15] If R is a uniquely clean ring with nil Jacobson radical, then $R[[x, \sigma]]$ is USC.

Proposition 5.1 [15] If the group ring RG is USC, then R is USC and G is a 2-group.

If R is a strongly regular or commutative strongly π-regular ring and if G is a locally finite group, then RG is strongly clean (strongly π-regular indeed) (see Theorem 2.5). Thus, a family of strongly clean group rings that are not USC are obtained by choosing G to be locally finite groups that are not 2-groups.

It was proved in [12] that if R is a uniquely clean ring and G is a locally finite 2-group, then the group ring RG is uniquely clean. We have been unable to prove the analog for USC group rings. But, there do exist examples of USC group rings that are not uniquely clean.

Example 5.2 [15] If R is a commutative uniquely clean ring and if G is an abelian 2-group, then $\mathbb{T}_n(R)G \cong \mathbb{T}_n(RG)$ is USC for all $n \geq 1$.

The following example indicates that the abelianness of the group G is not necessary for RG to be USC.

Example 5.3 [15] Let $D_4 = \langle h, g : h^2 = 1, g^4 = 1, hg = g^3 h\rangle$ be the dihedral group of order 8. Then $\mathbb{T}_2(\mathbb{Z}_2)D_4$ is a USC ring.

Acknowledgments

The work was supported by the National Natural Science Foundation of

China (No.10571026) and the Specialized Research Fund for the Doctoral Program of Higher Education (20060286006).

References

1. D. D. Anderson and V. P. Camillo, Commutative rings whose elements are a sum of a unit and an idempotent, *Comm. Algebra* **30**(2002), 3327-3336.
2. P. Ara, Strongly π-regular rings have stable range one, *Proc. Amer. Math. Soc.* **124**(1996), 3293-3298.
3. P. Ara, W. R. Goodearl, K. C. O'Meara and E. Pardo, Separative cancellation for projective modules over exchange rings, *Israel J. Math.* **105**(1998), 105-137.
4. F. Azarpanah, When is $C(X)$ clean ring? *Acta. Math. Hunger.* **94**(2002), 53-58.
5. G. Borooah, A. J. Diesl and T. J. Dorsey, Strongly clean matrix rings over commutative local rings, *J. Pure Appl. Algebra* **212**(2008), 281-296.
6. G. Borooah, A. J. Diesl and T. J. Dorsey, Strongly clean triangular matrix rings, *J. Algebra,* **312**(2007), 773-797.
7. W. D. Burgess and P. Menal, On strongly π-regular rings and homomorphisms into them, *Comm. Algebra* **16**(1988), 1701-1725.
8. V. P. Camillo and D. Khurana, A characterization of unit regular rings, *Comm. Algebra* **29**(2001), 2293-2295.
9. V. P. Camillo, D. Khurana, T. Y. Lam, W. K. Nicholson and Y. Zhou, Continuous modules are clean, *J. Algebra* **304**(2006), 94-111.
10. V. P. Camillo and J. J. Simón, The Nicholson-Varadarajan theorem on clean linear transformations, *Glasgow Math. J.* **44**(2002), 365-369.
11. V. P. Camilo and H. P. Yu, Exchange rings, units and idempotents, *Comm. Algebra* **22**(1994), 4737-4749.
12. J. L. Chen, W. K. Nicholson and Y. Q. Zhou, Group rings in which every element is uniquely the sum of a unit and an idempotent, *J. Algebra* **306**(2006), 453-460.
13. J. L. Chen, X. D. Yang and Y. Q. Zhou, When is the 2×2 matrix ring over a commutative local ring strongly clean? *J. Algebra* **301**(2006), 280-293.
14. J. L. Chen, X. D. Yang and Y. Q. Zhou, On strongly clean matrix and triangular matrix rings, *Comm. Algebra* **34**(2006), 3659-3674
15. J. L. Chen, Z. Wang and Y. Q. Zhou, Rings in which elements are uniquely the sum of an idempotent and a unit which commute, *J. Pure Appl. Algebra,* to appear.
16. J. L. Chen and Y. Q. Zhou, Strongly clean power series rings, *Proc. Edinburgh Math. Soc.* **50**(2007), 73-85.
17. A. Chin and H. Chen, On strongly π-regular group rings, *Southeast Asian Bull. Math.* **26**(2002), 387-390.
18. P. Crawely and B. Jónsson, Refinements for infinite direct decompositions of algebraic systems, *Pacific J. Math.* **14**(1964), 797-855.
19. M. F. Dischinger, Sur les anneaux fortement réguliers, *C.R. Acad. Sci. Paris* **283**(1976), 571-573.

20. K. R. Goodearl and R. B. Warfield, Algebras over zero-dimensional rings, *Math. Ann.* **223**(1976), 157-168.
21. A. W. Hager and C. M. Kimber, Clean rings of continuous functions, *Algebra Univers.* **56**(2007), 77-92.
22. J. Han and W. K. Nicholson, Extensions of clean rings, *Comm. Algebra* **29**(2001), 2589-2595.
23. D. Handelman, Perspectivity and cancellation in regular rings, *J. Algebra* **48**(1977), 1-16.
24. M. Henriksen, Two classes of rings generated by their units, *J. Algebra* **31**(1974), 182-193.
25. J. Levitzki, On the structure of algebraic algebras and related rings. *Trans. Amer. Math. Soc.* **74**(1953), 384-409.
26. Y. L. Li, Strongly clean matrix rings over local rings, *J. Algebra* **312**(2007), 397-404.
27. W. W. McGovern, Clean semiprime f-rings with bounded inversion, *Comm. Algebra* **31**(2003), 3295-3304.
28. W. W. McGovern, Neat rings, *J. Algebra* **205**(2005), 243-265.
29. W. K. Nicholson, Lifting idempotents and exchange rings, *Trans. Amer. Math. Soc.* **229**(1977), 269-278.
30. W. K. Nicholson, Strongly clean rings and Fitting's Lemma, *Comm. Algebra* **28**(1999), 3583-3592.
31. W. K. Nicholson and K. Varadarajan, Countable linear transformations are clean, *Proc. Amer. Math. Soc.* **126**(1998), 61-64.
32. W. K. Nicholson, K. Varadarajan and Y. Q. Zhou, Clean endomorphism rings, *Arch. Math.* **83**(2004), 340-343.
33. W. K. Nicholson and Y. Zhou, Rings in which elements are uniquely the sum of an idempotent and a unit, *Glasgow Math. J.* **46**(2004), 227-236.
34. W. K. Nicholson and Y. Zhou, Clean rings: a survey, *Advances in Ring Theory*, Proc. of the 4th China-Japan-Korea Inter. Conf. 2005, 181-198.
35. W. K. Nicholson and Y. Q. Zhou, Endomorphisms that are the sum of a unit and a root of a fixed polynomial, *Canad. Math. Bull.* **49**(2006), 265-269.
36. E. Sánchez Campos, On strongly clean rings, unpublished.
37. P. Vámos, 2-good rings, *Quart. J. Math.* **56**(2005), 417-430.
38. Z. Wang and J. L. Chen, On two open problems about strongly clean rings, *Bull. Austrl. Math, Soc.* **70**(2004), 279-282.
39. Z. Wang and J. L. Chen, A note on clean rings, *Algebra Colloq.* **14**(2007), 537-540.
40. Z. Wang and J. L. Chen, 2-clean rings, *Canad. Math. Bull.* to appear
41. R. B. Warfield, Exchange rings and decompositions of modules, *Math. Ann.* **22**(1972), 31-36.
42. K. G. Wolfson, An ideal theoretic characterization of the ring of all linear transformations, *Amer. J. Math.* **75**(1953), 358-386.
43. G. S. Xiao and W. T. Tong, n-clean rings and weakly unit stable range rings, *Comm. Algebra* **33**(2005), 1501-1517.
44. X. D. Yang and Y. Q. Zhou, Some families of strongly clean rings, *Linear Algebra and its Appl.* **425**(2007), 119-129.

45. D. Zelinsky, Every linear transformation is a sum of nonsingular ones, *Proc. Amer. Math. Soc.* **5**(1954) 627-630.

Auslander-type Conditions

ZHAOYONG HUANG

Department of Mathematics, Nanjing University, Nanjing 210093, P. R. China
E-mail: huangzy@nju.edu.cn

We survey some recent results on Noetherian rings satisfying the Auslander-type conditions, with emphasis on the homological behavior of such rings.

1. Introduction

Throughout this article, Λ is a left and right Noetherian ring (unless stated otherwise), $\operatorname{mod}\Lambda$ is the category of finitely generated left Λ-modules and

$$0 \to \Lambda \to I_0(\Lambda) \to I_1(\Lambda) \to \cdots \to I_i(\Lambda) \to \cdots$$

is the minimal injective resolution of Λ as a left Λ-module.

For a module $M \in \operatorname{mod}\Lambda$ and a non-negative integer n, recall that the *grade* of M, denoted by $\operatorname{grade} M$, is said to be at least n if $\operatorname{Ext}^i_\Lambda(M, \Lambda) = 0$ for any $0 \leq i < n$; and the *strong grade* of M, denoted by $\operatorname{s.grade} M$, is said to be at least n if $\operatorname{grade} X \geq n$ for any submodule X of M (see [3] and [7]). Bass in [8] proved the following result.

Theorem 1.1. *For a commutative Noetherian ring Λ, the following statements are equivalent:*

(1) The self-injective dimension of Λ is finite.
(2) The flat dimension of $I_i(\Lambda)$ is at most i for any $i \geq 0$.
(3) $\operatorname{grade}\operatorname{Ext}^i_\Lambda(M, \Lambda) \geq i$ for any $M \in \operatorname{mod}\Lambda$ and $i \geq 1$.

A commutative Noetherian ring Λ is called *Gorenstein* if it satisfies one of the above equivalent conditions. For the non-commutative case, Λ is also called *Gorenstein* if the left and right self-injective dimensions of Λ are finite. It was proved by [44, Lemma A] that the left and right self-injective dimensions of a Gorenstein ring are identical.

Theorem 1.2. *[15, Auslander's Theorem 3.7] The following statements are equivalent:*

(1) *The flat dimension of* $I_i(\Lambda)$ *is at most* i *for any* $0 \leq i \leq n-1$.
(2) $\operatorname{s.grade}\operatorname{Ext}^i_\Lambda(M,\Lambda) \geq i$ *for any* $M \in \operatorname{mod}\Lambda$ *and* $1 \leq i \leq n$.
(3) *The flat dimension of* $I_i(\Lambda^{op})$ *is at most* i *for any* $0 \leq i \leq n-1$.
(4) $\operatorname{s.grade}\operatorname{Ext}^i_\Lambda(N,\Lambda) \geq i$ *for any* $N \in \operatorname{mod}\Lambda^{op}$ *and* $1 \leq i \leq n$.

Λ is called *n-Gorenstein* if it satisfies one of the above equivalent conditions, and Λ is said to satisfy the *Auslander condition* if it is n-Gorenstein for all n. Theorem 1.2 means that the notion of n-Gorenstein rings (and hence that of the Auslander condition) is left-right symmetric. Motivated by Theorem 1.2, the notion of the Auslander-type conditions was introduced in [24] as follows.

Definition 1.1. [24] Let $n, k \geq 0$. We say that Λ is $G_n(k)$ if $\operatorname{s.grade}\operatorname{Ext}^{i+k}_\Lambda(M,\Lambda) \geq i$ for any $M \in \operatorname{mod}\Lambda$ and $1 \leq i \leq n$. Similarly, we say that Λ is $g_n(k)$ if $\operatorname{grade}\operatorname{Ext}^{i+k}_\Lambda(M,\Lambda) \geq i$ for any $M \in \operatorname{mod}\Lambda$ and $1 \leq i \leq n$. We say that Λ is $G_n(k)^{op}$ (resp. $g_n(k)^{op}$) if Λ^{op} is $G_n(k)$ (resp. $g_n(k)$). We call both $G_n(k)$ and $g_n(k)$ the *Auslander-type conditions.*

The following relations are obvious for any $n \geq n'$ and $k \leq k'$:

$$\begin{array}{ccc} G_n(k) & \Rightarrow & G_{n'}(k') \\ \Downarrow & & \Downarrow \\ g_n(k) & \Longrightarrow & g_{n'}(k') \end{array}$$

The Auslander-type conditions can be regarded as certain non-commutative analogs of commutative Gorenstein rings. Such conditions, especially dominant dimension and the n-Gorenstein ring, play a crucial role in representation theory and non-commutative algebraic geometry (e.g. [2, 6, 7, 10, 12, 13, 15, 16, 18, 24, 28, 31, 32, 33, 34, 35, 36, 38, 40, 41, 42]). They are also interesting from the viewpoint of some unsolved homological conjectures, e.g. the finitistic dimension conjecture, Nakayama conjecture, Gorenstein symmetry conjecture, and so on. In this article, we survey some recent results on Noetherian rings satisfying the Auslander-type conditions, with emphasis on the homological behavior of such rings. In Section 2, we give some equivalent characterizations of the conditions $G_n(k)$ and $g_n(k)$, respectively. In Section 3, we investigate the properties of rings being $G_n(k)$ or $g_n(k)$, especially for the case $k = 0, 1$. In Section 4, we give the definition of the (weak) (m, n)-condition, which is closely related to the conditions $G_n(k)$ and $g_n(k)$. Then we investigate the properties of rings satisfying certain (m, n)-condition.

2. Characterizations of the Auslander-type conditions

In this section, we give some equivalent characterizations of the Auslander-type conditions $G_n(k)$ and $g_n(k)$, respectively.

Put $(\ )^* = \mathrm{Hom}_\Lambda(\ ,\Lambda)$ and $\mathbb{E}_n = \mathrm{Ext}^n_\Lambda(\ ,\Lambda)$ for any $n \geq 0$. Let $M \in \mathrm{mod}\,\Lambda$ and $P_1 \to P_0 \to M \to 0$ be a projective resolution of M in $\mathrm{mod}\,\Lambda$. Then $\mathrm{Coker}(P_0^* \to P_1^*)$ is called the *transpose* of M, and denoted by $\mathrm{Tr}\,M$ (see [3]).

Definition 2.1. [3] Let $k \geq 1$ and M be a module in $\mathrm{mod}\,\Lambda$. M is called *k-torsionfree* if $\mathbb{E}_i(\mathrm{Tr}\,M) = 0$ for any $1 \leq i \leq k$; and M is called *k-syzygy* if there exists an exact sequence $0 \to M \to P_{k-1} \to \cdots \to P_1 \to P_0$ in $\mathrm{mod}\,\Lambda$ with all P_i projective.

We use $\mathcal{F}_k$ to denote the subcategory of $\mathrm{mod}\,\Lambda$ consisting of k-torsionfree modules, and $\Omega^k(\mathrm{mod}\,\Lambda)$ to denote the subcategory of $\mathrm{mod}\,\Lambda$ consisting of k-syzygy modules. It is well known that that $\mathcal{F}_k \subseteq \Omega^k(\mathrm{mod}\,\Lambda)$ for any $k \geq 1$ (see [3, Theorem 2.17]).

For subcategories $\mathcal{C}_i$ $(i = 1, 2)$ of $\mathrm{mod}\,\Lambda$, we use $\mathcal{E}(\mathcal{C}_1, \mathcal{C}_2)$ to denote the subcategory of $\mathrm{mod}\,\Lambda$ consisting of $C \in \mathrm{mod}\,\Lambda$ such that there exists an exact sequence $0 \to C_2 \to C \to C_1 \to 0$ with $C_i \in \mathcal{C}_i$ $(i = 1, 2)$. For a left Λ-module M, we use $\mathrm{pd}\,M$, $\mathrm{fd}\,M$ and $\mathrm{id}\,M$ to denote the projective dimension, flat dimension and injective dimension of M, respectively.

The following result gives some equivalent characterizations of $G_n(k)$.

Theorem 2.1. *[24, 2.4(7) and Theorem 3.5] The conditions (1)–(3) are equivalent for any $n, k \geq 0$. If $k \geq 1$, then (1)–(4) are equivalent:*

(1) Λ is $G_n(k)$.
(2) $\mathrm{fd}\,I_i(\Lambda^{op}) \leq i + k$ for any $0 \leq i \leq n-1$.
(3) For any exact sequence $0 \to A \xrightarrow{f} B \to C \to 0$ with $C \in \Omega^k(\mathrm{mod}\,\Lambda)$, $\mathbb{E}_i\mathbb{E}_i(f)$ is a monomorphism for any $0 \leq i \leq n-1$.
(4) $\mathcal{E}(\Omega^{i+k}(\mathrm{mod}\,\Lambda), \Omega^{i+k}(\mathrm{mod}\,\Lambda)) \subseteq \mathcal{F}_{i+1}$ for any $0 \leq i \leq n-1$.

Remark 2.1. $G_n(0)$ (resp. $G_\infty(0)$) is just the n-Gorenstein ring (resp. the Auslander condition).

Let $k \geq 1$. We denote by

$$\mathcal{W}_k = \{M \in \mathrm{mod}\,\Lambda \mid \mathbb{E}_i(M) = 0 \text{ for any } 1 \leq i \leq k\}$$

and

$$\mathcal{P}_k = \{M \in \mathrm{mod}\,\Lambda \mid \mathrm{pd}\,M < k\}.$$

For a module $M \in \operatorname{mod}\Lambda$, $\Omega^k(M)$ denotes a k-syzygy module of M.

The following result gives some equivalent characterizations of $g_n(k)$, where (1) $\Leftrightarrow$ (3) for the case $k = 1$ is [3, Proposition 2.26].

Theorem 2.2. *[24, Theorem 3.4] The conditions (1) and (2) are equivalent for any $n, k \geq 0$. If $k \geq 1$, then (1)–(5) are equivalent:*

(1) Λ is $g_n(k)$.

(2) For any monomorphism $A \xrightarrow{f} B$ with $A, B \in \Omega^{k+1}(\operatorname{mod}\Lambda)$, $\mathbb{E}_i\mathbb{E}_i(f)$ is a monomorphism for any $0 \leq i \leq n-1$.

(3) $\Omega^{i+k}(\operatorname{mod}\Lambda) \subseteq \mathcal{F}_{i+1}$ for any $1 \leq i \leq n$.

(4) For any $C \in \operatorname{mod}\Lambda$ and $0 \leq i \leq n$, there exists an exact sequence $0 \to Y \to X \to \Omega^{k-1}(C) \to 0$ with $X \in \mathcal{W}_{i+1}$ and $Y \in \mathcal{P}_{i+1}$.

(5) For any $C \in \operatorname{mod} R$ and $0 \leq i \leq n$, there exists an exact sequence $0 \to \Omega^k(C) \to Y' \to X' \to 0$ with $X' \in \mathcal{W}_{i+1}$ and $Y' \in \mathcal{P}_{i+1}$.

Now we concentrate on the conditions $G_n(k)$ and $g_n(k)$ for the case $k = 0, 1$.

In [24, Theorem 4.1], we gave a quick proof of the following remarkable left-right symmetry of $G_n(k)$ and $g_n(k)$ for the case $k = 0, 1$, where (1) is in Theorem 1.2, (2) is in [18, Theorem 4.7] and [23, Theorem 2.4], and (3) is in [7, Theorem 0.1] and [18, Theorem 4.1].

Theorem 2.3. *(Symmetry)*

(1) $G_n(0) \Leftrightarrow G_n(0)^{op}$.

(2) $g_n(1) \Leftrightarrow g_n(1)^{op}$.

(3) $G_n(1) \Leftrightarrow g_n(0)^{op}$.

Question 2.1. [24, Question 4.1.1] It is natural to ask for the existence of a common generalization of the conditions $G_n(k)$ and $g_n(k)$ satisfying certain "left-right symmetry". For example, does there exist some natural condition $G_n(k,l)$ for each triple (n,k,l) of non-negative integers with the following properties?

(i) $G_n(k,0) = G_n(k)$, and $G_n(k,1) = g_n(k)$.

(ii) $G_n(k,l) \Leftrightarrow G_n(l,k)^{op}$.

(iii) $G_n(k,l) \Rightarrow G_{n'}(k',l')$ if $n \geq n'$, $k \leq k'$ and $l \leq l'$.

Combining [6, Proposition 3.4] with Theorems 1.2 and 2.1, we have the following equivalent characterizations of $G_n(0)$.

Theorem 2.4. *The conditions (1)–(3) and their opposite versions are equivalent. If Λ is an Artinian algebra, then all conditions are equivalent:*

(1) Λ *is* $G_n(0)$.
(2) $\mathrm{fd}\, I_i(\Lambda) \leq i$ *for any* $0 \leq i \leq n-1$
(3) $\mathbb{E}_i\mathbb{E}_i$ *preserves monomorphisms in* $\mathrm{mod}\,\Lambda$ *for any* $0 \leq i \leq n-1$.
(4) *All simple composition factors of* $\mathbb{E}_i(S)$ *have grade at least* i *for any simple* Λ*-module* S *and* $1 \leq i \leq n$.
(5) *Opposite side version of* (i) $(1 \leq i \leq 4)$.

Remark 2.2. It was proved in [29, Theorem 8] that if Λ is a left and right Artinian ring and k is a positive integer, then Λ is $G_n(0)$ if and only if a (lower) triangular matrix ring of degree k over Λ is also $G_n(0)$. Note that this is a generalization of [15, Theorem 3.10] where the case $k = 2$ was established.

By Theorems 2.2 and 2.3, we have the following equivalent characterizations of $g_n(1)$.

Theorem 2.5. *The following statements are equivalent:*

(1) Λ *is* $g_n(1)$.
(2) *For any monomorphism* $A \xrightarrow{f} B$ *with* $A, B \in \Omega^2(\mathrm{mod}\,\Lambda)$, $\mathbb{E}_i\mathbb{E}_i(f)$ *is a monomorphism for any* $0 \leq i \leq n-1$.
(3) $\Omega^i(\mathrm{mod}\,\Lambda) = \mathcal{F}_i$ *holds for any* $1 \leq i \leq n+1$.
(4) *For any* $C \in \mathrm{mod}\,\Lambda$ *and* $0 \leq i \leq n$, *there exists an exact sequence* $0 \to Y \to X \to C \to 0$ *with* $X \in \mathcal{W}_{i+1}$ *and* $Y \in \mathcal{P}_{i+1}$.
(5) *For any* $C \in \mathrm{mod}\,\Lambda$ *and* $0 \leq i \leq n$, *there exists an exact sequence* $0 \to \Omega^1(C) \to Y \to X \to 0$ *with* $X \in \mathcal{W}_{i+1}$ *and* $Y \in \mathcal{P}_{i+1}$.
(6) *Opposite side version of* (i) $(1 \leq i \leq 5)$.

Let $\mathcal{D}$ be a full subcategory of mod Λ. Recall that $\mathcal{D}$ is said to be *closed under extensions* if the middle term B of any short sequence $0 \to A \to B \to C \to 0$ is in $\mathcal{D}$ provided that the end terms A and C are in $\mathcal{D}$. We use $\mathrm{add}\,\mathcal{D}$ to denote the subcategory of $\mathrm{mod}\,\Lambda$ consisting of all Λ-modules isomorphic to direct summands of finite direct sums of modules in $\mathcal{D}$. For any $k \geq 1$, we denote by $\mathcal{I}_k = \{M \in \mathrm{mod}\,\Lambda \mid \mathrm{id}\, M < k\}$.

The following result gives some equivalent characterizations of $G_n(1)$, where (1) $\Leftrightarrow$ (2) $\Leftrightarrow$ (3) $\Leftrightarrow$ (4) are in Theorem 2.1, (1) $\Leftrightarrow$ (8) $\Leftrightarrow$ (9) follow from Theorems 2.3 and 2.2, (1) $\Leftrightarrow$ (5) $\Leftrightarrow$ (6) $\Leftrightarrow$ (7) are in [7, Theorem 0.1] and [19, Theorem 3.3], and (2) + (5) $\Rightarrow$ (10) $\Rightarrow$ (6) are in [24, Theorem 4.4].

Theorem 2.6. *The following conditions (1)–(9) are equivalent. If* Λ *is an Artinian algebra, then (10) is also equivalent:*

(1) Λ *is* $G_n(1)$.
(2) $\mathrm{fd}\, I_i(\Lambda^{op}) \leq i+1$ *holds for any* $0 \leq i \leq n-1$.
(3) *For any exact sequence* $0 \to A \xrightarrow{f} B \to C \to 0$ *with* $C \in \Omega^1(\mathrm{mod}\,\Lambda)$, $\mathbb{E}_i\mathbb{E}_i(f)$ *is a monomorphism for any* $0 \leq i \leq n-1$.
(4) $\mathcal{E}(\Omega^i(\mathrm{mod}\,\Lambda), \Omega^i(\mathrm{mod}\,\Lambda)) \subseteq \mathcal{F}_i$ *holds for any* $1 \leq i \leq n$.
(5) $\Omega^i(\mathrm{mod}\,\Lambda)$ *is closed under extensions for any* $1 \leq i \leq n$.
(6) $\mathrm{add}\,\Omega^i(\mathrm{mod}\,\Lambda)$ *is closed under extensions for any* $1 \leq i \leq n$.
(7) $\mathcal{F}_i$ *is closed under extensions for any* $1 \leq i \leq n$.
(8) Λ *is* $\mathcal{G}_n(0)^{op}$.
(9) *For any monomorphism* $A \xrightarrow{f} B$ *with* $A, B \in \Omega^1(\mathrm{mod}\,\Lambda^{op})$, $\mathbb{E}_i\mathbb{E}_i(f)$ *is a monomorphism for any* $0 \leq i \leq n-1$.
(10) *For any* $C \in \mathrm{mod}\,\Lambda$ *and* $1 \leq i \leq n$, *there exist exact sequences* $0 \to Y \to X \to C \to 0$ *and* $0 \to C \to Y' \to X' \to 0$ *with* $X, X' \in \Omega^i(\mathrm{mod}\,\Lambda)$ *and* $Y, Y' \in \mathcal{I}_{i+1}$.

Example 2.1. Contrary to the condition $G_n(0)$, the condition $G_n(1)$ is not left-right symmetric. Consider the following example. Let K be a field and Δ the quiver:

$$1 \underset{\beta}{\overset{\alpha}{\rightleftarrows}} 2 \xrightarrow{\gamma} 3.$$

(1) If $\Lambda = K\Delta/(\alpha\beta\alpha)$, then $\mathrm{fd}\, I_0(\Lambda) = 1$ and $\mathrm{fd}\, I_0(\Lambda^{op}) \geq 2$. (2) If $\Lambda = K\Delta/(\gamma\alpha, \beta\alpha)$, then $\mathrm{fd}\, I_0(\Lambda) = 2$ and $\mathrm{fd}\, I_0(\Lambda^{op}) = 1$.

By Theorems 1.1, 2.4 and 2.6, we have the following result about commutative Gorenstein rings.

Corollary 2.1. *If* Λ *is commutative, then the following statements are equivalent:*

(1) Λ *is Gorenstein.*
(2) Λ *is* $G_\infty(0)$.
(3) Λ *is* $G_\infty(1)$.

3. Properties of rings satisfying the Auslander-type conditions

In this section, we investigate the properties of rings satisfying the Auslander-type conditions. These properties involve duality theory, the socle of modules, homological dimensions, homological finiteness of certain subcategories, cotorsion pairs, Evans-Griffith presentations, and so on.

For a non-negative integer l, put $\mathcal{C}_l(\Lambda) = \{X \in \operatorname{mod}\Lambda \mid X = \mathbb{E}_l(Y)$ for some $Y \in \operatorname{mod}\Lambda^{op}$ and $\operatorname{grade} Y \geq l\}$. The following result is a duality between $\mathcal{C}_l(\Lambda)$ and $\mathcal{C}_l(\Lambda^{op})$, which generalizes results in [32, 6.2] and [26, Theorem 4].

Theorem 3.1. *[30, Theorem 1.2] Let Λ be $G_n(0)$ and $0 \leq l \leq n-1$. Then $\mathbb{E}_l$ gives a duality between $\mathcal{C}_l(\Lambda)$ and $\mathcal{C}_l(\Lambda^{op})$, and $\mathbb{E}_l\mathbb{E}_l$ is isomorphic to the identity functor.*

The following result is a duality between simple Λ-modules and simple Λ^{op}-modules.

Theorem 3.2. *[30, Theorem 1.3] Let Λ be a Noetherian algebra which is $G_n(0)$ and $0 \leq l \leq n-1$. Then $\mathbb{F}_l := \operatorname{Soc}\mathbb{E}_l$ gives a duality between simple Λ-modules X with* $\operatorname{grade} X = l$ *and that of Λ^{op}, and $\mathbb{F}_l\mathbb{F}_l$ is isomorphic to the identity functor. Moreover,* $\operatorname{grade}\mathbb{E}_l(X)/\mathbb{F}_l(X) > l$.

Let Λ be an Artinian algebra. Recall that the Nakayama conjecture states that Λ is self-injective if $I_i(\Lambda)$ is projective for any $i \geq 0$ (see [37] or [43]), and the generalized Nakayama conjecture states that each indecomposable injective Λ-module occurs as the direct summand of some $I_i(\Lambda)$ (see [4]). In view of Theorems 1.1 and 1.2, Auslander and Reiten conjectured in [6] that Λ is Gorenstein if it is $G_\infty(0)$. This conjecture is situated between the Nakayama conjecture and the generalized Nakayama conjecture. The following result is related to this conjecture. It means that the Gorenstein symmetry conjecture holds true for a left and right Artinian ring which is $G_\infty(1)$. Recall from [9] that the Gorenstein symmetry conjecture states that the left and right self-injective dimensions of any Artinian algebra are identical.

Proposition 3.1. *[21, Proposition 4.6] If a left and right Artinian ring Λ is $G_\infty(1)$, then* $\operatorname{id}\Lambda = \operatorname{id}\Lambda^{op}$.

Observe that Auslander and Reiten proved in [6, Corollary 5.5(b)] that if an Artinian algebra Λ is $G_\infty(0)$, then $\operatorname{id}\Lambda = \operatorname{id}\Lambda^{op}$. Proposition 3.1 is a generalization of this result.

Theorem 3.3. *(1) [16, Proposition 1.1] If Λ is $G_n(0)$ with* $\operatorname{id}\Lambda = \operatorname{id}\Lambda^{op} = n$, *then* $\operatorname{pd} I_n(\Lambda) = \operatorname{fd} I_n(\Lambda) = n$ *and so Λ is $G_\infty(0)$.*

(2) [27, Theorem 2] If Λ is $G_\infty(0)$ with $\operatorname{id}\Lambda = \operatorname{id}\Lambda^{op} = n$, *then any injective indecomposable Λ-module E with* $\operatorname{fd} E = n$ *is isomorphic to a direct summand of $I_n(\Lambda)$ and is isomorphic to the injective envelope of a*

simple Λ*-module. Thus if* M *is a* Λ*-module with* $\operatorname{id} M = n$*, then* $I_n(M)$ *has an essential socle, where* $I_n(M)$ *is the* $(n+1)$*-st term in a minimal injective resolution of* M*.*

As an immediate consequence of Theorem 3.3, we have the following result.

Corollary 3.1. *[28, Theorem 6] If* Λ *is* $G_\infty(0)$ *with* $\operatorname{id}\Lambda = \operatorname{id}\Lambda^{op} = n$*, then* $I_n(\Lambda)$ *has an essential socle.*

Recall that the *finitistic dimension* of Λ, denoted by $\operatorname{fin.dim}\Lambda$, is defined as $\sup\{\operatorname{pd} X \mid X \in \operatorname{mod}\Lambda \text{ and } \operatorname{pd} X < \infty\}$. By using Theorem 2.2, it is not difficult to get the following result.

Lemma 3.1. *[24, Lemma 5.1] Assume that* Λ *is* $g_{n+1}(k)$ *with* $n \geq 0$ *and* $k \geq 1$*. If* $\operatorname{fin.dim}\Lambda = n$*, then* $n \leq \operatorname{id}\Lambda \leq n + k$*.*

In the following result, the case for $k \geq 1$ follows from Lemma 3.1, and the case for $k = 0$ is in [23, Corollary 2.15].

Theorem 3.4. *[24, Theorem 5.2] If* Λ *is* $g_\infty(k)$ *with* $k \geq 0$*, then* $\operatorname{fin.dim}\Lambda \leq \operatorname{id}\Lambda \leq \operatorname{fin.dim}\Lambda + k$*.*

As an application of Theorem 3.4, we have the following result, where (1) and (2) follow from the symmetry of $G_\infty(0)$ and the fact that $G_\infty(1) \Leftrightarrow g_\infty(0)^{op}$ (see Theorem 2.3), respectively.

Corollary 3.2. *(1) If* Λ *is* $G_\infty(0)$*, then* $\operatorname{fin.dim}\Lambda = \operatorname{id}\Lambda$ *and* $\operatorname{fin.dim}\Lambda^{op} = \operatorname{id}\Lambda^{op}$*.*

(2) If Λ *is* $G_\infty(1)$*, then* $\operatorname{fin.dim}\Lambda \leq \operatorname{id}\Lambda \leq \operatorname{fin.dim}\Lambda + 1$ *and* $\operatorname{fin.dim}\Lambda^{op} = \operatorname{id}\Lambda^{op}$*.*

Definition 3.1. [5] Assume that $\mathcal{D}$ is a full subcategory of $\operatorname{mod}\Lambda$ and $C \in \operatorname{mod}\Lambda$, $D \in \mathcal{D}$. A morphism $f : D \to C$ is said to be a *right* $\mathcal{D}$*-approximation* of C if $\operatorname{Hom}_\Lambda(X, f) : \operatorname{Hom}_\Lambda(X, D) \to \operatorname{Hom}_\Lambda(X, C) \to 0$ is exact for any $X \in \mathcal{D}$. A right $\mathcal{D}$-approximation $f : D \to C$ is called *minimal* if an endomorphism $g : D \to D$ is an automorphism whenever $f = fg$. The subcategory $\mathcal{D}$ is said to be *contravariantly finite* in $\operatorname{mod}\Lambda$ if any module in $\operatorname{mod}\Lambda$ has a right $\mathcal{D}$-approximation. Dually, we define the notions of *(minimal) left* $\mathcal{D}$*-approximations* and *covariantly finite subcategories*. The subcategory of $\operatorname{mod}\Lambda$ is said to be *functorially finite* in $\operatorname{mod}\Lambda$ if it is both contravariantly finite and covariantly finite in $\operatorname{mod}\Lambda$.

As applications of Theorems 2.5 and 2.6, we have the following result about homological finiteness of $\mathcal{P}_k$.

Corollary 3.3. *[24, Corollary 4.7]*

(1) *If Λ is $g_n(1)$, then $\mathcal{P}_{i+1}$ is covariantly finite in* $\operatorname{mod}\Lambda$ *for any $0 \leq i \leq n$.*
(2) *[22, Theorem 3.6] If an Artinian algebra Λ is $g_n(0)$, then $\mathcal{P}_{i+1}$ is functorially finite in* $\operatorname{mod}\Lambda$ *for any $0 \leq i \leq n$.*

Definition 3.2. (1) [39] A pair $(\mathcal{C}, \mathcal{D})$ of full subcategories of $\operatorname{mod}\Lambda$ is called a *cotorsion pair* if

$$\mathcal{C} = \{C \in \operatorname{mod}\Lambda \mid \operatorname{Ext}^1_\Lambda(C, \mathcal{D}) = 0\} \text{ and } \mathcal{D} = \{D \in \operatorname{mod}\Lambda \mid \operatorname{Ext}^1_\Lambda(\mathcal{C}, D) = 0\}.$$

(2) [24] For an Artinian algebra Λ, a cotorsion pair $(\mathcal{C}, \mathcal{D})$ is called *functorially finite* if the following equivalent conditions are satisfied:

(i) $\mathcal{C}$ is contravariantly finite in $\operatorname{mod}\Lambda$.

(ii) $\mathcal{D}$ is covariantly finite in $\operatorname{mod}\Lambda$.

(iii) For any $C \in \operatorname{mod}\Lambda$, there exists an exact sequence $0 \to Y \to X \xrightarrow{f} C \to 0$ with $X \in \mathcal{C}$ and $Y \in \mathcal{D}$.

(iv) For any $C \in \operatorname{mod}\Lambda$, there exists an exact sequence $0 \to C \xrightarrow{g} Y' \to X' \to 0$ with $X' \in \mathcal{C}$ and $Y' \in \mathcal{D}$.

For any $m, n \geq 0$, put $\mathcal{X}_{n,m} = \mathcal{W}_n \cap \mathcal{F}_m$ and $\mathcal{Y}_{n,m} = \operatorname{add}\mathcal{E}(\mathcal{I}_m, \mathcal{P}_n)$. When Λ is an Artinian algebra over R, we denote by $\mathbb{D} : \operatorname{mod}\Lambda \to \operatorname{mod}\Lambda^{op}$ the duality induced by the Matlis duality of R.

Theorem 3.5. *[24, Corollary 4.9 and Theorem 1.3] Let Λ be an Artinian algebra which is $G_\infty(1)$ and $G_\infty(1)^{op}$ (In particular, let Λ be an Artinian algebra which is $G_\infty(0)$). Then $(\mathcal{X}_{i,j-1}, \mathcal{Y}_{i,j})$ $(i \geq 0,\ j \geq 1)$ and $(\mathcal{Y}_{i,j}, \mathbb{D}\,\mathcal{X}^{op}_{j,i-1})$ $(i \geq 1,\ j \geq 0)$ form functorially finite cotorsion pairs.*

By Theorem 3.5, $(\mathcal{W}_i, \mathcal{Y}_{i,1})$ $(j := 1)$ and $(\mathcal{F}_{j-1}, \mathcal{I}_j)$ $(i := 0)$ form functorially finite cotorsion pairs. In addition, as an immediate consequence of Theorem 3.5, we have the following result.

Corollary 3.4. *Under the assumption of Theorem 3.5, $\mathcal{W}_1 \supseteq \mathcal{W}_2 \supseteq \cdots \supseteq \mathcal{W}_i \supseteq \cdots$ is a chain of contravariantly finite subcategories of* $\operatorname{mod}\Lambda$.

Let Λ be a commutative Noetherian ring and let n be a non-negative integer and $M \in \Omega^n(\operatorname{mod}\Lambda)$. Recall from [14] that an *Evans-Griffith presentation* of M is an exact sequence in $\operatorname{mod}\Lambda$:

$$0 \to S \to B \to M \to 0,$$

where B is an n-syzygy module of $\mathbb{E}_{n+1}(\operatorname{Tr} M)$ and $S \in \Omega^{n+2}(\operatorname{mod} \Lambda)$. In the case Λ is not necessarily commutative, we also call such an exact sequence an *Evans-Griffith presentation* of M.

Theorem 3.6. *[20, Proposition 4.4] Let Λ be $G_n(1)$. Then, for any $0 \leq d \leq n-1$, each module in $\Omega^d(\operatorname{mod} \Lambda)$ has an Evans-Griffith presentation.*

By Theorem 3.6, we have the following result.

Corollary 3.5. *[20, Corollary 4.5] If Λ is $G_\infty(1)$, then for any non-negative integer d, each module in $\Omega^d(\operatorname{mod} \Lambda)$ has an Evans-Griffith presentation.*

Combining Corollaries 3.5 and 2.1, we immediately have the following result.

Corollary 3.6. *[20, Corollary 4.6] If Λ is a commutative Gorenstein ring, then for any non-negative integer d, each module in $\Omega^d(\operatorname{mod} \Lambda)$ has an Evans-Griffith presentation.*

Observe that a special instance of Corollary 3.6 was already considered by Evans and Griffith in [14, Theorem 2.1]. They showed that if Λ is a commutative Noetherian local ring with finite global dimension and contains a field then each non-free d-syzygy of rank d has an Evans-Griffith presentation. Corollary 3.6 generalizes this result to much more general setting.

4. (m, n)-conditions

In this section, we give the definition of the (weak) (m, n)-condition, and then investigate the properties of rings satisfying certain (m, n)-condition.

Definition 4.1. [30] Let $m, n \geq 1$. Λ is said to satisfy the *(m, n)-condition* (or Λ^{op} satisfies the *$(m, n)^{op}$-condition*) if $\operatorname{s.grade} \mathbb{E}_m(N) \geq n$ for any $N \in \operatorname{mod} \Lambda^{op}$. Similarly, Λ is said to satisfy the *weak (m, n)-condition* (or Λ^{op} satisfies the *weak $(m, n)^{op}$-condition*) if $\operatorname{grade} \mathbb{E}_m(N) \geq n$ for any $N \in \operatorname{mod} \Lambda^{op}$.

Remark 4.1. (1) By [32, 6.1], Λ satisfies the (m, n)-condition if and only if $\operatorname{fd} I_i(\Lambda) \leq m-1$ for any $0 \leq i \leq n-1$.

(2) It is easy to see that Λ is $G_n(k)$ if and only if Λ satisfies the $(k+i, i)^{op}$-condition for any $1 \leq i \leq n$, and Λ is $g_n(k)$ if and only if Λ satisfies the weak $(k+i, i)^{op}$-condition for any $1 \leq i \leq n$.

(3) For a module $M \in \operatorname{mod}\Lambda$ and $n \geq 1$, recall that the *dominant dimension* of M, denoted by $\operatorname{dom.dim} M$, is said to be at least n if the first n terms in a minimal injective resolution of M are flat. So, $\operatorname{dom.dim}\Lambda \geq n$ if and only if Λ satisfies the $(1, n)$-condition. It was proved in [17, Theorem] that $\operatorname{dom.dim}\Lambda = \operatorname{dom.dim}\Lambda^{op}$.

The following result gives some relations between different (weak) (m, n)-conditions.

Proposition 4.1. *[24, Lemma 5.4] and [30, Lemma 2.3]*

(1) (m, l) + *weak* $(l, n) \Rightarrow (m, n)$.
(2) *weak* (m, l) + *weak* $(l, n) \Rightarrow$ *weak* (m, n).
(3) (m, l) + *weak* $(l, n)^{op} \Rightarrow (m, n)$.
(4) *weak* (m, l) + *weak* $(l, n)^{op} \Rightarrow$ *weak* (m, n).

It is known that Λ is $G_n(0)$ if and only if so is Λ^{op} (see Theorem 2.3). However, the (i, i)-condition does not possess such a symmetric property in general. For example, the finite dimensional algebra given by the quiver $1 \xrightarrow{\alpha} 2 \xrightarrow{\beta} 3 \leftarrow 4$ modulo the ideal $\beta\alpha$ satisfies exactly one of the $(2, 2)$ and $(2, 2)^{op}$-conditions.

The following result gives a sufficient condition that the (i, i)-condition implies the $(i, i)^{op}$-condition.

Proposition 4.2. *[24, Corollary 5.7]* $G_{n-1}(1) + (n, n) \Rightarrow (n, n)^{op}$.

In particular, putting $n = 3$ in Proposition 4.2, we get the following result.

Corollary 4.1. $(2, 2)^{op} + (3, 3) \Rightarrow (3, 3)^{op}$.

In [24], we gave an example satisfying the conditions in Corollary 4.1 as follows.

Example 4.1. Let Λ be a finite dimensional algebra given by the quiver:

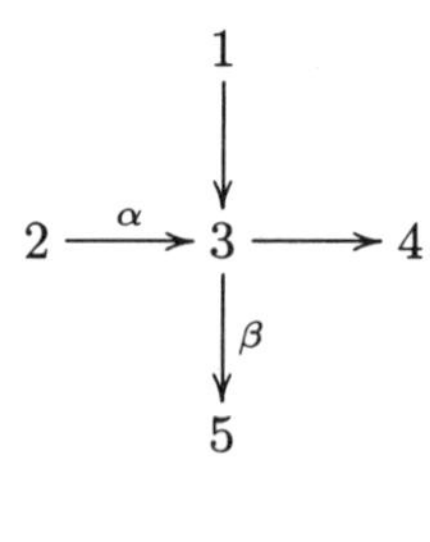

modulo the ideal $\beta\alpha$. Then $\operatorname{fd} I_0(\Lambda) = \operatorname{fd} I_1(\Lambda) = \operatorname{fd} I_0(\Lambda^{op}) = \operatorname{fd} I_1(\Lambda^{op}) = 1$, and $\operatorname{fd} I_2(\Lambda) = \operatorname{fd} I_2(\Lambda^{op}) = 2$.

Recall from [10] that a module $M \in \operatorname{mod}\Lambda$ is called *pure* if $\operatorname{grade} X = \operatorname{grade} M$ for any non-zero submodule X of M. Björk in [10, p.144] raised a question: If Λ is $G_\infty(0)$ with finite left and right self-injective dimensions, is $\mathbb{E}_{\operatorname{grade} M}(M)$ pure for any $M \in \operatorname{mod}\Lambda$? Björk and Ekström in [11, Proposition 2.11] gave a positive answer to this question, and then Iyama proved in [30] that the answer to it is positive in more general case.

Proposition 4.3. *[30, Proposition 2.9] If Λ satisfies the $(n,n)^{op}$-condition, then for any $M \in \operatorname{mod}\Lambda$ with $\operatorname{grade} M = n$, $\mathbb{E}_n(M)$ is pure with $\operatorname{grade}\mathbb{E}_n(M) = n$.*

For a positive integer n, we denote $\mathcal{E}_n(\Lambda_\Lambda) = \{M \in \operatorname{mod}\Lambda \mid M = \mathbb{E}_n(N)$ for some $N \in \operatorname{mod}\Lambda^{op}\}$. Auslander showed in [1, Proposition 3.3] that any direct summand of a module in $\mathcal{E}_1(\Lambda_\Lambda)$ is still in $\mathcal{E}_1(\Lambda_\Lambda)$. He then asked whether any submodule of a module in $\mathcal{E}_1(\Lambda_\Lambda)$ is still in $\mathcal{E}_1(\Lambda_\Lambda)$. Recall that a full subcategory $\mathcal{X}$ of $\operatorname{mod}\Lambda$ is said to be *submodule closed* if any non-zero submodule of a module in $\mathcal{X}$ is also in $\mathcal{X}$. Then the above Auslander's question is equivalent to the following question: Is $\mathcal{E}_1(\Lambda_\Lambda)$ submodule closed?

Proposition 4.4. *[21, Corollaries 3.9 and 3.14]*

(1) If Λ is $G_\infty(0)$ with $\operatorname{id}\Lambda = \operatorname{id}\Lambda^{op} = n$, then $\mathcal{E}_n(\Lambda_\Lambda)$ is submodule closed.
(2) If $\operatorname{id}\Lambda = \operatorname{id}\Lambda^{op} = 1$, then $\mathcal{E}_1(\Lambda_\Lambda)$ is submodule closed if and only if Λ satisfies the $(1,1)$-condition.
(3) If $\operatorname{id}\Lambda = \operatorname{id}\Lambda^{op} = 2$, then $\mathcal{E}_2(\Lambda_\Lambda)$ is submodule closed if and only if Λ satisfies the $(2,2)$-condition.

As an application of Proposition 4.4, the following examples were constructed in [21] to illustrate that neither $\mathcal{E}_1(\Lambda_\Lambda)$ nor $\mathcal{E}_2(\Lambda_\Lambda)$ are submodule closed in general, by which the above Auslander's question is answered negatively.

Example 4.2. (1) Let Λ be a finite dimensional algebra given by the quiver:

$$2 \longleftarrow 1 \longrightarrow 3.$$

Then $\operatorname{id}\Lambda = \operatorname{id}\Lambda^{op} = 1$ and $\operatorname{fd} I_0(\Lambda) = 1$. By Proposition 4.4(2), $\mathcal{E}_1(\Lambda_\Lambda)$ is not submodule closed.

(2) Let Λ be a finite dimensional algebra given by the quiver:

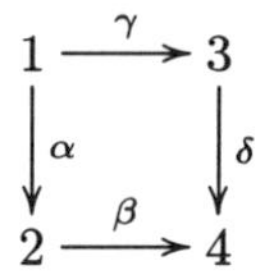

modulo the ideal $\beta\alpha$. Then $\operatorname{id}\Lambda = \operatorname{id}\Lambda^{op} = 2$ and $\operatorname{fd} I_0(\Lambda) = 2$. By Proposition 4.4(3), $\mathcal{E}_2(\Lambda_\Lambda)$ is not submodule closed.

It is clear that $\operatorname{mod}\Lambda \supseteq \mathcal{E}_1(\Lambda_\Lambda) \supseteq \mathcal{E}_2(\Lambda_\Lambda) \supseteq \cdots \supseteq \mathcal{E}_i(\Lambda_\Lambda) \supseteq \cdots$. For any positive integer n, $\mathcal{E}_n(\Lambda_\Lambda)$ is submodule closed for Λ being $G_\infty(0)$ with $\operatorname{id}\Lambda = \operatorname{id}\Lambda^{op} = n$ by Proposition 4.4(1), and neither $\mathcal{E}_1(\Lambda_\Lambda)$ nor $\mathcal{E}_2(\Lambda_\Lambda)$ are submodule closed in general by Example 4.2. It is interesting to know whether $\mathcal{E}_n(\Lambda_\Lambda)$ (where $n \geq 3$) is submodule closed in general.

Acknowledgments

The author was partially supported by the Specialized Research Fund for the Doctoral Program of Higher Education (Grant No. 20060284002), NSFC (Grant No. 10771095) and NSF of Jiangsu Province of China (Grant No. BK2007517).

References

1. M. Auslander, Comments on the functor Ext. *Topology* 1969, **8**: 151–166.
2. M. Auslander, Representation dimension of Artin algebras, Lecture Notes; Queen Mary College: London, 1971.
3. M. Auslander and M. Bridger, Stable Module Theory, Memoirs Amer. Math. Soc. 94; Amer. Math. Soc.: Providence, RI, 1969.
4. M. Auslander and I. Reiten, On a generalized version of the Nakayama conjecture. *Proc. Amer. Math. Soc.* 1975, **52**: 69–74.
5. M. Auslander and I. Reiten, Applications of contravariantly finite subcategories. *Adv. Math.* 1991, **86**: 111–152.
6. M. Auslander and I. Reiten, k-Gorenstein algebras and syzygy modules. *J. Pure Appl. Algebra* 1994, **92**: 1–27.
7. M. Auslander and I. Reiten, Syzygy modules for Noetherian rings. *J. Algebra* 1996, **183**: 167–185.
8. H. Bass, On the ubiquity of Gorenstein rings. *Math. Z.* 1963, **82**: 8–28.
9. A. Beligiannis and I. Reiten, Homological and homotopical aspects of torsion theories, Memoirs Amer. Math. Soc. 188; Amer. Math. Soc.: Providence, RI, 2007.

10. J. E. Björk, The Auslander condition on Noetherian rings, In: Seminaire d'Algebre Paul Dubreil et Marie-Paul Malliavin, 39eme Annee (Paris, 1987/1988), Lecture Notes in Math. 1404; Springer-Verlag: Berlin, 1989, 137–173.
11. J. E. Björk and E. K. Ekström, Filtered Auslander-Gorenstein rings, In: Operator Algebras, Unitary Representations, Enveloping Algebras, and Invariant Theory, Paris, 1989, Progr. Math. 92; Birkhäuser: Boston, 1990, 425–448.
12. J. Clark, Auslander-Gorenstein rings for beginners, In: International Symposium on Ring Theory (Kyongju, 1999), Trends Math.; Birkhäuser: Boston, 2001, 95–115.
13. K. Erdmann, T. Holm, O. Iyama and J. Schröer, Radical embeddings and representation dimension. *Adv. Math.* 2004, **185**: 159–177.
14. E. G. Evans and P. Griffith, Syzygies of critical rank. *Quart. J. Math. Oxford* 1984, **35**: 393–402.
15. R. M. Fossum, P. Griffith and I. Reiten, Trivial Extensions of Abelian Categories, Homological Algebra of Trivial Extensions of Abelian Categories with Applications to Ring Theory, Lecture Notes in Math. 456; Springer-Verlag: Berlin, 1975.
16. K. R. Fuller and Y. Iwanaga, On n-Gorenstein rings and Auslander rings of low injective dimension, In: Representations of algebras (Ottawa, ON, 1992), CMS Conf. Proc. 14; Amer. Math. Soc.: Providence, RI, 1993, 175–183.
17. M. Hoshino, On dominant dimension of Noetherian rings. *Osaka J. Math.* 1989, **26**: 275–280.
18. M. Hoshino and K. Nishida, A generalization of the Auslander formula, In: Representations of Algebras and Related Topics, Fields Institute Communications 45; Amer. Math. Soc.: Providence, RI, 2005, 175–186.
19. Z. Y. Huang, Extension closure of k-torsionfree modules. *Comm. Algebra* 1999, **27**: 1457–1464.
20. Z. Y. Huang, Syzygy modules for quasi k-Gorenstein rings. *J. Algebra* 2006, **299**: 21–32.
21. Z. Y. Huang, Generalized tilting modules with finite injective dimension. *J. Algebra* 2007, **311**: 619–634.
22. Z. Y. Huang, Approximation presentations of modules and homological conjectures. *Comm. Algebra* 2008, **36**: 546–563.
23. Z. Y. Huang, On the grade of modules over noetherian rings. *Comm. Algebra* (to appear).
24. Z. Y. Huang and O. Iyama, Auslander-type conditions and cotorsion pairs. *J. Algebra* 2007, **318**: 93–110.
25. K. Igusa, S. O. Smalϕ and G. Todorov, Finite projectivity and contravariant finiteness. *Proc. Amer. Math. Soc.* 1990, **109**: 937–941.
26. Y. Iwanaga, Duality over Auslander-Gorenstein rings. *Math. Scand.* 1997, **81**: 184–190.
27. Y. Iwanaga and J. I. Miyachi, Modules of the highest homological dimension over a Gorenstein ring, In: Trends in the representation theory of finite-dimensional algebras (Seattle, WA, 1997), Contemp. Math. 229; Amer. Math. Soc.: Providence, RI, 1998, 193–199.

28. Y. Iwanaga and H. Sato, On Auslander's n-Gorenstein rings. *J. Pure Appl. Algebra* 1996, **106**: 61–76.
29. Y. Iwanaga and T. Wakamatsu, Auslander-Gorenstein property of triangular matrix rings. *Comm. Algebra* 1995, **23**: 3601–3614.
30. O. Iyama, Symmetry and duality on n-Gorenstein rings. *J. Algebra* 2003, **269**: 528–535.
31. O. Iyama, The relationship between homological properties and representation theoretic realization of Artin algebras. *Trans. Amer. Math. Soc.* 2005, **357**: 709–734.
32. O. Iyama, τ-categories III, Auslander orders and Auslander-Reiten quivers. *Algebr. Represent. Theory* 2005, **8**: 601–619.
33. O. Iyama, Finiteness of Representation dimension. *Proc. Amer. Math. Soc.* 2003, **131**: 1011–1014.
34. O. Iyama, Higher-dimensional Auslander-Reiten theory on maximal orthogonal subcategories. *Adv. Math.* 2007, **210**: 22–50.
35. O. Iyama, Auslander correspondence. *Adv. Math.* 2007, **210**: 51–82.
36. J. I. Miyachi, Injective resolutions of Noetherian rings and cogenerators. *Proc. Amer. Math. Soc.* 2000, **128**: 2233–2242.
37. T. Nakayama, On algebras with complete homology. *Abh. Math. Sem. Univ. Hamburg* 1958, **22**: 300–307.
38. R. Rouquier, Representation dimension of exterior algebras. *Invent. Math.* 2006, **165**: 357–367.
39. L. Salce, Cotorsion theories for abelian groups, In: Symposia Mathematica **XXIII**, (Conf. Abelian Groups and their Relationship to the Theory of Modules, INDAM, Rome, 1977); Academic Press: London-New York, 1979, 11–32.
40. S. P. Smith, Some finite-dimensional algebras related to elliptic curves, In: Representation theory of algebras and related topics (Mexico City, 1994), CMS Conf. Proc. 19; Amer. Math. Soc.: Providence, RI, 1996, 315–348.
41. H. Tachikawa, Quasi-Frobenius Rings and Generalizations, QF-3 Rings and QF-1 Rings, Lecture Notes in Math. 351; Springer-Verlag: Berlin, 1973.
42. T. Wakamatsu, Tilting modules and Auslander's Gorenstein property. *J. Algebra* 2004, **275**: 3–39.
43. K. Yamagata, Frobenius algebras, In: Handbook of Algebra 1; North-Holland Publishing Co.: Amsterdam, 1996, 841–887.
44. A. Zaks, Injective dimension of semiprimary rings. *J. Algebra* 1969, **13**: 73–86.

Koszul differential graded algebras and modules

J.-W. He

Department of Mathematics, Shaoxing college of Arts and Sciences, Shaoxing Zhejiang 312000, China
E-mail: jwhe@zscas.edu.cn

Q.-S. Wu

Institute of Mathematics, Fudan University, Shanghai 200433, China
E-mail: qswu@fudan.edu.cn

This is a survey of our recent work [10–13] on the koszulity of differential graded algebras and modules.

Keywords: Differential graded algebra, Derived category, Koszul algebra, BGG correspondence

1. Introduction

Our work is motivated by the following two problems: one is presented by Manin [1] — how to generalize the koszulity of graded algebras to differential graded (DG, for short) algebras; and the other is to generalize the Bernstein-Gelfand-Gelfand (BGG) correspondence [2] to the DG setting.

The categories of DG modules over DG algebras, or more generally, DG categories [3,4] and their derived categories are important tools in algebraic topology and algebraic geometry [5–8]. Attempts to extend the koszulity to the DG setting have been made by several authors [5,8]. In their terminology, a DG algebra is said to be koszul if the underlying graded algebra is koszul. Koszul DG algebras in their sense are applied to discuss configuration spaces. However, we take a different point of view. Let k be a field. By a connected DG algebra over k we mean a positively graded k-algebra $A = \oplus_{n\geq 0}A^n$ with $A^0 = k$ such that there is a differential $d : A \to A$ of degree 1 which is also a graded derivation. A connected DG algebra A is said to be a *koszul DG algebra* if the minimal semifree resolution of the trivial DG module ${}_Ak$ has a semifree basis consisting of homogeneous elements of degree zero. Our definition of koszul DG algebra is a natural generalization of the usual koszul algebra. As we will see, a connected graded algebra regarded as a DG algebra with zero differential is a koszul DG algebra if and only if it is a koszul algebra in the usual sense. Examples of koszul DG algebras can be found in various fields. For example, let M be a connected n-dimensional C^∞ manifold, and let $(\mathcal{A}^*(M) = \bigoplus_{i=0}^n \mathcal{A}^i(M), d)$ be the de Rham complex of M, then $(\mathcal{A}^*(M), d)$ is a commutative DG algebra and by de Rham theorem [9] the 0-*th* cohomology group $H^0(\mathcal{A}^*(M)) \cong \mathbb{R}$. Hence the DG algebra $\mathcal{A}^*(M)$ has a minimal model [7] or Sullivan model [6] A, which is certainly a connected DG algebra. If the manifold M has some further properties (e.g., $M = T^n$ the n-dimensional torus), then the de Rham cohomology algebra $H(\mathcal{A}^*(M))$ is a koszul algebra. Hence the cohomology algebra of its minimal model A is koszul as A is quasi-isomorphic to $\mathcal{A}^*(M)$. Then A is a koszul DG algebra, see Ref. [10].

The concept of koszul DG modules is introduced in Ref. [11]. Let A be a connected DG algebra, $\mathcal{D} = \mathcal{D}^+_{dg}(A)$ be the derived category of bounded below DG modules. Then there is a natural t-structure $(\mathcal{D}^{\geq 0}, \mathcal{D}^{\leq 0})$ on the triangulated category $\mathcal{D}$ [7]. Koszul DG modules over A are just the objects in the heart of the t-structure $(\mathcal{D}^{\geq 0}, \mathcal{D}^{\leq 0})$. Similar to the usual koszul algebra case, there is a duality between some suitable subcategory of the derived category of a koszul DG algebra A and some suitable subcategory of the derived category of the Ext-algebra $E = \mathrm{Ext}^*_A(k, k)$. Using the duality theorem, we give a description of the heart of the t-structure given above.

Associated to any DG module X_A over a connected DG algebra A, we define a koszul DG complex $K(X)$, which is a minimal semifree DG module [12]. The koszul DG complex of the trivial module k_A is called the koszul DG complex of the DG algebra A, and is denoted by $K(A)$. There is also a so called koszul complex $K(X, Y)$ associated two any DG modules X_A and ${}_AY$. Koszul DG complexes reflect some properties of the associated DG modules. Especially, if the DG algebra A is koszul, then we may apply suitable koszul DG complex to compute the Ext- and Tor-groups and the Hochschild cohomologies of DG modules. Usual koszul algebras [14–16] fit well in our koszul DG setting. If we view a usual koszul algebra R as a special koszul DG algebra A, then our koszul DG complex $K(A)$ coincides with the usual koszul complex of R [14,15].

Bernstein-Gelfand-Gelfand established an equivalence [2] between the stable category of finitely generated graded modules over the exterior algebra $\bigwedge V$ with $V = kx_0 \oplus kx_1 \oplus \cdots \oplus kx_n$, and the bounded derived category of coherent sheaves on the projective space $\mathbb{P}^n$. This equivalence is now called the *BGG correspondence*. BGG correspondence has been generalized to noncommutative projective geometry by several authors. Let R be a (noncommutative) koszul algebra. If R is AS-regular, Jorgensen [17] proved that there is an equivalence between the stable category over the graded Frobenius algebra $E(R) = \mathrm{Ext}^*_R(k,k)$ and the derived category of the noncommutative analogue $\mathrm{QGr}(R)$ of the quasi-coherent sheaves over R; Martínez Villa-Saorín [18] proved that the stable category of the finite dimensional modules over $E(R)$ is equivalent to the bounded derived category of the noncommutative analogue $\mathrm{qgr}R$ of the coherent sheaves over R. Mori [19] proved a similar version under a more general condition. We prove a similar correspondence in our DG setting [10]. In some special case, the DG version of the BGG correspondence coincides with the classical one [2,18].

2. Preliminaries

Throughout, k is a field and all algebras are k-algebras; unadorned $\otimes$ means $\otimes_k$ and Hom means Hom_k.

By a *graded algebra* we mean a $\mathbb{Z}$-graded algebra. An *augmented* graded algebra is a graded algebra A with an augmentation map $\varepsilon : A \to k$ which is a graded algebra morphism. A positively graded algebra $A = \bigoplus_{n\geq 0} A_n$ with $A_0 = k$ is called a *connected* graded algebra. Let M and N be graded A-modules. $\underline{\mathrm{Hom}}_A(M,N)$ is the set of all graded A-module morphisms. If L is a graded vector space, $L^\# = \underline{\mathrm{Hom}}(L,k)$ is the graded vector space dual.

By a (cochain) *DG algebra* we mean a graded algebra $A = \bigoplus_{n\in\mathbb{Z}} A^n$ with a differential $d : A \to A$ of degree 1, which is also a graded derivation. An *augmented DG algebra* is a DG algebra A such that the underlying graded algebra is augmented with augmentation map $\varepsilon : A \to k$ satisfying $\varepsilon \circ d = 0$. $\ker \varepsilon$ is called the augmented ideal of A. A *connected* DG algebra is a DG algebra such that the underlying graded algebra is connected. Any graded algebra can be viewed as a DG algebra with differential $d = 0$; in this case it is called a DG algebra with *trivial* differential.

Let (A, d_A) be a DG algebra. A *left differential graded module* over A (DG A-module for short) is a left graded A-module M with a differential $d_M : M \to M$ of degree 1 such that d_M satisfies the graded Leibnitz rule

$$d_M(a\,m) = d_A(a)\,m + (-1)^{|a|} a\, d_M(m)$$

for all graded elements $a \in A$, $m \in M$.

A right DG module over A is defined similarly. We denote A^{op} as the opposite DG algebra of A, whose product is defined as $a \cdot b = (-1)^{|a|\cdot|b|} ba$

for all graded elements $a, b \in A$. Right DG modules over A can be identified with DG A^{op}-modules.

Dually, by a (cochain) *DG coalgebra* we mean a graded coalgebra $C = \bigoplus_{n\in\mathbb{Z}} C^n$ with a differential $d: C \to C$ of degree 1, which is also a graded coderivation. A *coaugmented DG coalgebra* is a DG coalgebra C with a graded coalgebra map $\eta : k \longrightarrow C$, called coaugmentation map, such that $d \circ \eta = 0$. If C is a coaugmented DG coalgebra, then C has a decomposition $C = k \oplus \bar{C}$, where $\bar{C}$ is the kernel of the counit ε_C, which is isomorphic to the cokernel $\widetilde{C}$ of η. There is a coproduct $\bar{\Delta} : \bar{C} \to \bar{C} \otimes \bar{C}$ defined by $\bar{\Delta}(c) = \Delta(c) - 1 \otimes c - c \otimes 1$, such that $(\bar{C}, \bar{\Delta})$ is a coalgebra without counit. Δ induces a coproduct $\widetilde{\Delta}$ over $\widetilde{C}$. $(\bar{C}, \bar{\Delta})$ and $(\widetilde{C}, \widetilde{\Delta})$ are isomorphic as coalgebras. A coaugmented DG coalgebra C is *cocomplete* if, for any homogeneous element $x \in \bar{C}$, there is an integer n such that $\bar{\Delta}^n(x) = (\bar{\Delta} \otimes 1^{\otimes n-1}) \circ \cdots \circ (\bar{\Delta} \otimes 1) \circ \bar{\Delta}(x) = 0$. A right DG C-comodule N is a graded right C-comodule with a graded coderivation d_N (i.e. $\rho_N d_N = (d_N \otimes 1 + 1 \otimes d_C)\rho_N$) of degree 1. A cocomplete right DG C-comodule is defined similarly [20].

For the standard facts about DG modules, semifree modules and semifree resolutions of DG modules, etc, see Refs. [6] and [21]. A DG A-module M is said to be *bounded below* if $M^n = 0$ for $n \ll 0$. Let A be a DG algebra, M and N be left DG A-modules, W be a right DG A-module. Following Refs. 7] and [22], for $n \in \mathbb{Z}$, the n-th differential Ext and Tor are defined as

$$\mathrm{Ext}^n_A(M, N) = H^n(\mathrm{RHom}_A(M, N)) \quad \text{and} \quad \mathrm{Tor}^n_A(W, M) = H^n(W \otimes^L_A M).$$

2.1. *Bar and cobar construction*

Let s be the suspension (shifting) map with $(sX)^n = X^{n-1}$ for any cochain complex X. Then $s^i : X \to s^i X$ is of degree i for any $i \in \mathbb{Z}$.

Let A be an augmented DG algebra with differential d. The *bar construction* $(B(A), \delta)$ of A is a coaugmented DG coalgebra. For any right DG A-module (M, d_M), the *bar construction* $B(M; A) = M \otimes B(A)$ *of* M is a right DG $B(A)$-comodule [6,10].

Let C be a coaugmented DG coalgebra with differential d. The *cobar construction* $(\Omega(C), \partial)$ of C is an augmented DG algebra. For any right DG C-comodule (M, ρ, d_M), The *cobar construction* $\Omega(M; C) = M \otimes \Omega(C)$ of M is a right DG $\Omega(C)$-module [6,10].

2.2. *Twisting cochain*

A *twisting cochain* [20,23,24] is a graded linear map $\tau : C \to B$, where B is an augmented DG algebra with multiplication m and augmentation map ε_B, and C is a coaugmented DG coalgebra with coproduct Δ and coaugmentation map η_C, such that

$$\varepsilon_B \circ \tau \circ \eta_C = 0, \text{ and}$$
$$m \circ (\tau \otimes \tau) \circ \Delta + d_B \circ \tau + \tau \circ d_C = 0.$$

The *canonical twisting cochain* from C to $\Omega(C)$ is the composition

$$\tau_0 : C \xrightarrow{\pi} \overline{C} \xrightarrow{s} \Omega(C),$$

where $\overline{C} = \operatorname{coker} \eta_C$, π is the natural projection and s is the shift map.

Let $\tau : C \to B$ be a twisting cochain, (M^C, ρ) a right DG C-comodule. The *twisted tensor product* $N \otimes_\tau B$]13,20,24] is a right DG B-module.

Let $f : B \to B'$ be a morphism of DG algebras. Then $\tau' = f \circ \tau : C \to B'$ is a twisting cochain. We have a right DG B-module morphism

$$N \otimes f : N \otimes_\tau B \longrightarrow N \otimes_{\tau'} B'.$$

Dually, if M_B is a right DG module the *twisted tensor product* $M \otimes_\tau C$ is a right DG C-comodule.

2.3. *Some notations.*

Let A be an augmented DG algebra. Let $\mathcal{C}(A)$ be the category of left DG A-modules, $\mathcal{K}(A)$ be the homotopy category of $\mathcal{C}(A)$, and $\mathcal{K}^0(A)$ be the full subcategory of $\mathcal{K}(A)$ consisting of semifree DG modules with a semifree basis concentrated in degree zero. $\mathcal{D}_{dg}(A)$ stands for the derived category of left DG A-modules and $\mathcal{D}_{dg}(A^{op})$ for the derived category of right DG A-modules; $\mathcal{D}^c(A)$ (resp. $\mathcal{D}^c(A^{op})$) stands for the full triangulated subcategory of $\mathcal{D}_{dg}(A)$ (resp. $\mathcal{D}_{dg}(A^{op})$) consisting of all the compact objects [3, Sect. 5]. If A is a connected DG algebra, then $\mathcal{D}^c(A)$ (resp. $\mathcal{D}^c(A^{op})$) is equivalent to the full triangulated subcategory $\langle {}_AA \rangle$ (resp. $\langle A_A \rangle$) generated by the object ${}_AA$ (resp. A_A), that is, the smallest full triangulated subcategory containing ${}_AA$ (resp. A_A) as an object and closed under isomorphisms.

Let E be an algebra. The notation $\mathcal{D}^*(E)$, where $* = +, -, b$, stands for the derived category of, respectively, bounded below, bounded above and bounded cochain complexes of left E-modules. $\mathcal{D}^*(E^{op})$ stands for the right version of $\mathcal{D}^*(E)$.

3. Koszul complexes of DG algebras and modules

In the homology theory of rings and algebras, the koszul complex plays an important role in computing homology groups [14,15,22]. Usually, the koszul complex provides varies koszul resolutions for certain modules. For instance, if R is a noetherian commutative ring, the koszul complex of an R-regular sequence provides a free resolution of the module R/I, where I is the ideal of R generated by the elements in the regular sequence [22]. Also if Λ is a koszul connected graded k-algebra, then the koszul complex associated to Λ provides a free resolution to the trivial module k [14,15]. Motivated by Lefèvre-Hasegawa's work [20], we introduce the concept of koszul DG complexes. Usual koszul algebras [14–16] fit well in our koszul DG setting. If we view a usual koszul algebra Λ as a special koszul DG algebra A, then our koszul DG complex $K(A)$ coincides with the usual koszul complex of Λ [14,15].

3.1. *Filtered coalgebras and comodules*

In this subsection, let C be an ordinary coaugmented coalgebra (considered as a DG coalgebra concentrated in degree zero) and $A = \Omega(C)$ be the cobar construction of C. Then A is a connected DG algebra. Recall that $\tau_0 : C \longrightarrow A$ is the canonical twisting cochain.

Lemma 3.1. *(Ref. 12) For any $X \in \mathcal{K}^0(A^{op})$, $H^i(X \otimes_{\tau_0} C) = 0$ if $i \neq 0$.*

Following this lemma, we can form an additive functor Φ which is the composition

$$\Phi : \mathcal{K}^0(A^{op}) \overset{-\otimes_{\tau_0} C}{\longrightarrow} \mathcal{C}(\text{Com-}C) \overset{\mathrm{H}^0}{\longrightarrow} \text{Com-}C, \tag{1}$$

where Com-C is the category of right C-comodules and $\mathcal{C}$(Com-C) is the category of cochain complexes of right C-comodules.

Let $(C, \triangle, \varepsilon, \eta)$ be a coaugmented coalgebra with counit ε and coaugmentation map η. We say C is *filtered* if there is a filtration $0 \subseteq F^0(C) \subseteq F^1(C) \subseteq \cdots$ such that

$$F^0(C) = k, \ \cup_{i\geq 0} F^i(C) = C, \text{ and } \Delta(F^n(C)) \subseteq \sum_{i+j=n} F^i(C) \otimes F^j(C).$$

Let $\pi : C \to \overline{C}$ be the projection map, where $\overline{C}$ is the cokernel of η. Let

$$F^0(C) = k, \ F^i(C) = \{c \in C | \pi^{\otimes i} \Delta^i(c) = 0\} \quad \text{for } i \geq 1, \tag{2}$$

where $\Delta^i = (\Delta \otimes id^{\otimes i-1}) \cdots (\Delta \otimes id)\Delta$. Then C is a filtered coalgebra if and only if $\cup_{i \geq 0} F^i(C) = C$. In this case, C is called a *cocomplete* coalgebra [20], and we say the filtration (2) is the *associated filtration* of the cocomplete coalgebra C.

Let M^C be a right C-comodule with comodule action ρ. We say M^C is *filtered* if there is a filtration $0 \subseteq F^0(M) \subseteq F^1(M) \subseteq \cdots \subseteq F^n(M) \subseteq \cdots$ such that $\cup_{i\geq 0} F^i(M) = M$, $F^0(M) = \{m \in M | \rho(m) = m \otimes 1\}$ and $\Delta(F^n(M)) \subseteq \sum_{i+j=n} F^i(M) \otimes F^j(C)$.

For a filtered comodule M^C, let $P = M \otimes_{\tau_0} A$. Then P is a semifree DG A-module with a semifree filtration $P(i) = F^i(M) \otimes_{\tau_0} A$ for $i \geq 0$.

Now let C be a cocomplete coalgebra with the associated filtration (2), and M^C a comodule. Define

$$F^i(M) = \{m \in M | (id \otimes \pi^{\otimes i+1})\rho^{i+1}(c) = 0\} \quad \text{for } i \geq 0, \tag{3}$$

where $\rho^{i+1} = (\rho \otimes id^{\otimes i}) \cdots (\rho \otimes id)\rho$. Then one can see that M^C, with the filtration (3), is a filtered comodule. We call the filtration (3) the *induced filtration* on M. Therefore, given a comodule M^C, we get a semifree DG module $P = M \otimes_{\tau_0} A$. Moreover the semifree filtration $P(i) = F^i(M) \otimes_{\tau_0} A$ $(i \geq 0)$ is standard. Hence we get an additive functor

$$\Psi = - \otimes_{\tau_0} A : \text{Com-}C \longrightarrow \mathcal{K}^0(A^{op}). \tag{4}$$

Proposition 3.1. *(Ref. [12]) Let C be a cocomplete coalgebra. Then the additive functors (Φ, Ψ) are a pair of quasi-inverse equivalences of additive categories:*

$$\mathcal{K}^0(A^{op}) \underset{\Psi}{\overset{\Phi}{\rightleftarrows}} Com\text{-}C.$$

In particular, $\mathcal{K}^0(A^{op})$ is an abelian category.

3.2. *Koszul DG complexes*

In this subsection, A is a connected DG algebra. Let $B(A)$ be the bar construction of A. Then $B(A)$ is a coaugmented DG coalgebra concentrated in nonnegative degrees. As we know, $\mathrm{Ext}^*_A(k_A, k_A)$ is a graded algebra concentrated in nonpositive degrees and $\mathrm{Tor}^*_A(k_A, {}_Ak)$ is a graded coalgebra concentrated in nonnegative degrees. Moreover $\mathrm{Ext}^*_A(k_A, k_A)$ is the dual algebra of $\mathrm{Tor}^*_A(k_A, {}_Ak)$. Let $C = \mathrm{Tor}^0_A(k_A, {}_Ak)$. Since k_A has a semifree resolution $B(A, A) = B(A) \otimes A \longrightarrow k_A$, it follows $\mathrm{Tor}^0_A(k_A, {}_Ak) = \mathrm{H}^0(B(A)) = Z^0(B(A))$, the 0-th cocycles of $B(A)$. Therefore $C = \mathrm{Tor}^0_A(k_A, {}_Ak)$ is a coaugmented cocomplete coalgebra. The inclusion map $C \to B(A)$ induces a morphism of DG algebras $\Omega(C) \longrightarrow \Omega(B(A))$. By [6, Ex.2, p.272], there is a quasi-isomorphism of DG algebras

$$\zeta : \Omega(B(A)) \longrightarrow A. \tag{5}$$

Therefore we get a morphism of DG algebras

$$\varphi : \Omega(C) \longrightarrow A. \tag{6}$$

As before $\tau_0 : C \to \Omega(C)$ is the canonical twisting cochain. Let

$$\tau = \varphi \circ \tau_0. \tag{7}$$

Then it is a twisting cochain from C to A.

Definition 3.1. (Ref. 12) Let A and C be as above. We call the right DG A-module $K(A) = C \otimes_\tau A$ *the koszul DG complex* of A.

Since C is cocomplete, we have seen in the previous subsection that C is a filtered coalgebra with the associated filtration (2). The koszul DG complex $K(A)$ is a semifree DG module with a standard semifree filtration $K(A)(i) = F^i(C) \otimes_\tau A$ for $i \geq 0$.

Similarly, we can define koszul complexes on DG modules. Let X_A be a right DG A-module. Then $\mathrm{Tor}_A^*(X, k)$ is a right graded $\mathrm{Tor}_A^*(k, k)$-comodule. In particular, $\mathrm{Tor}_A^0(X, k)$ is a right $C = \mathrm{Tor}_A^0(k, k)$ comodule. Let $\tau : C \longrightarrow A$ be the twisting cochain in (7) in the preceding subsection.

Definition 3.2. (Ref. 12) Let X_A be a right DG module. We call the DG module $K(X) = \mathrm{Tor}_A^0(X, k) \otimes_\tau A$ *the koszul DG complex* of X.

Since $C = \mathrm{Tor}_A^0(k, k)$ is a cocomplete coalgebra, $M = \mathrm{Tor}_A^0(X, k)$ is a filtered comodule with the induced filtration (3). Hence koszul complex $K(X) = M \otimes_\tau A$ is a semifree DG module.

Definition 3.3. (Ref. 12) Let ${}_AY$ be a left DG A-module, X_A a right DG A-module. Define a complex $K(X, Y) = X \otimes_\tau C \otimes_\tau Y$, that is, as vector spaces $K(X, Y) = X \otimes C \otimes Y$, and the differential is defined by

$$\begin{aligned} d(x \otimes c \otimes y) = d_X(x) \otimes c \otimes y &+ \sum_{(c)} (-1)^{|x|} x \otimes c_{(1)} \otimes \tau(c_{(2)}) y \\ &+ (-1)^{|x|} x \otimes c \otimes d_Y(y) - \sum_{(c)} (-1)^{|x|} x\tau(c_{(1)}) \otimes c_{(2)} \otimes y \end{aligned}$$

where $x \in X$ and $y \in Y$ are homogeneous elements. We call $K(X, Y)$ the *koszul DG complex associated to X_A and ${}_AY$.*

Dual to Definition 3.3, we have cokoszul DG complexes. Let ${}_AU_A$ be a DG A-A-bimodule. Define a complex $\overline{K}(U)$ associated to U as follows. As a graded space $\overline{K}(U) = \underline{\mathrm{Hom}}(C, U)$, and the differential is given by

$$d(f)(c) = d_U(f(c)) + \sum_{(c)} \tau(c_{(1)})f(c_{(2)}) - \sum_{(c)} (-1)^{|f|} f(c_{(1)})\tau(c_{(2)}),$$

where $f \in \underline{\mathrm{Hom}}(C, U)$ and $c \in C$ are homogeneous elements. We call the complex $\overline{K}(U)$ *the cokoszul DG complex* [12] of the DG bimodule U. In fact, one may see that as complexes $\overline{K}(U) \cong \underline{\mathrm{Hom}}_{A^e}(K(A, A), U)$, where $A^e = A \otimes A^{op}$.

Remark 3.1. If we assume in addition that the connected DG algebra A admits an extra grading (usually called the Adams grading [7]), that is, $A = \bigoplus_{i,j\geq 0} A^i_j$ such that $A^0_0 = k$, $A^0_j = 0$ and the differential preserves the lower grading, then everything discussed in the preceding sections also works well. However, in this case, everything admits an additional grading. For instance, $C = \mathrm{Tor}^0_A(k, k)$ is a graded cocomplete coalgebra and the koszul complex $K(A)$ of A is a right DG module with an additional grading, and the twisting cochain $\tau : C \to A$ is of degree $(1, 0)$ in which the first degree corresponding to the upper grading and the second to the lower grading.

Let $R = k \oplus R_1 \oplus R_2 \oplus \cdots$ be a quadratic algebra. Let $A = \bigoplus_{i,j\geq 0} A^i_j$ be a DG algebra such that $A^i_i = R_i$ and $A^i_j = 0$ for $i \neq j$. In this case, a DG A-module can be regarded as a complex of graded R-modules. One may check that the koszul DG complexes $K(A) = C \otimes_\tau A$ is just the usual koszul complex [14,15,25] if we identify a DG module over A with a complex of graded R-modules.

3.3. *Koszul DG algebras and modules*

Definition 3.4. (Refs. 10,11) Let A be a connected DG algebra. A DG module M_A is *koszul* if it admits a semifree resolution with a semifree basis concentrated in degree zero. If the trivial module k_A is koszul then we say A is a *left koszul* DG algebra.

Right koszul DG algebra is defined similarly. Proposition 2.2 in Ref.10 says that a connected DG algebra is left koszul if and only it is right koszul.

If A and A' are quasi-isomorphic connected DG algebras, then the derived categories $\mathcal{D}_{dg}(A)$ and $\mathcal{D}_{dg}(A')$ are equivalent as triangulated categories [7]. So A is koszul if and only if A' is.

Our definition of koszul DG algebras is a natural generalization of the classical koszul algebras. Recall that a connected graded algebra R is called a *koszul algebra* [14,15] if the trivial module ${}_Rk$ has a linear free resolution

$$\cdots \to Q_n \to \cdots \to Q_1 \to Q_0 \to {}_Rk \to 0.$$

Proposition 3.2. *(Ref. 10) Let A be a connected DG algebra. If the cohomology algebra $H(A)$ is a koszul algebra, then A is a koszul DG algebra.*

Example 3.1. Let A be the graded algebra $k\langle x,y\rangle/(y^2,yx)$, where $|x| = |y| = 1$. Let $d(x) = xy$ and $d(y) = 0$. Then d induces a differential d over A and A is a DG algebra. It is not hard to check that $H(A) = k \oplus ky$, which is a koszul algebra. Hence by Proposition 2.2, A is a koszul DG algebra.

Remark 3.2. Each koszul algebra R is the cohomology algebra of a certain koszul DG algebra with nontrivial differential. In fact, R can be viewed as a connected DG algebra with a trivial differential. Then $\Omega B(R)$ is quasi-isomorphic to R as DG algebras. Hence $H(\Omega B(R)) \cong H(R) \cong R$. Clearly $\Omega B(R)$ is a connected DG algebra with a nontrivial differential. By Proposition 2.2, $\Omega B(R)$ is a koszul DG algebra.

3.4. *Koszul DG complexes and Hochschild cohomology*

Let A be a koszul DG algebra, X_A and ${}_AY$ are DG modules. In this case, the koszul DG complexes $K(A)$, $K(X)$ and $K(X,Y)$ have nice properties.

Theorem 3.1. *(Ref. 12) Let A be a connected DG algebra. Then A is a koszul DG algebra if and only if the koszul DG complex $K(A)$ is exact except at 0-th position, and $H^0(K(A)) = k$.*

In this case, $K(A)$ is a semifree resolution of the trivial DG module k_A.

Similarly, we have

Proposition 3.3. *(Ref. 12) If A is a koszul DG algebra, then a bounded below DG module X_A is koszul if and only if the koszul DG complex $K(X)$ is a minimal semifree resolution of X.*

The (co)koszul DG complexes associated to DG modules may be applied to compute the Ext- and Tor-groups and Hochschild cohomologies of DG modules. Let ${}_AU_A$ be a DG A-A-bimodule. Recall that the Hochschild cohomology of U is defined to be

$$\mathrm{HH}^i(U) = \mathrm{Ext}^i_{A^e}(A,U), \text{ for } i \in \mathbb{Z}.$$

Theorem 3.2. *(Ref. 12) Let A be a connected DG algebra. The following are equivalent.*

(i) A is a koszul DG algebra.

(ii) For any bounded below DG modules X_A and ${}_AY$, and $i \in \mathbb{Z}$,

$$\mathrm{Tor}_A^i(X,Y) \cong H^i K(X,Y).$$

(iii) For any bounded below A-A-bimodule U, and $i \in \mathbb{Z}$,

$$HH^i(U) \cong H^i \overline{K}(U).$$

(iv) For any bounded below DG modules X_A and X'_A, and $i \in \mathbb{Z}$,

$$\mathrm{Ext}_A^i(X,X') \cong H^i \overline{K}(\underline{\mathrm{Hom}}(X,X')).$$

4. Koszul DG algebras

In this section, we give a duality theorem between a koszul DG algebra A and its Ext-algebra $E(A) = \mathrm{Ext}_A^*(k,k)$. If the trivial module ${}_Ak$ lies in $\mathcal{D}^c(A)$, then $E(A)$ is finite dimensional. By using results in Ref. 20, we may deduce a duality theorem on the derived categories $\mathcal{D}^c(A)$ and $\mathcal{D}^b(E(A))$. However, if ${}_Ak$ is not in $\mathcal{D}^c(A)$, then $E(A)$ is of infinite dimension. In this case, by using the Foxby duality, we may also deduce a duality theorem on $\mathcal{D}^c(A)$ and $\mathcal{D}^b(E(A))$.

4.1. *The Ext-algebra of a koszul DG algebra*

Let A be a koszul DG algebra. Let $E = E(A) = \oplus_{i\in\mathbb{Z}} \mathrm{Ext}_A^i({}_Ak, {}_Ak)$. Since A is koszul, $E = \mathrm{Ext}_A^0({}_Ak, {}_Ak)$.

Let $R = T(V)/(U)$ be a quadratic algebra, where V is a finite dimensional vector space, and $T(V) = k \oplus V \oplus V^{\otimes 2} \oplus \cdots$ is the tensor algebra over V, and $U \subseteq V \otimes V$ is the subspace spanned by the relations. Recall that the *quadratic dual* $R^!$ of R is defined as $T(V^*)/(U^\perp)$, where V^* is the dual vector space of V and $U^\perp \subseteq (V \otimes V)^* \cong V^* \otimes V^*$ is the orthogonal complement of U.

Theorem 4.1. *(Refs. 10,13) Let A be a koszul DG algebra. Then*

(i) The Ext-algebra E of A is a local algebra with $E/J = k$, where J is the Jacobson radical of E;

(ii) If $H(A)$ is a koszul algebra, then $gr(E) = (H(A))^!$, where $gr(E)$ is the graded algebra associated with the radical filtration of the local algebra E.

There is a duality theorem between A and E.

Theorem 4.2 (Koszul duality on Ext-algebra). *(Refs. 10,13) If* ${}_Ak \in \mathcal{D}^c(A)$ *or* $\dim H(A) < \infty$, *then* $\mathrm{Ext}^*_E(k,k) \cong H(A)$.

4.2. *Koszul Duality*

Let (B, m_B, d_B) be an augmented DG algebra with an augmentation map $\varepsilon_B : B \to k$, and (C, Δ, d_C) be a coaugmented DG coalgebra with a coaugmentation map $\eta_C : k \to C$. Assume $\tau : C \to B$ is a twisting cochain.

Let DGmod-B be the category of right DG B-modules and DGcom-C be the category of right DG C-comodules. Then there is a pair of adjoint functors (L, R) [20,26]:

$$\text{DGcom-}C \underset{R=-\otimes_\tau C}{\overset{L=-\otimes_\tau B}{\rightleftarrows}} \text{DGmod-}B.$$

Let DGcomc-C be the category of cocomplete right DG C-comodules. For any $M, N \in$ DGcomc-C, a DG comodule morphism $f : M \to N$ is called a *weak equivalence related to* τ [20,26] if $L(f) : LM \to LN$ is a quasi-isomorphism. Note that a weak equivalence related to the canonical twisting cochain $\tau_0 : C \to \Omega(C)$ (see Section 3.2) is a quasi-isomorphism. But the converse is not true in general [26]. Let $\mathcal{K}(C)$ be the homotopy category of DGcomc-C. Equipped with the natural exact triangles, $\mathcal{K}(C)$ is a triangulated category. Let $\mathcal{W}$ be the class of weak equivalences in the category $\mathcal{K}(C)$. Then $\mathcal{W}$ is a multiplicative system. The coderived category $\mathcal{D}_{dg}(C)$ of C is defined to be $\mathcal{K}(C)[\mathcal{W}^{-1}]$, the localization of $\mathcal{K}(C)$ at the class $\mathcal{W}$ of weak equivalences [20,26]. Let $\mathcal{D}_{dg}(B^{op})$ be the derived category of right DG B-modules. The following theorem is proved by Lefèvre-Hasegawa in [20, Ch.2], and also can be found in Ref. 26.

Theorem 4.3. *Let C be a cocomplete DG coalgebra, B an augmented DG algebra and $\tau : C \to B$ be a twisting cochain. Then the following are equivalent:*

(i) The map τ induces a quasi-isomorphism $\Omega(C) \to B$;
(ii) The adjunction map

$$B \otimes_\tau C \otimes_\tau B \to B$$

is a quasi-isomorphism;
(iii) The functors L and R induce an equivalence of triangulated categories (also denoted by L and R)

$$\mathcal{D}_{dg}(C) \underset{R}{\overset{L}{\rightleftarrows}} \mathcal{D}_{dg}(B^{op}).$$

Now let A be a koszul DG algebra. Suppose ${}_Ak \in \mathcal{D}^c(A)$. By Theorem 4.1, its Ext-algebra E is a finite dimensional local algebra with the residue field k. Hence the vector space dual E^* is a coaugmented coalgebra which is of course cocomplete. Regard E^* as a coaugmented DG coalgebra concentrated in degree zero. Then all the DG E^*-comodules are cocomplete. Let $C = E^*$ and $B = \Omega(C)$. Then B is a connected DG algebra, and the canonical twisting cochain $\tau_0 : C \to \Omega(C)$ satisfies the condition (i) in the Theorem 4.3. Hence we have the following equivalence of triangulated categories

$$\mathcal{D}_{dg}(E^*) \underset{R}{\overset{L}{\rightleftarrows}} \mathcal{D}_{dg}(\Omega(E^*)^{op}).$$

Clearly, the functors L and R send bounded below objects to bounded below objects. Then the above equivalence induces the following equivalence of triangulated categories:

$$\mathcal{D}^+_{dg}(E^*) \underset{R}{\overset{L}{\rightleftarrows}} \mathcal{D}^+_{dg}(\Omega(E^*)^{op}).$$

Since the DG coalgebra E^* is concentrated in degree zero, the coderived DG category $\mathcal{D}^+_{dg}(E^*)$ is just the derived categories $\mathcal{D}^+(E^*)$ of bounded below complexes of comodules over the coalgebra E^*. Hence we get an equivalence

$$\mathcal{D}^+(E^*) \underset{R}{\overset{L}{\rightleftarrows}} \mathcal{D}^+_{dg}(\Omega(E^*)^{op}).$$

Since E is a finite dimensional algebra, the category of left E-modules is isomorphic to the category of right E^*-comodules [27, 1.6.4]. Hence there is an equivalence of triangulated categories

$$\mathcal{D}^+(E) \underset{G}{\overset{F}{\rightleftarrows}} \mathcal{D}^+(E^*),$$

where $\mathcal{D}^+(E)$ is the derived category of bounded below cochain complexes of left E-modules.

Since A is koszul, there is a quasi-isomorphism of DG algebras $\varphi : \Omega(E^*) \longrightarrow A$ [10[. Hence by [7, Proposition 4.2], the following gives an equivalence of triangulated categories

$$\mathcal{D}^+(A^{op}) \underset{-\otimes^L_{\Omega(E^*)}A}{\overset{\varphi^*}{\rightleftarrows}} \mathcal{D}^+(\Omega(E^*)^{op}).$$

Let $\Phi = (-\otimes^L_{\Omega(E^*)} A)\circ L\circ F$ and $\Psi = G\circ R\circ\varphi^*$. Thus we have the following theorem.

Theorem 4.4 (Koszul Equivalence). *(Ref. 10) Let A be a koszul DG algebra and E be its Ext-algebra. If ${}_Ak \in \mathcal{D}^c(A)$, then we have an equivalence of triangulated categories*

$$\mathcal{D}^+(E) \underset{\Psi}{\overset{\Phi}{\rightleftarrows}} \mathcal{D}^+_{dg}(A^{op}).$$

It is easy to see that $\Phi({}_Ek) = L(k^{E^*}) \otimes^L_{\Omega(E^*)} A = \Omega(E^*) \otimes^L_{\Omega(E^*)} A = A_A$. Temporarily write $\langle {}_Ek \rangle$ the full triangulated subcategory of $\mathcal{D}^+(E)$ generated by ${}_Ek$. By restricting Φ and Ψ, we get an equivalence of triangulated categories

$$\langle {}_Ek \rangle \underset{\Psi_{\mathbf{res}}}{\overset{\Phi_{\mathbf{res}}}{\rightleftarrows}} \mathcal{D}^c(A^{op}).$$

Since E is a finite dimensional local algebra, $\langle {}_Ek \rangle = \mathcal{D}^b(\text{mod-}E)$, where mod-$E$ is the category of finitely generated left E-modules. Thus we have an equivalence of triangulated categories

$$\mathcal{D}^b(\text{mod-}E) \underset{\Psi_{\mathbf{res}}}{\overset{\Phi_{\mathbf{res}}}{\rightleftarrows}} \mathcal{D}^c(A^{op}).$$

Now, we are able to state a DG version of the koszul duality.

Theorem 4.5 (Koszul Duality). *(Ref. 10) Let A be a koszul DG algebra and E be its Ext-algebra. Suppose ${}_Ak \in \mathcal{D}^c(A)$. Then there is a duality of triangulated categories*

$$\mathcal{D}^b(\textit{mod-}E^{op}) \underset{\mathcal{G}}{\overset{\mathcal{F}}{\rightleftarrows}} \mathcal{D}^c(A^{op}).$$

4.3. *Koszul Duality in non-compact case*

The Koszul Duality Theorem 4.5 is valid only under the condition that the trivial module ${}_Ak$ is compact. The DG algebras A arising form algebraic topology and geometry usually have the property that the cohomology algebra $H(A)$ is finite dimensional. If A is a connected DG algebra with $H(A)$ finite dimensional, then the trivial module ${}_Ak$ must not be compact [13,28]. We wonder whether the koszul duality theorem holds for this kind of koszul DG algebras.

Let A be a koszul DG algebra. Let $\mathcal{I}$ be a K-injective resolution of ${}_Ak$. Let $B = \underline{\mathrm{Hom}}_A(\mathcal{I}, \mathcal{I})$. Then B is a DG algebra and $\mathcal{I}$ is a DG $A \otimes B$-module. Since A is koszul, $H^i(B) = 0$ for $i \neq 0$. We have the following truncation

$$B' := \cdots B^{-n} \longrightarrow B^{-n+1} \longrightarrow \cdots \longrightarrow B^{-1} \longrightarrow Z^0(B) \longrightarrow 0,$$

where $Z^0(B)$ is the 0-th cocycles of B. Then one can check that B' is a DG subalgebra of B. Of course the inclusion map $B' \hookrightarrow B$ is a quasi-isomorphism. Hence $\mathcal{I}$ is a DG $A \otimes B'$-module. Now let

$$F = \mathrm{R}\underline{\mathrm{Hom}}_A(-,\mathcal{I}) : \mathcal{D}(A) \to \mathcal{D}(B') \text{ and } G = \mathrm{R}\underline{\mathrm{Hom}}_{B'}(-,\mathcal{I}) : \mathcal{D}(B') \to \mathcal{D}(A)$$

Then F and G is a pair of adjoint contravariant functors. Let

$$\mathcal{A}(A) = \{M \in \mathcal{D}(A) | \ M \cong GFM \} \text{ and } \mathcal{B}(B') = \{N \in \mathcal{D}(B') | \ N \cong FGN \}$$

be the Auslander class and Bass class respectively.

Lemma 4.1 (Foxby duality). *If (F, G) be a pair of adjoint contravariant triangulated functors between triangulated categories $\mathcal{C}$ and $\mathcal{D}$, then*

(i) The Auslander class and the Bass class are full triangulated subcategories,

(ii) F and G induce a pair of dualities on the Auslander class and the Bass class.

Since the Ext-algebra E of A is a local algebra, we may deduce the following koszul duality theorem.

Theorem 4.6 (Koszul duality). *(Ref. 13) Let A be a koszul DG algebra, E be its Ext-algebra. If* $\dim H(A) < \infty$ *and E is noetherian, then we have a pair of dualities of triangulated categories*

$$\theta : \mathcal{D}_{fd}(A) \longrightarrow \mathcal{D}^b(\textit{mod-}E) \text{ and } \phi : \mathcal{D}^b(\textit{mod-}E) \longrightarrow \mathcal{D}_{fd}(A),$$

where $\mathcal{D}^b(\textit{mod-}E)$ is the bounded derived category of finitely generated left E-modules, and $\mathcal{D}_{fd}(A)$ is the full triangulated subcategory of $\mathcal{D}(A)$ consisting of DG modules M such that $\dim H(M) < \infty$

Moreover, under this duality, $\theta({}_A A) = {}_E k$ and $\theta({}_A k) = {}_E E$.

Remark 4.1. Koszul duality in the A_∞ setting may be found in Ref. 29.

5. Koszul DG modules

Let A be a connected DG algebra, M_A be a bounded below DG module. If M has bounded Ext-groups , then it must contain a koszul DG submodule (up to a shift) when A is a koszul DG algebra. Moreover M can be "approximated" by koszul DG modules. The triangulated category $\mathcal{D}^+_{dg}(A^{op})$ has a natural t-structure [7]. This t-structure induces a t-structure on its subcategory $\mathcal{D}^c(A^{op})$. Moreover, if the cohomology algebra $H(A)$ is finite dimensional and A is koszul, the triangulated category $\mathcal{D}_{fd}(A)$ also has a t-structure. By using the Koszul Duality Theorems in the previous section, we give a description of the hearts of the above t-structures.

5.1. *DG modules with bounded Ext-groups*

We say a DG module M_A has *bounded Ext-groups* if there are integers $s > t$ such that $\mathrm{Ext}^i_A(M,k) = 0$ for $i \leq t$ or $i \geq s$. The purpose of this subsection is to prove that a DG module over a koszul DG algebra has bounded Ext-groups if and only if it can be approximated by a sequence of koszul DG modules (up to shift). The results here can be regarded as natural generalizations of some results in Ref. 30.

Lemma 5.1. *(Ref. 11) Let M_A a bounded below DG module. If there is an integer t such that $H^iM = 0$ for all $i \geq t$, then $\mathrm{Ext}^i_A(M,k) = 0$ for all $i \leq -t$.*

This lemma shows that the vanishing of the cohomology groups of a bounded below DG module M implies the vanishing of the Ext-groups of M. If the cohomology algebra $H(A)$ is finite dimensional, then the bounded property of the cohomology groups of M is equivalent to the bounded property of the Ext-groups of M. The following theorem tells us that any DG A-module with bounded Ext-group admits a koszul DG submodule (up to a shift).

Theorem 5.1. *(Ref. 11) Let M_A be a bounded below DG module such that $\mathrm{Ext}^i_A(M,k) = 0$ for $i \leq -t$. Then $\tau^{\geq t}M[t]$ is a koszul DG module.*

Corollary 5.1. *(i) For any DG module $M_A \in \mathcal{D}^c(A)$, there is an integer t such that $\tau^{\geq t}M[t]$ is a koszul DG module.*

(ii) For any koszul DG module M_A, and any integer $t \geq 0$, $\tau^{\geq t}M[t]$ is koszul.

Theorem 5.2. *(Ref. 11) A bounded below DG module M_A has bounded Ext-groups if and only if there is a filtration of DG submodules*

$$0 = F_0M \subseteq F_1M \subseteq \cdots \subseteq F_nM = M$$

such that $F_iM/F_{i-1}M$ is a koszul DG module (up to a shift) for $1 \leq i \leq n$.

5.2. *Hearts of t-structures*

Let $\mathcal{D} = \mathcal{D}^+_{dg}(A^{op})$, and

$$\mathcal{D}^{\geq 0} = \{M \in \mathcal{D} \mid \mathrm{Ext}^i_A(M,k) = 0 \text{ for } i > 0\};$$

$$\mathcal{D}^{\leq 0} = \{M \in \mathcal{D} \mid \mathrm{Ext}^i_A(M,k) = 0 \text{ for } i < 0\}.$$

Then $(\mathcal{D}^{\geq 0}, \mathcal{D}^{\leq 0})$ is a t-structure on the triangulated category $\mathcal{D}$ [7]. This induces a t-structure $(\overline{\mathcal{D}}^{\geq 0}, \overline{\mathcal{D}}^{\leq 0})$ on the triangulated subcategory $\mathcal{D}^c(A^{op})$, where

$$\overline{\mathcal{D}}^{\geq 0} = \{M \in \mathcal{D}^c(A^{op}) \mid \operatorname{Ext}_A^i(M,k) = 0 \text{ for } i > 0\};$$

$$\overline{\mathcal{D}}^{\leq 0} = \{M \in \mathcal{D}^c(A^{op}) \mid \operatorname{Ext}_A^i(M,k) = 0 \text{ for } i < 0\}.$$

Let $\mathcal{A} = \overline{\mathcal{D}}^{\geq 0} \cap \overline{\mathcal{D}}^{\leq 0}$ be the heart.

Now let A be a koszul DG algebra, $E = \operatorname{Ext}_A^0({}_Ak, {}_Ak)$ be its Ext-algebra. Let $(\mathcal{T}^{\geq 0}, \mathcal{T}^{\leq 0})$ be the t-structure on $\mathcal{D}^b(\text{mod-}E)$ defined by the truncations, that is,

$$\mathcal{T}^{\geq 0} = \{M \in \mathcal{D}^b(\text{mod-}E) \mid H^iM = 0 \text{ for } i < 0\};$$

$$\mathcal{T}^{\leq 0} = \{M \in \mathcal{D}^b(\text{mod-}E) \mid H^iM = 0 \text{ for } i > 0\}.$$

Proposition 5.1. *(Ref. 11) Suppose that A is a koszul algebra such that ${}_Ak \in \mathcal{D}^c(A)$. The functors $\mathcal{F}$ and $\mathcal{G}$ in Theorem 4.5 are t-contra-exact functors, that is,*

$$\mathcal{F}(\mathcal{T}^{\geq 0}) \subseteq \overline{\mathcal{D}}^{\leq 0}, \quad \mathcal{F}(\mathcal{T}^{\leq 0}) \subseteq \overline{\mathcal{D}}^{\geq 0}; \quad \textit{and} \quad \mathcal{G}(\overline{\mathcal{D}}^{\geq 0}) \subseteq \mathcal{T}^{\leq 0}, \quad \mathcal{G}(\overline{\mathcal{D}}^{\leq 0}) \subseteq \mathcal{T}^{\geq 0}.$$

Notice that $E^{op} = \operatorname{Ext}_A^0({}_Ak, {}_Ak)^{op} = \operatorname{Ext}_A^0(k_A, k_A)$. Hence a left $\operatorname{Ext}_A^0(k_A, k_A)$-module may be viewed as a right E-module. Then we have

Theorem 5.3. *(Ref. 11) Let A be a koszul DG algebra, $(\overline{\mathcal{D}}^{\geq 0}, \overline{\mathcal{D}}^{\leq 0})$ be the t-structure on $\mathcal{D}^c(A^{op})$, and $\mathcal{A} = \overline{\mathcal{D}}^{\geq 0} \cap \overline{\mathcal{D}}^{\leq 0}$ be the heart. Let E be the Ext-algebra of A. If $k_A \in \mathcal{A}$, then*

$$\operatorname{Ext}_A^0(-, k_A) \cong \operatorname{Hom}_{\mathcal{A}}(-, k_A) : \mathcal{A} \longrightarrow \textit{mod-}E$$

gives a duality between the abelian categories.

Now let $\mathcal{K} = \mathcal{D}_{fd}(A)$, and

$$\mathcal{K}^{\geq 0} = \{{}_AN \in \mathcal{K} \mid \operatorname{Ext}_A^i(N,k) = 0 \text{ for } i > 0\},$$

$$\mathcal{K}^{\leq 0} = \{{}_AN \in \mathcal{K} \mid \operatorname{Ext}_A^i(N,k) = 0 \text{ for } i < 0\}.$$

In general, the pairing $(\mathcal{K}^{\geq 0}, \mathcal{K}^{\leq 0})$ may not be a t-structure on $\mathcal{K}$.

By Lemma 5.1 and the Koszul Duality Theorem 5.1, we have

Proposition 5.2. *(Ref. 13) Let A be a koszul DG algebra such that $H(A)$ is of finite dimension and its Ext-algebra E is noetherian. Then the pairing*

$(\mathcal{K}^{\geq 0}, \mathcal{K}^{\leq 0})$ *defines a t-structure on* $\mathcal{D}_{fd}(A)$*, and the functors* θ *and* ϕ *in Theorem 5.1 are t-contra-exact functors, that is,*

$$\phi(\mathcal{T}^{\geq 0}) \subseteq \mathcal{K}^{\leq 0}, \quad \phi(\mathcal{T}^{\leq 0}) \subseteq \mathcal{K}^{\geq 0}; \quad \textit{and} \quad \theta(\mathcal{K}^{\geq 0}) \subseteq \mathcal{T}^{\leq 0}, \quad \theta(\mathcal{K}^{\leq 0}) \subseteq \mathcal{T}^{\geq 0}.$$

Let $\mathcal{H} = \mathcal{K}^{\geq 0} \bigcap \mathcal{K}^{\leq 0}$ be the heart of the t-structure. Then we have

Theorem 5.4. *(Ref. 13) Let* A *and* E *be as in previous proposition. Then* $\mathrm{Ext}^0_A(-, k) : \mathcal{H} \longrightarrow$ *mod-*E *is an anti-equivalence between the abelian categories.*

6. BGG correspondence

Bernstein-Gelfand-Gelfand [2] established an equivalence of categories

$$\overline{\text{grmod-}}\Lambda(V) \cong \mathcal{D}^b(\mathrm{Coh}\mathbb{P}^n),$$

where $\overline{\text{grmod-}}\Lambda(V)$ is the stable category of finitely generated graded modules over the exterior algebra $\Lambda(V)$ of an $(n+1)$-dimensional space $V = kx_0 \oplus kx_1 \oplus \cdots \oplus kx_n$, and $\mathcal{D}^b(\mathrm{Coh}\mathbb{P}^n)$ is the bounded derived category of coherent sheaves over the n-dimensional projective space $\mathbb{P}^n$. This equivalence is now called the *BGG correspondence* in literature. A sketch of the proof of the BGG correspondence can be found also in [31, P.273, Ex.1]. The BGG correspondence has been generalized to noncommutative projective geometry by several authors [17–19]. Let R be a koszul noetherian AS-Gorenstein algebra with finite global dimension. Then its Ext-algebra $E(R)$ is a Frobenius algebra [25]. A version of the noncommutative BGG correspondence was proved in Ref. 18, which was stated as

$$\overline{\text{grmod-}}E(R) \cong \mathcal{D}^b(\mathrm{qgr}R^{op}),$$

where $\mathrm{qgr}R^{op}$ is the quotient category grmod-$R^{op}/\mathrm{tors}R^{op}$ [32]. Let $\mathcal{D}^b_{fd}(\text{grmod-}R^{op})$ be the full subcategory of $\mathcal{D}^b(\text{grmod-}R^{op})$ consisting of objects X with finite dimensional cohomology groups. It is well known that [33]

$$\mathcal{D}^b(\mathrm{qgr}R^{op}) = \mathcal{D}^b(\text{grmod-}R^{op})/\mathcal{D}^b_{fd}(\text{grmod-}R^{op}).$$

Hence the above BGG correspondence can be stated as

$$\overline{\text{grmod-}}\,E(R) \cong \mathcal{D}^b(\text{grmod-}R^{op})/\mathcal{D}^b_{fd}(\text{grmod-}R^{op}). \tag{8}$$

In this section, we deduce a correspondence similar to (8) in our koszul DG setting.

Recall that a connected DG algebra A is *right AS-Gorenstein* (AS stands for Artin-Schelter) if $\mathrm{RHom}_{A^{op}}(k, A) \cong k[n]$ for some integer n [34,35];

A is *right AS-regular* if A is right AS-Gorenstein and $k_A \in \mathcal{D}^c(A^{op})$. A connected DG algebra A is *Frobenius* [35] if there is a quasi-isomorphism ${}_AA \longrightarrow {}_AA^\sharp[l]$ for some integer l.

6.1. *BGG correspondence for Koszul AS-regular DG algebras*

From Koszul Duality Theorem 4.5, we can deduce the following proposition.

Proposition 6.1. *Let A be a koszul DG algebra with Ext-algebra $E = \mathrm{Ext}^*_A(k,k)$. Then A is right AS-regular if and only if E is Frobenius.*

Let $\overline{\text{mod-}E^{op}}$ be the stable category of mod-E^{op}. If E is Frobenius, then $\overline{\text{mod-}E^{op}}$ is a triangulated category.

Theorem 6.1 (BGG Correspondence). *(Ref. 10) Let A be a koszul DG AS-regular algebra with Ext-algebra E. Then there is a duality of triangulated categories*

$$\overline{\textit{mod-}E^{op}} \rightleftarrows \mathcal{D}^c(A^{op})/\mathcal{D}_{fd}(A^{op}).$$

6.2. *BGG correspondence for Koszul Frobenius DG algebras*

Recall that a noetherian local algebra R with residue field k is said to be *Gorenstein* if there is an integer $d \geq 0$ such that

$$\mathrm{Ext}^n_R(k,R) = \begin{cases} 0, & n \neq d; \\ k, & n = d. \end{cases}$$

Let A be a koszul DG algebra such that the cohomology algebra $H(A)$ is finite dimensional. Let E be its Ext-algebra. Assume that E is noetherian. From the Koszul Duality Theorem 5.2, we can deduce the following proposition.

Proposition 6.2. *(Ref. 13) E is Gorenstein if and only if A is Frobenius.*

Let J be the Jacobson radical of E. An E-module M is called a J-*torsion* module if for any element $m \in M$ there is an integer n such that $J^n m = 0$. Let $\mathrm{tor}\, E$ be the full subcategory of $\mathrm{mod}\, E$ consisting of all the J-torsion modules. Since E is noetherian, $\mathrm{tor}\, E$ is a thick abelian subcategory of $\mathrm{mod}\, E$. Write $\mathrm{qmod}\, E$ to be the quotient category $\mathrm{mod}\, E/\mathrm{tor}\, E$. Since E is noetherian, $\mathrm{tor}\, E$ is exactly the category of all finite dimensional E-modules.

Theorem 6.2 (BGG correspondence). *(Ref. 13) Let A be a koszul DG algebra with Ext-algebra E. If E is noetherian, then we have anti-equivalence of triangulated categories*

$$\mathcal{D}^b(qmod\, E) \longrightarrow \mathcal{D}_{fd}(A)/\mathcal{D}^c(A).$$

Remark 6.1. BGG correspondence in A_∞-setting may be found in Refs. 29 and 36.

6.3. *BGG correspondence for graded koszul algebra revisited*

As the discussions in Remark 3.1, we now consider the bigraded DG algebras. We show that the BGG correspondence obtained above implies the classical one on graded koszul algebras.

A bigraded DG algebra (the extra graded is usually called the Adams grading) A is said to be *Adams connected* if (1) $A^i_j = 0$ for $i < 0$ or $j < 0$, and (2) $A^0_0 = k$, $A^0_j = 0$ and $A^i_0 = 0$ for $i, j \neq 0$. The Ext-algebra $\mathcal{E}xt^{*,*}_A(k,k)$ is also bigraded.

Let A be an Adams connected DG algebra. It is called a koszul Adams DG algebra if $\mathcal{E}xt^{i,*}_A(k,k) = \bigoplus_{j\in\mathbb{Z}} \mathcal{E}xt^{i,j}_A(k,k) = 0$ for all $i \neq 0$. A is called an *AS-Gorenstein* Adams DG algebra if $\mathrm{RHom}_{A^{op}}(k, A) \cong k[r](s)$. Moreover if $\mathcal{E}xt^{*,*}_A(k,k)$ is finite dimensional, then A is called an *AS-regular* Adams DG algebra.

For a koszul Adams connected DG algebra A, its Ext-algebra $\mathcal{E} = \mathcal{E}xt^{*,*}_A(k,k)$ has the property that $\mathcal{E}^i_j = 0$ for $i \neq 0$ or $j > 0$. Now let $S_j = \mathcal{E}^0_{-j}$. Then $S = \oplus_{j\geq 0} S_j$ is a connected graded algebra.

Let A be a koszul Adams connected DG algebra. Let $\mathcal{AD}_{dg}(A^{op})$ be the derived category of bigraded DG right A-modules. Denote $\mathcal{AD}^c(A^{op})$ as the full triangulated subcategory of $\mathcal{AD}_{dg}(A^{op})$ generated A_A. Then the BGG correspondence in Theorem 6.1 reads as follows.

Theorem 6.3. *Let A be a koszul AS-regular Adams DG algebra and S be the connected graded such that $S_j = \mathcal{E}xt^{0,-j}_A(k,k)$. Then there is an duality of triangulated categories*

$$\overline{grmod\text{-}S^{op}} \cong \mathcal{AD}^c(A^{op})/\mathcal{AD}_{fd}(A^{op}),$$

where $\overline{grmod\text{-}S^{op}}$ is the stable category of graded right S-modules, and $\mathcal{AD}_{fd}(A^{op})$ is the full triangulated subcategory of $\mathcal{AD}^c(A^{op})$ consisting of objects M such that $\dim H(M) < \infty$.

Now let R be a noetherian connected graded algebra. Let A be the bigraded DG algebra with trivial differential by taking $A^i_i = R_i$ and $A^i_j = 0$ if $i \neq j$. If R is a koszul algebra, then it is not hard to see that A is a koszul Adams connected DG algebra. Moreover, $\mathrm{Ext}^{0,-j}_A(k,k) = R^!_j$ for all $j \geq 0$, i.e., $S = R^! = E(R) = \mathrm{Ext}^*_R(k,k)$. Suppose that gl. dim $R < \infty$. Then $S = E(R)$ is finite dimensional. Since $E(R)$ is finite dimensional grmod-$E(R)^{op}$ is dual to grmod-$E(R)$. Hence $\mathcal{D}^b(\text{grmod-}S^{op}) = \mathcal{D}^b(\text{grmod-}E(R)^{op})$ is dual to $\mathcal{D}^b(\text{grmod-}E(R))$. Let us inspect the category $\mathcal{AD}^c(\mathcal{A}^{op})$ in Theorem 6.3. Since the differential of A is trivial and A is concentrated in the diagonal of the first quadrant, the triangulated category $\mathcal{AD}_{dg}(\mathcal{A}^{op})$ is naturally equivalent to the derived category $\mathcal{D}(\text{Grmod-}R^{op})$ of the category Grmod-R of right graded R-modules. Under this equivalence, A_A is corresponding to R_R in $\mathcal{D}(\text{Grmod-}R^{op})$. Hence $\mathcal{AD}^c(\mathcal{A}^{op})$ is equivalent to the full triangulated subcategory of $\mathcal{D}(\text{Grmod-}R^{op})$ generated by R_R (closed under the shifts on the grading of R_R), which is equivalent to $\mathcal{D}^b(\mathrm{proj}\, R^{op})$, the bounded derived category of finitely generated graded projective right R-modules. Since R is noetherian and has finite global dimension, $\mathcal{D}^b(\mathrm{proj} R^{op})$ is equivalent to $\mathcal{D}^b(\text{grmod-}R^{op})$, the bounded derived category of finitely generated graded right R-modules. In summary we have the equivalence (which is established in Ref. 15) of triangulated categories if R is noetherian and of finite global dimension

$$\mathcal{D}^b(\text{grmod-}\, E(R)) \cong \mathcal{D}^b(\text{grmod-}\, R^{op}). \tag{9}$$

Moreover, we assume that R is a noetherian koszul AS-regular algebra. Then the Adams connected DG algebra A is koszul Adams AS-regular DG algebra. Hence in the left hand of (9), $\overline{\text{grmod-}S^{op}}$ is dual to $\overline{\text{grmod-}\, E(R)}$. Since $\mathcal{AD}^c(\mathcal{A}^{op})$ is equivalent to $\mathcal{D}^b(\text{grmod-}R^{op})$, the full triangulated subcategory $\mathcal{AD}_{fd}(\mathcal{A}^{op})$ is equivalent to $\mathcal{D}^b_{fd}(\text{grmod-}R^{op})$, the triangulated subcategory consisting of objects X such that HX is finite dimensional. Hence in the right hand of (9),

$$\mathcal{AD}^c(\mathcal{A}^{op})/\mathcal{AD}_{fd}(\mathcal{A}^{op}) \cong \mathcal{D}^b(\text{grmod-}R^{op})/\mathcal{D}^b_{fd}(\text{grmod-}R^{op})$$

which is equivalent to $\mathcal{D}^b(\mathrm{qgr} R^{op})$ [33]. In summary we get the BGG correspondence established in Ref. [18]

$$\overline{\text{grmod-}\, E(R)} \cong \mathcal{D}^b(\mathrm{qgr} R^{op}).$$

Remark 6.2. The classical BGG correspondence also can be deduced from Theorem 6.2 (see Ref. 13).

Acknowledgments

The second author is supported by the NSFC (key project 10731070) and supported by the Doctorate Foundation (No. 20060246003), Ministry of Education of China.

References

1. Y. I. Manin, *Quantum Groups and Noncommutative Geometry*, Université de Montréal, Centre de Recherches Mathématiques, Montreal, QC, 1988.
2. I. N. Bernstein, I. M. Gelfand and S. I. Gelfand, Algebraic bundles over $\mathbb{P}^n$ and problems in linear algebra, Funct. Anal. Appl. **12** (1979), 212-214.
3. B. Keller, Deriving DG categories, Ann. scient. Éc. Norm. Sup., 4 série, **27** (1994), 63-102.
4. B. Keller, On differential graded categories, arXiv:math.KT/0601185.
5. R. Bezrukavnikov, Koszul DG-algebras arising from configuration spaces, Geom. Funct. Anal, **4** (1994), 119-135.
6. Y. Félix, S. Halperin and J.-C. Thomas, *Rational Homotopy Theory*, GTM **205**, Springer-Verlag, New York, 2001.
7. I. Kříž and J. P. May, Operads, algebras, modules and motives, Astérisque **233**, 1995.
8. A. Polishchuk and L. Positselski, *Quadratic Algebras*, University Lecture Series **37**, American Mathematical Society, Providence, RI, 2005.
9. S. Morita, *Geometry of Differetial Forms*, Transl. Math. Monographs **201**, American Mathematical Society, Providence, RI, 2001.
10. J.-W. He and Q.-S. Wu, Koszul differential graded algebras and BGG correspondence, arXiv:math.RA/0712.1324.
11. J.-W. He and Q.-S. Wu, Koszul differential graded modules, preprint 2007.
12. J.-W. He and Q.-S. Wu, Koszul differential graded complexes, preprint 2007.
13. J.-W. He and Q.-S. Wu, Koszul differential graded algebras and BGG correspondence II, preprint 2007.
14. S. Priddy, Koszul resolutions, Trans. Amer. Math. Soc. **152** (1970), 39-60.
15. A. A. Beilinson, V. Ginzburg and W. Soergel, Koszul duality patterns in representation theory, J. Amer. Math. Soc. **9** (1996), 473-527.
16. C. Löfwall, On the subalgebra generated by the one-dimensional elements in the Yoneda Ext-algebra, Algebra, Algebraic Topology and Their Interactions (Stockholm, 1983), 291-338, LNM **1183**, Springer-Verlag, Berlin-New York, 1986.
17. P. Jorgensen, A noncommutative BGG correspondence, Pacific J. Math. **218** (2005), 357-377.
18. R. Martínez Villa and M. Saorín, Koszul equivalence and dualities, Pacific J. Math. **214** (2004), 359-378.
19. I. Mori, Riemann-Roch like theorem for triangulated categories, J. Pure Appl. Algebra **193** (2004), 263-285.
20. K. Lefèvre-Hasegawa, *Sur les A_∞-Catégories*, Université Paris 7, Thése de Doctorat, 2003.

21. L. L. Avramov, H.-B. Foxby and S. Halperin, *Differetial Homological algebra*, manuscript.
22. C. Weibel, *An Introduction to Homological Algebra*, Cambridge University Press, Cambridge, 1994.
23. J. Huebschmann, Berikashvilis functor $\mathcal{D}$ and the deformation equation, Festschrift in honor of N. Berikashvilis 70th birthday, Proceedings of A. Razmadze Institute **119**, 59-72, 1999.
24. D. Husemoller, J. C. Moore and J. Stasheff, Differential homological algebra and homogeneous spaces, J. Pure Appl. Algebra **5** (1974), 113-185.
25. S. P. Smith, Some finite dimensional algebras related to elliptic curves, CMS Conf. Proc. **19** (1996), 315-348.
26. B. Keller, Koszul duality and coderived categories, preprint, 2003.
27. S. Montgomery, *Hopf algebras and their actions on rings*, CBMS Regional Conference Series in Mathematics **82**, Providence, RI, 1993.
28. X.-F. Mao and Q.-S. Wu, Homological Invariants for Connected DG Algebras, to appear in Comm. Algebra.
29. D.-M. Lu, J. H. Palmieri, Q.-S. Wu, J.J. Zhang, Koszul Equivalences in A_∞-algebras, arXiv:math.RA/0710.5492.
30. R. Martinez-Villa, D. Zacharia, Approximations with modules having linear resolutions, J. Algebra, Vol. 266, 671-697, 2003.
31. S. I. Gelfand and Y. I. Manin, *Methods of Homological algebra*, Springer-Verlag, 1997.
32. M. Artin and J. J. Zhang, Noncommutative projective schemes, Adv. Math. **109** (1994), 228-287.
33. J.I. Miyachi, Localization of triangulated categories and derived categories, J. Algebra **141** (1991), 463-483.
34. Y. Félix, S. Halperin and J.-C. Thomas, Gorenstein spaces, Adv. Math. **71** (1988), 92-112.
35. D.-M Lu, J. H. Palmieri, Q.-S. Wu and J. J. Zhang, A_∞-algebras for ring theorists, Alg. Colloq. **11** (2004), 91-128.
36. V. Baranovsky, BGG correspondence for toric complete intersections, arXiv:math.RT/07061398.
37. A. Beligiannis, The homological theory of contravariantly finite subcategories: Auslander-Buchweitz contexts, Gorenstein categories and (co-)stablization, Comm. Algebra **28** (2000), 4547-4596.
38. E. L. Green and R. Martínez Villa, Koszul and Yoneda algebras, Canad. Math. Soc. Conference Proceedings **18** (1996), 247-297.
39. D.-M. Lu, J. H. Palmieri, Q.-S. Wu, J.J. Zhang, Artin-Schelter Regular Algebras of Dimension 4 and Their -Ext-Algebras, Duke Math. J. **137** (2007), 537-584.
40. R. Martínez Villa, Koszul algebras and sheaves over projective space, arXiv:math.RT/0405538.

QUASI-CONTINUOUS RINGS SATISFYING CERTAIN CHAIN CONDITIONS*

Jin Yong Kim
Department of Mathematics and Institute of Natural Sciences, Kyung Hee University
Suwon 449–701, South Korea
E-mail: jykim@khu.ac.kr

Jaekyung Doh and Jae Keol Park*
Department of Mathematics, Busan National University
Busan 609–735, South Korea
**E-mail: jkpark@pusan.ac.kr*

In this paper, two-sided quasi-continuous rings or left CS rings with certain chain conditions are QF. For example, it is shown that two-sided quasi-continuous left perfect rings with DCC on left annihilators are QF. Also it is proved that left CS left GP-injective rings with ACC on left annihilators are QF.

Keywords: QF ring, quasi-continuous ring, GP-injective ring, left perfect ring, left CS ring

1. Introduction

A well-known result of Faith [10] asserts that a left self-injective ring is quasi-Frobenius (simply, QF) provided it has ACC on left annihilators. Armendariz and Park [3] proved that a left self-injective ring is QF if the factor ring modulo its left socle is left Goldie. In [20] it was also shown that a left Artinian two-sided quasi-continuous ring is QF. In this paper we extend these results to the case of left perfect rings with DCC on left annihilators and to that of left CS left GP-injective rings with ACC on left annihilators.

Throughout this paper all rings considered are associative with identity and all modules are unitary. For a ring R, we write $J(R)$, $Z({}_RR)$, $\mathrm{Soc}({}_RR)$,

---

*2000 Mathematics Subject Classification. 16L60, 16P60, 16D50.

and $\mathrm{Soc}(R_R)$ for the Jacobson radical, the left singular ideal, the socle of ${}_RR$, and the socle of R_R, respectively. For a nonempty subset X of a ring R, $\ell_R(X)$ and $r_R(X)$ represent the left annihilator and the right annihilator of X in R, respectively. Recall that a ring R is *left continuous* [24] if it satisfies the following:

(C1) every left ideal of R is essential in a direct summand of ${}_RR$; and
(C2) every left ideal isomorphic to a direct summand of ${}_RR$ is itself a direct summand of ${}_RR$.

A ring R is said to be *left CS*, if it satisfies condition (C1) only. A left CS-ring R is said to be *left quasi-continuous* if $Re \cap Rf = 0$ where e and f are idempotents in R, then $Re + Rf$ is a direct summand of ${}_RR$. Similarly, right quasi-continuous rings can be defined. Both left and right quasi-continuous rings are called *two-sided quasi-continuous.* Recall that R is called *left P-injective*, if every left R-homomorphism from a principal left ideal of R into R extends to an endomorphism of ${}_RR$. A ring R is said to be *generalized left principally injective* (simply, *left GP-injective*) if, for any nonzero a in R, there exists a positive integer n such that $a^n \neq 0$ and any left R-homomorphism from Ra^n to R extends to an endomorphism of ${}_RR$. In [18, Theorem 2], it was shown that $J(R) = Z({}_RR)$ if R is left GP-injective. Recall that a ring R is called a *left perfect* if $J(R)$ is left T-nilpotent and the factor ring $R/J(R)$ is semisimple Artinian.

2. QF-Rings via Two-Sided Quasi-continuous Rings

In this section, by using a result derived from supplying an affirmative answer to a conjecture in [4, Remark 1] to the case of left continuous rings, QF rings are characterized in terms of two-sided quasi-continuous rings with certain chain conditions. As byproduct, we can show that a left Artinian two-sided quasi-continuous ring is QF, which is the main result in [20].

According to Azumaya [4] an R-module A is called a *finite*, (resp. *single*), *extension* of its submodule B if the factor module B/A is finitely generated (resp. cyclic). Also a submodule B of A is said to be *finitely*, (resp. *singly*), *split* in A if, for every submodule A' of A which is a finite (resp. single) extension of B, B is a direct summand of A'. Also a module M is said to be *finitely projective*, (resp. *singly projective*), if every epimorphism onto M is finitely split (resp. singly split). It was conjectured [4, Remark 1] that every flat left R-module is finitely projective if and only if the ascending chain $\ell_R(a_1) \subseteq \ell_R(a_1a_2) \subseteq \ell_R(a_1a_2a_3) \subseteq \cdots$ terminates for every infinite sequence of elements $a_1, a_2, a_3, \ldots$ of R. In general this conjecture is known to be false [25]. But [9, Corollary 3.5] gave an affirmative answer to this conjecture for GP-injective rings. In the following we also give a positive answer to the conjecture for the case of left continuous rings.

Proposition 2.1. *Let R be a left continuous ring. Then the following statements are equivalent:*

(1) *R is left perfect.*
(2) *Every flat left R-module is singly projective.*
(3) *For every infinite sequence $a_1, a_2, \ldots$ in R, the ascending chain $\ell_R(a_1) \subseteq \ell_R(a_1a_2) \subseteq \ell_R(a_1a_2a_3) \subseteq \cdots$ terminates.*

Proof. (1)$\Rightarrow$(2) is clear and (2)$\Rightarrow$(3) follows from [4, Corollary 25]. (3)$\Rightarrow$(1) Since R is left continuous, we have $J = J(R) = Z(_RR)$ and R/J is regular. For a sequence $a_1, a_2, \ldots$ in $J(R)$, there exists a positive integer n such that $\ell_R(a_1a_2\cdots a_n) = \ell_R(a_1a_2\cdots a_na_{n+1})$ by hypothesis. Hence $Ra_1a_2\cdots a_n \cap \ell_R(a_{n+1}) = 0$. Since $a_{n+1} \in J(R) = Z(_RR)$, $a_1a_2\cdots a_n = 0$. Hence $J(R)$ is left T-nilpotent. It remains to show that $R/J(R)$ is orthogonally finite. Let $\{f_1, f_2, \ldots, f_n, \ldots\}$ be a set of countably infinite nonzero orthogonal idempotents in $R/J(R)$. Then there exists nonzero orthogonal idempotents $e_1, e_2, \ldots$ in R such that $f_i = e_i + J$. Let $a_i = 1 - (e_1 + e_2 + \cdots + e_i)$, where $i = 1, 2, \ldots$. Then $a_{i+1} = a_i - a_ie_{i+1}a_i$, $e_{i+1}a_i = e_{i+1} \neq 0$ and $e_{i+1}a_{i+1} = 0$. Hence $\ell_R(a_i) \subsetneq \ell_R(a_{i+1})$, where $i = 1, 2, \ldots$. Let $b_i = 1 - e_i$ for $i = 1, 2, \ldots$. Then $a_i = b_1b_2\cdots b_i$, and so $\ell_R(b_1b_2\cdots b_i) \subsetneq \ell_R(b_1b_2\cdots b_{i+1})$. Thus we have the following strictly ascending chain $\ell_R(b_1) \subsetneq \ell_R(b_1b_2) \subsetneq \cdots$. It is a contradiction. Hence $R/J(R)$ is orthogonally finite. Therefore $R/J(R)$ is semisimple Artinian and so R is left perfect. □

Corollary 2.1. *If R is a left continuous ring with ACC on left annihilators, then R is semiprimary.*

Proof. Note that $J(R) = Z(_RR)$. By Proposition 2.1, R is left perfect. Since R has ACC on left annihilators, $J(R)$ is nilpotent by [1, Proposition 29.1] and hence R is semiprimary. □

Lemma 2.1. *Let R be a two-sided continuous ring. Then the following are equivalent:*

(1) *R is QF.*
(2) *R has ACC on right annihilators.*

(3) $R/Soc(R_R)$ is right Goldie.

Proof. See [2, Corollary 2.7] or [8, Theorem 1]. □

Theorem 2.1. *The following statements are equivalent for a ring R:*

(1) R is QF.
(2) R is a two-sided quasi-continuous and left perfect ring with DCC on left annihilators.
(3) R is a two-sided quasi-continuous and left perfect ring with the factor ring modulo its right socle right Goldie.
(4) R is a two-sided quasi-continuous and right perfect ring with the factor ring modulo its right socle right Goldie.

Proof. We only need to show that (2)⇒(1), (3)⇒(1), and (4)⇒(1). For these, it is enough to see that R is two-sided continuous by Lemma 2.1.
(2)⇒(1) Since R is left perfect right quasi-continuous, R is a right continuous ring by [7, Theorem 12]. From the right hand version of Corollary 2.1, R is semiprimary. Again by [7, Theorem 12], R is left continuous.
(3)⇒(1) By [7, Theorem 12], R is right continuous. Also by [2, Theorem 2.2 and Proposition 1.4], R is semiprimary. Hence R is a two-sided continuous ring.
(4)⇒(1) Let $S = \mathrm{Soc}(R_R)$. Since $J = J(R)$ is right T-nilpotent, the ideal $(J + S)/S$ of the ring R/S is also right T-nilpotent. By [1, Proposition 29.1], $(J + S)/S$ is nilpotent. So there exists a positive integer n such that $J^n \subseteq S$. Thus $J^{n+1} \subseteq JS = 0$, hence J is nilpotent. So R is semiprimary. An application of [7, Theorem 12] ensures that R is a two-sided continuous ring. □

Immediately we have the following from Theorem 2.1.

Corollary 2.2 ([20, Theorem 1]). *A left Artinian two-sided quasi-continuous ring is QF.*

Example 2.1.

(1) There is a semiprimary ring R (hence left and right perfect) such that $R/\mathrm{Soc}(R_R)$ (resp. $R/\mathrm{Soc}(_RR)$) is right (resp. left) Goldie and R satisfies DCC on right annihilators, but R is not two-sided quasi-continuous. For a field F, let V be an infinite dimensional vector

space over F and $R = \begin{pmatrix} F & V \\ 0 & F \end{pmatrix}$, which is a generalized upper triangular matrix ring. Then R is semiprimary which is neither left nor right Artinian. Also $\mathrm{Soc}({}_RR) = \begin{pmatrix} F & V \\ 0 & 0 \end{pmatrix}$ and $\mathrm{Soc}(R_R) = \begin{pmatrix} 0 & V \\ 0 & F \end{pmatrix}$. Thus $R/\mathrm{Soc}({}_RR) \cong F \cong R/\mathrm{Soc}(R_R)$.

(2) Let R be a commutative local domain which is not a field. Then R is two-sided quasi-continuous, R has ACC on left annihilators, and $R/\mathrm{Soc}(R_R)$ is right Goldie, but R is not QF. Thus Lemma 2.1 cannot be extended to the case of two-sided quasi-continuous rings. Also Theorem 2.1 cannot be extended to the case of local (hence semiperfect) rings.

(3) ([6, Remarks 2]) There is a two-sided quasi-continuous and left (right) perfect ring R such that $R/\mathrm{Soc}(R_R)$ is not right Goldie and R does not satisfy DCC on left annihilators. Let R be the ring of polynomials in countably many indeterminates $x_1, x_2, \ldots$ over the field $\mathbb{Z}_2$ with two elements, where the following conditions are imposed: (α) $x_k^3 = 0$ for all k; (β) $x_k x_j = 0$ for $k \neq j$; and (γ) $x_k^2 = x_j^2$ for all k, j. Then by [6], R is commutative, semiprimary and local, but not self-injective. Also from [19, Example, p.330], the ring R is continuous. Since R is not QF, by Lemma 2.1 or by Theorem 2.1, it follows that $R/\mathrm{Soc}(R_R)$ is not right Goldie and R does not satisfy DCC on left annihilators.

3. QF-Rings via Left CS Left GP-Injective Rings

In this section, QF rings are characterized by using left CS property and left GP-injectivity. Also we give an alternative proof of the result [13, Theorem 3.4] related to Johns rings. Ikeda and Nakayama[14] considered the following conditions on a ring R:

(a) every principal right ideal is a right annihilator.

(b) $r_R(I_1 \cap I_2) = r_R(I_1) + r_R(I_2)$ for each pair of left ideals I_1 and I_2.

It was proved in [14] that a left self-injective ring satisfies (a) and (b). However, the converse is not true [24, Example 3]. Also it was shown in [14] that a ring R is left P-injective if and only if R satisfies (a). Recently, in [7], a ring satisfying the condition (b) is called a *left Ikeda-Nakayama ring*, (simply, *left IN-ring*). It was shown that every left IN-ring is left quasi-continuous [7, Theorem 5]. Also, it follows from [5, Theorem 4.1] that a ring which satisfies (a) and (b), and ACC on left annihilators must be QF.

Also Johns proved [16, Theorem 2] that a ring which satisfies (a) and (b), and ACC on right annihilators is QF. A ring R is called *left Johns* if R is left Noetherian ring in which every left ideal is a left annihilator. His proof in [16] is known to be incorrect because he used the false fact that a left Johns ring is left Artinian. However, it is well known [13, Theorem 3.4] that the theorem is true. In the following we give another proof of this theorem.

Theorem 3.1. *Assume that a ring R satisfies (a) and (b) with ACC on right annihilators. Then R is QF.*

Proof. It is enough to show that R is right Artinian. Now R is a left P-injective ring, hence R satisfies the condition (C2) by [21, Theorem 1.2]. By [7, Theorem 5], R is left quasi-continuous. Therefore R is a left continuous ring. So the factor ring $R/J(R)$ is von Neumann regular left continuous. Also, by [16, Lemma 5], every finitely generated right ideal is a right annihilator. Thus R is right Noetherian, so the factor ring $R/J(R)$ is semisimple Artinian. Now note that $J(R)$ is nilpotent by [16, Lemma 1]. Hence R is semiprimary. By Hopkins-Levitzki Theorem, R is right Artinian. □

A ring R is called *left Kasch* [12] if every simple left R-module embeds in R, equivalently if $r_R(M) \neq 0$ for every maximal left ideal M of R, or equivalently if every maximal left ideal is a left annihilator of R. Right Kasch rings are defined analogously. Left and right Kasch rings are called *two-sided Kasch* ring. Pardo and Asensio [22, Corollary 2.7] proved that if R is a left CS and left Kasch ring then $\text{Soc}({}_RR)$ is a finitely generated essential left ideal.

Proposition 3.1. *For a ring R we have the following:*

(1) If R is a left GP-injective ring with ACC on left annihilators, then R is a right Artinian and two-sided Kasch ring.
(2) R is semisimple Artinian if and only if R is semiprime (or left nonsingular) left GP-injective ring with ACC on left annihilators.
(3) If R is a left CS left GP-injective ring with ACC on left annihilators then R is a two-sided Artinian and two-sided Kasch ring.

Proof. (1) By [9, Theorem 3.7], R is right Artinian. Thus R is semiprimary, and so $\text{Soc}({}_RR)$ is essential in ${}_RR$. Hence R is left and right Kasch by [2, Theorem 2.5].
(2) It is obvious from the part (1).
(3) See [9, Theorems 2.5 and 3.7] and [22, Corollary 2.7). □

Recently Pardo and Yousif [23, Theorem 2.2] showed that if R is a left CS and right Kasch ring then R is left continuous ring. Hence the ring R in Proposition 3.1 (3) is a two-sided Artinian and left continuous ring. There is a two-sided Artinian ring with one sided continuity that is not necessarily continuous on the other side [11, Example 7.11′, p.338] (see also [15]). Using these results, we can establish the following result which extends some results in [2] and [10].

Theorem 3.2. *The following statements are equivalent for a ring R:*

(1) R is QF.
(2) R is a left CS left GP-injective ring with ACC on left annihilators.
(3) R is a left continuous left GP-injective ring with the factor ring modulo its left socle left Goldie.
(4) R is a left GP-injective left IN-ring with ACC on left annihilators.

Proof. We only need to prove (2)⇒(1), (3)⇒(1), and (4)⇒(1).
(2)⇒(1) By Proposition 3.1 (3), R is a left Artinian and right Kasch ring. So the factor ring modulo its left socle is left Goldie. Now R is a left CS and right Kasch ring, hence R is left continuous [23, Theorem 2.2]. Thus by [19, Theorem 1], it remains only to show that $K = r_R\ell_R(K)$ for every minimal right ideal of R. Let $K = aR$ be a minimal right ideal of R. Since R is left GP-injective, there exists a positive integer n such that a^nR is a nonzero right annihilator by [17, Lemma 3]. So $K = aR = a^nR = r_R\ell_R(a^nR) = r_R\ell_R(K)$ for every minimal right ideal of R.
(3)⇒(1) Since R is left GP-injective, we have $K = r_R\ell_R(K)$ for every minimal right ideal of R by the same proof in (2)⇒(1). Then [19, Theorem 1] can be applied.
(4)⇒(1) Since every left IN-ring is left quasi-continuous by [7, Theorem 5], R is QF by (2)⇒(1). □

Example 3.1.

(1) Let $R = \prod_{n=1}^{\infty} F_n$, where $F_n = F$ is a field for all n. Then R is left self-injective von Neumann regular (hence left CS and GP-injective). But R does not satisfy ACC on left annihilators and $R/\mathrm{Soc}({}_RR)$ is not left Goldie.
(2) By [11, Example 7.11′.2, p.338] and [15], there is a local left Artinian ring R with only three left ideals which are 0, $J(R)$ and R. Also every left ideal is a left annihilator and hence R is right P-injective (hence right GP-injective), R is left CS, but not right CS.

Also R is a left IN-ring, but R is not left GP-injective.

(3) Let R be the 2-by-2 upper triangular matrix ring over a field F. Then R is left CS and left Artinian. But in this case R is not left GP-injective. In fact, let $\alpha = \begin{pmatrix} 0 & 1 \\ 0 & 0 \end{pmatrix} \in R$. Note $\alpha^2 = 0$. Thus for R to be left GP-injective, the left R-homomorphism $f : {}_R R\alpha \to {}_R R$ defined by $f(\alpha) = \begin{pmatrix} 1 & 0 \\ 0 & 0 \end{pmatrix}$ has to be extended to an R-endomorphism of ${}_R R$. But this is impossible. So R is not left GP-injective.

(4) By Theorem 3.2 (4), a left P-injective, left IN-ring with ACC on left annihilators is QF. Note that a left P-injective ring satisfies C2 condition. The ring in (2) is a left C2, left IN ring with ACC on left annihilators, but it is not QF.

Acknowledgement

The first author was supported by grant No.R05-2002-000-00715-0 from the Basic Research Program of the Korea Science & Engineering Foundation.

Added in Proof

Recently, the authors found that the Theorem 2.1 was announced by J.Chen, L. Shen and Y. Zhou in Characterizations of QF Rings, Comm. in Algebra 35(2007), 281-288.

References

1. F. W. Anderson and K. R. Fuller, *Rings and Categories of Modules*, Springer-Verlag, Berlin-Heidelberg-New York, 1974.
2. P. Ara and J. K. Park, *On continuous semiprimary rings*, Comm. Algebra **19** (1991), 1945–1957.
3. E. P. Armendariz and J. K. Park, *Self-injective rings with restricted chain conditions*, Arch. Math. **58** (1992), 24–33.
4. G. Azumaya, *Finite splitness and finite projectivity*, J. Algebra **106** (1987), 114–134.
5. J-E. Björk, *Rings satisfying certain chain conditions*, J. Reine Angew. Math. **245** (1970), 63–73.
6. V. P. Camillo, *Commutative rings whose principal ideals are annihilators*, Portugal. Math. **46** (1989), Fasc 1, 33–37.
7. V. P. Camillo, W. K. Nicholson, and M. F. Yousif, *Ikeda-Nakayama rings*, J. Algebra **226** (2000), 1001–1010.
8. V. P. Camillo and M. F. Yousif, *Continuous rings with ACC on annihilators*, Canad. Math. Bull. **34** (1991), 462–464.

9. J. L. Chen and N. Q. Ding, *On general principally injective rings*, Comm. Algebra **27** (1999), 2097–2116.
10. C. Faith, *Rings with ascending chain condition on annihilators*, Nagoya Math. J. **27** (1966), 179–191.
11. C. Faith, *Algebra: Rings, Modules, and Categories I*, Springer-Verlag, Berlin-Heidel-berg-New York, 1973.
12. C. Faith, *Injective Modules and Injective Quotient Rings*, Lecture Notes in Pure and Applied Math. **72**, 1982.
13. C. Faith and P. Menal, *A counter example to a conjecture of Johns*, Proc. Amer. Math. Soc. **116** (1992), 21–26.
14. M. Ikeda and T. Nakayama, *On some characteristic properties of quasi-Frobenius and regular rings*, Proc. Amer. Math. Soc. **5** (1954), 15–19.
15. S. K. Jain, S. R. Lopez-Permouth, and S. T. Rizvi, *Continuous rings with acc on essentials are artinian*, Proc. Amer. Math. Soc. **108** (1990), 583–586
16. B. Johns, *Annihilator conditions in Noetherian rings*, J. Algebra **49** (1977), 222-224.
17. R. Y. C. Ming, O*n regular rings and Artinian rings (II)*, Riv. Math. Univ. Parma **11** (1985), 101–109.
18. S. B. Nam, N. K. Kim, and J. Y. Kim, *On simple GP-injective modules*, Comm. Algebra **23** (1995), 5437–5444.
19. W. K. Nicholson and M. F. Yousif, *Continuous rings with chain conditions*, J. Pure and Appl. Algebra **97** (1994), 325–332.
20. W. K. Nicholson and M. F. Yousif, *On quasi-continuous rings*, Proc. Amer. Math. Soc. **120** (1994), 1049–1051.
21. W. K. Nicholson and M. F. Yousif, *Principally injective rings*, J. Algebra **174** (1995), 77–93.
22. J. L. G. Pardo and P. A. G. Asensio, *Rings with finite essential socle*, Proc. Amer. Math. Soc. **125** (1997), 971–977.
23. J. L. G. Pardo and M. F. Yousif, *Semiperfect min-CS rings*, Glasgow Math. J. **41** (1999), 231–238.
24. Y. Utumi, *On continuous regular rings and semisimple self-injective rings*, Canad. J. Math. **12** (1960), 597–605.
25. S. Zhu, *On rings over which every flat left module is finitely projective*, J. Algebra **139** (1991), 311–321.

Morita duality and recent development

KAZUTOSHI KOIKE

Okinawa National College of Technology
Nago City, Okinawa 905-2192, Japan
E-mail: koike@okinawa-ct.ac.jp

We give a survey of recent development of Morita duality. In particular, we will focus on self-duality of (quasi-)Harada rings and applications to Azumaya's conjecture ("every exact ring has a self-duality") and related self-duality of locally distributive rings.

1. Introduction and Notation

Morita duality plays an important role in ring theory and has been investigated by many ring theorists. For the theory of Morita duality, there are good literature, the book of Xue [29] and the survey papers of Müller [18] and Ánh [3]. In this paper, we shall mainly focus on Morita duality and self-duality of certain artinian rings and give a survey of recent development of these area. In Section 2 we give the definition and fundamental results of Morita duality. In Section 3 we introduce the concepts of several types of self-dualities (weakly symmetric self-duality, good self-duality and almost self-duality) and characterize rings with these self-dualities. We also present examples of rings without these self-dualities. We devote Section 4 to review of the self-duality of serial rings. In Sections 5 and 6 we discuss the existences of several types of self-dualities of Harada rings and quasi-Harada rings. In Section 7, as applications of study of Harada rings and quasi-Harada rings, we present several results about Azumaya's conjecture ("every exact ring has a self-duality") and related problems. Particularly we focus on the self-duality of locally distributive rings. In the final section, we present some problems about the self-duality of locally distributive rings. Many parts of material are taken from the author's papers [12–14].

Throughout this paper, all rings have identity and all modules are unitary. The symbols R and S always denote rings. For a ring R, we denote the category of right (resp. left) R-modules by Mod-R (resp. R-Mod). The injective hull, the Jacobson radical, the socle and the top of a module X are denoted by $E(X)$, $\mathrm{rad}(X)$, $\mathrm{soc}(X)$ and $\mathrm{top}(X)$, respectively. Here the top of X means $X/\mathrm{rad}(X)$. For notation, definitions and known results in ring theory, see Anderson and Fuller [2] and Lam [16]. Particularly, for results about Morita duality, see Xue [29].

2. Definition of Morita duality

We begin with the definition of Morita duality. For an (S, R)-bimodule U, we denote U-dual functors by

$$(-)^* = \mathrm{Hom}_R(-, U) : \text{Mod-}R \to S\text{-Mod},$$
$$(-)^* = \mathrm{Hom}_S(-, U) : S\text{-Mod} \to \text{Mod-}R.$$

For a right R-module X, there is the *evaluation map* $X \to X^{**}$ defined by $x \mapsto [f \mapsto f(x)]$ for each $x \in X$ and $f \in X^*$. For a left S-module Y, the evaluation map $Y \to Y^{**}$ is defined similarly. A right R-module X_R (resp. a left S-module $_SY$) is said to be *U-reflexive* if the evaluation map $X \to X^{**}$ (resp. $Y \to Y^{**}$) is an isomorphism. The pair of functors $(-)^* : \text{Mod-}R \rightleftarrows S\text{-Mod} : (-)^*$ induces a duality (contravariant category equivalence) between the full subcategories of U-reflexive right R- and left S-modules.

As the following shows, the definition of Morita duality requires that the classes of U-reflexive modules contain generators and are closed under taking submodules and factor modules (Morita [17]).

Definition 2.1. An (S, R)-bimodule $_SU_R$ defines a *Morita duality* if the following conditions hold:

(1) R_R and $_SS$ are U-reflexive.
(2) Every submodule and every factor module of a U-reflexive module is U-reflexive.

Any duality (contravariant category equivalence) between full subcategories of Mod-R and S-Mod that contain R_R and $_SS$ can be represented by the U-dual functor for some bimodule $_SU_R$. Thus, in the definition of Morita duality, we can concentrate on the U-dual functors.

We say that a ring R has a *right Morita duality* and R is *right Morita dual* to S if there exists a bimodule $_SU_R$ that defines a Morita duality. In the case $R = S$, we say that $_RU_R$ defines a *self-duality.*

The following is a basic characterization of Morita duality due to Morita [17].

Theorem 2.2. *For a bimodule $_SU_R$, the following conditions are equivalent:*

(1) *$_SU_R$ defines a Morita duality.*
(2) *Every factor module of R_R, $_SS$, U_R and $_SU$ is U-reflexive.*
(3) *U_R and $_SU$ are injective cogenerators and $_SU_R$ is faithfully balanced.*

Most typical examples of rings with self-duality are:

Example 2.3. (1) Every finite dimensional algebra R over a field K has a self-duality. The usual K-duality can be regarded as the self-duality induced by the (R, R)-bimodule $\mathrm{Hom}_K(R, K)$.

(2) An artinian ring R is *quasi-Frobenius (QF)* in case R_R and $_RR$ are injective. Any QF ring R has a self-duality defined by the regular bimodule $_RR_R$.

An origin of Morita duality is in the examples above. Before Morita [17] established the definition of Morita duality for general rings, several ring theorists investigated duality between categories of modules over artinian rings. In particular, Tachikawa [27] showed that if a right artinian ring R has a finitely generated injective cogenerator U_R, then $S = \mathrm{End}(U_R)$ is left artinian and $(-)^*$ induces a duality between the categories of finitely generated right R- and left S-modules. With the converse, the following result is obtained in Morita [17] and Azumaya [5] independently.

Theorem 2.4. *A right artinian ring R has a right Morita duality if and only if a minimal injective cogenerator U_R is finitely generated. In this case, $S = \mathrm{End}(U_R)$ is left artinian and $_SU_R$ induces a duality between the categories of finitely generated right R- and left S-modules.*

3. Several types of self-dualities

What kind of rings have a self-duality? This is one of the most important and interesting problems in the theory of Morita duality. To study the problem, it seems available to consider several types of self-dualities, weakly symmetric self-duality, good self-duality and almost self-duality.

Definition 3.1. A bimodule $_RU_R$ defines a *weakly symmetric self-duality* if $_RU_R$ defines a self-duality and $(\mathrm{top}(eR))^* \cong \mathrm{top}(Re)$ for each primitive idempotent e of R.

Remark 3.2. (1) Let R be a basic artinian ring with right Morita duality induced by a minimal injective cogenerator U_R. Then R has a weakly symmetric self-duality if and only if there exists a ring isomorphism $\sigma : R \to \text{End}(U_R)$ such that $\text{soc}(\sigma(e)U) \cong \text{top}(eR)$ for each primitive idempotent e of R. This ring isomorphism σ is just a *Nakayama isomorphism* of Kado and Oshiro [11] essentially.

(2) Let R be a basic QF ring. Then R is a *weakly symmetric QF ring* if the regular bimodule ${}_RR_R$ defines a weakly symmetric self-duality. R has a weakly symmetric self-duality if and only if there exists a ring automorphism σ of R such that $\text{soc}(\sigma(e)R) \cong \text{top}(eR)$ for each primitive idempotent e of R. Such a ring automorphism is called a *Nakayama automorphism.* Thus a weakly symmetric QF ring R is just a QF ring such that the identity map is a Nakayama automorphism.

If a bimodule ${}_RU_R$ defines a self-duality, then the composite of annihilators $l_R l_U(-)$ gives a lattice automorphism of the lattice of two-sided ideals of R. Good self-duality requires that the lattice automorphism is identical.

Definition 3.3. A bimodule ${}_RU_R$ defines a *good self-duality* if ${}_RU_R$ defines a self-duality and $l_U l_R(I) = I$ for each two-sided ideal I of R.

Remark 3.4. If ${}_RU_R$ defines a good self-duality, then ${}_RU_R$ defines a weakly symmetric self-duality. For locally distributive rings (serial rings), the converse holds. (Remark 7.4(2).)

Typical examples of rings with good self-duality are:

Example 3.5. (1) Finite dimensional algebras R over a field K have a good self-duality induced by ${}_R\text{Hom}_K(R, K)_R$.

(2) Commutative artinian rings R have a good self-duality induced by $E(R/\operatorname{rad}(R))$.

Simson [26] introduced the concept of almost self-duality, which can be regarded as a generalization of self-duality.

Definition 3.6. R has an *almost self-duality* if there is a series of rings $R_1 = R, R_2, \ldots, R_n, R_{n+1} = R$ such that each R_i is right Morita dual to R_{i+1}.

The author [13] characterized rings with good self-duality, weakly symmetric self-duality and almost self-duality by using QF rings as follows. (The good self-duality version of (1) can be proved by a similar way of the weakly symmetric version [13, Theorem 1.6].)

Theorem 3.7. *Let R be an artinian ring.*

(1) R has a weakly symmetric self-duality (resp. good self-duality) if and only if there exist a QF ring S with weakly symmetric self-duality (resp. good self-duality) and an idempotent e of S such that $R \cong eSe$.

(2) R has an almost self-duality if and only if there exist a QF ring S and an idempotent e of S such that $R \cong eSe$.

Remark 3.8. For self-duality, a similar characterization does not hold. That is, there exist a QF ring S and an idempotent e of S such that eSe does not have a self-duality. In view of Theorem 3.7 (2) above, this means that there exists an artinian ring without self-duality but with almost self-duality. We shall present an example of such a ring in Example 3.9.

It should be noted that we can construct all of rings

(1) an artinian ring without self-duality but with almost self-duality,
(2) an artinian ring without weakly symmetric self-duality but with self-duality,
(3) a QF ring without weakly symmetric self-duality (Nakayama automorphism)

from any one of the three rings above. We present the constructions.

(1) $\Rightarrow$ (2). Let R be a ring without self-duality but with almost self-duality. Let $R_1 = R, R_2, \ldots, R_n, R_{n+1} = R$ be a series of rings such that each R_i is right Morita dual to R_{i+1}. Then the ring product $R_1 \times R_2 \times \cdots \times R_n$ does not have a weakly symmetric self-duality but has a self-duality. (See Kraemer [15, Proposition 6.5 and Remarks 6.1 and 6.2].)

(2) $\Rightarrow$ (3). Let R be an artinian ring without weakly symmetric self-duality but with self-duality. Let ${}_RU_R$ be a bimodule that defines a self-duality. Then the matrix ring $\begin{bmatrix} R & U \\ U & R \end{bmatrix}$ is a QF ring without weakly symmetric self-duality, where the multiplication is defined by $\begin{bmatrix} 0 & U \\ U & 0 \end{bmatrix} \times \begin{bmatrix} 0 & U \\ U & 0 \end{bmatrix} = 0$. (See [13, Theorem 1.6].)

(3) $\Rightarrow$ (1). Let R be a QF ring without weakly symmetric self-duality. We may assume that R is basic. Let $\{e_1, \ldots, e_m\}$ be a complete set of orthogonal primitive idempotents of R. For each $1 \leq i, j \leq m$, we define

sets S_{ij} of $i \times j$ matrices by

$$S_{ij} = M_{i\times j}(e_iRe_j) \text{ if } i \neq j,$$
$$S_{ii} = \{[a_{kl}] \in M_{i\times i}(e_iRe_i) \mid a_{kl} \in \mathrm{rad}(e_iRe_i) \text{ if } k > l\} \text{ if } i = j,$$

where $M_{i\times j}$ denote the sets of $(i \times j)$-matrices. Then S_{ii} are rings and S_{ij} are (S_{ii}, S_{jj})-bimodules. We define a matrix ring (indeed a $(1+2+\cdots+n, 1+2+\cdots+n)$-matrix ring) S by

$$S = \begin{bmatrix} S_{11} & \cdots & S_{1m} \\ \vdots & & \vdots \\ S_{m1} & \cdots & S_{mm} \end{bmatrix}.$$

Then S is an artinian ring without self-duality but with almost self-duality. Indeed R is a left Harada ring. (See Section 5.) This construction is due to Kado and Oshiro [11]. See also [14, Lemma 5.6].

By the remark above, if we have an example of rings without weakly symmetric self-duality but with almost self-duality (a ring of (1)), then we obtain examples of (2) and (3).

Example 3.9. We present an example of a ring without self-duality but with almost self-duality.

Schofield [25] constructed a division ring extension $F \geq G$ with $\dim({}_GF) = 2, \dim(F_G) = 3$ and $\dim({}_F\mathrm{End}(F_G)) = \dim(\mathrm{End}(F_G)_F) = 3$ for the left multiplication $F \to \mathrm{End}(F_G)$. This example answered Artin's problem "does there exist a division ring extension $F \geq G$ such that $\dim({}_GF) < \infty$, $\dim(F_G) < \infty$ and $\dim({}_GF) \neq \dim(G_F)$?" negatively. The example of Schofield was required in the results of Dowbor, Ringel and Simson [9]. Then by [9] the ring $R = \begin{bmatrix} F & F \\ 0 & G \end{bmatrix}$ is a hereditary artinian ring of finite representation type and R does not have a self-duality but has an almost self-duality with cycle 5. That is, there exist rings $R_1 = R, R_2, \ldots, R_5, R_6 = R$ such that each R_i is right Morita dual to R_{i+1} and $R_1, R_2, \ldots, R_5$ are pairwise non-isomorphic. See Kraemer [15, Section 6] for the details.

4. Serial rings

From now on, we shall focus our attention on certain artinian rings with Morita duality. Particularly, we shall treat serial rings, Harada rings, quasi-Harada rings, Azumaya's exact rings and locally distributive rings.

(Artinian) serial rings are fundamental artinian rings as well as QF rings. The self-duality of serial rings is the most successful case and has an interesting history.

Definition 4.1. An artinian ring R is a *serial ring* (*Nakayama ring*) if eR_R and $_R Re$ are uniserial (the lattices of submodules are linearly ordered) for each primitive idempotent e of R.

The self-duality of serial rings has been investigated by several ring theorists. Haack is one of them who tried the problem. Though he could not prove the self-duality of serial rings, in 1979 he [10] showed the following result, which is important in our point of view as we shall state later.

Theorem 4.2. *Every serial QF ring has a weakly symmetric self-duality (Nakayama automorphism).*

In 1984 Dischinger and Müller [8] succeeded in proving the weakly symmetric self-duality of serial rings in general.

Theorem 4.3. *Every serial ring has a weakly symmetric self-duality.*

However, in 1986 Waschbüsch [28] pointed out independently that the self-duality of serial rings had been claimed earlier by Amdal and Ringdal [1] in 1968 and he gave a proof of the self-duality of serial rings.

Several authors provided another proofs of the self-duality of serial rings. We should note that the proofs of the following are based on the weakly symmetric self-duality of serial QF rings (Theorem 4.2). Kado and Oshiro [11] pointed out that the self-duality of serial rings follows from the result of self-duality of Harada rings (Theorem 5.4). Ánh [4] also gave another proof recently. As we shall see in Theorem 7.9, the author obtained an improvement of the weakly symmetric self-duality of serial rings.

5. Harada rings

Oshiro [20–24] named certain (two-sided) QF-3 right QF-2 artinian rings left Harada rings and investigated these rings deeply.

Definition 5.1. An artinian ring R is said to be a *left Harada ring* if R has a basic set of orthogonal primitive idempotents $\{e_{ij} \mid 1 \le i \le m, 1 \le j \le n(i)\}$ such that

(1) $e_{i1}R_R$ is injective for each $i = 1, 2, \ldots, m$;
(2) $e_{ij}R \cong \mathrm{rad}(e_{i,j-1}R)$ for each $i = 1, 2, \ldots, m$ and $j = 2, 3, \ldots, n(i)$.

Remark 5.2. (1) Indeed, the definition above is that of *right co Harada rings*. Oshiro named two certain classes of rings (defined by some conditions about right modules) *right Harada rings* and *right co-Harada rings*. He proved that the classes of *left* Harada rings and *right* co-Harada rings coincide. So the condition of Definition 5.1 is often adopted as the definition of left Harada rings recently.

(2) Oshiro proved that every left Harada ring can be represented as a factor ring of a suitable *extension* of a QF ring. (See Example 5.3.) This fact is a central tool of study of self-duality of left Harada rings.

(3) Left Harada rings R are special QF-3 rings. eR is a minimal faithful right R-module, where $e = e_{11} + e_{21} + \cdots + e_{m1}$. We note that eRe becomes a left Harada ring again. By iteration taking the endomorphism rings of minimal faithful right modules for left Harada rings, we reach a QF ring.

We illustrate left Harada rings with a simple example.

Example 5.3. Let K be a field, Q the quiver $1 \underset{\beta}{\overset{\alpha}{\rightleftarrows}} 2$, and $R = KQ/\langle\alpha\beta\rangle$ the factor algebra of the path algebra KQ. Since e_1R is injective and $e_2R \cong \operatorname{rad}(e_1R)$, R is a left Harada ring (indeed a serial ring), where e_i denotes the primitive idempotent corresponding to the vertex "i" for $i = 1, 2$. R has the representation

$$R \cong \begin{bmatrix} A & A/\operatorname{soc}(A) \\ \operatorname{rad}(A) & A/\operatorname{soc}(A) \end{bmatrix} = \begin{bmatrix} A & A \\ \operatorname{rad}(A) & A \end{bmatrix} / \begin{bmatrix} 0 & \operatorname{soc}(A) \\ 0 & \operatorname{soc}(A) \end{bmatrix},$$

where $A = e_1Re_1$. The matrix ring $S = \begin{bmatrix} A & A \\ \operatorname{rad}(A) & A \end{bmatrix}$ is an "extension" of the QF ring A. Thus the left Harada ring R is represented as a factor ring of an *extension* S of the QF ring $A = e_1Re_1$.

The class of left Harada rings contains QF rings and serial rings, which have the self-duality. Thus it is natural to ask whether every left Harada ring has a self-duality. Kado and Oshiro [11] investigated the problem and obtain the following theorem.

Theorem 5.4. *The following are equivalent:*

(1) Every left Harada ring has a self-duality.
(2) Every left Harada ring has a weakly symmetric self-duality.
(3) Every QF ring has a weakly symmetric self-duality (Nakayama automorphism).

The author [12] pointed out that Kraemer [15] constructed examples of QF rings without weakly symmetric self-duality and constructed concrete examples of left Harada rings without self-duality (see Remark 3.8 and Example 3.9).

Thus the conditions of Theorem 5.4 are not true. However the theorem is very important. For example, as we stated in Section 4, the restriction of Theorem 5.4 to serial rings together with the weakly symmetric self-duality of serial QF rings shows the self-duality of serial rings. The proof is also useful. We can use the technique in various cases (e.g., the implication $(3) \not\rightarrow (1)$ in Remark 3.8 and Theorems 7.11 and 7.12).

Left Harada rings do not have a self-duality in general. However the author [13] obtained the following.

Theorem 5.5. *Every left Harada ring has an almost self-duality.*

6. Quasi-Harada rings

Baba and Iwase [7] generalized the concept of Harada rings as quasi-Harada rings.

Definition 6.1. An artinian ring R is said to be a *left quasi-Harada ring* if eR_R is quasi-injective for each primitive idempotent e of R.

Remark 6.2. (1) Left quasi-Harada rings are not QF-3 in general. But left quasi-Harada rings R are right QF-2 (i.e., $\mathrm{soc}(eR)$ is simple for each primitive idempotent e of R).

(2) Let R be a left quasi-Harada ring with basic set $\{e_1, e_2, \ldots, e_n\}$ of orthogonal primitive idempotents. Set

$$f = \sum \{e_i \mid \mathrm{Hom}_R(\mathrm{top}(e_i R), R) \neq 0\}.$$

Then fRf is a left quasi-Harada ring again. As is similar to left Harada rings (Remark 5.2), we reach a QF ring by iteration taking such the ring "fRf". We can reconstruct the left Harada ring R from the QF ring ([14, Theorem 3.13]). This fact allows us to reduce certain problems about left quasi-Harada rings to problems about QF rings.

Example 6.3. Let K be a field, Q the quiver $\alpha \circlearrowleft 1 \xrightarrow[\beta]{} 2$, and $R = KQ/\langle \alpha^2, \beta\alpha \rangle$ the factor algebra of the path algebra KQ. Then R_R and ${}_RR$ have the Loewy factors

$$R_R = \begin{smallmatrix} 1 \\ 1 \end{smallmatrix} \bigoplus \begin{smallmatrix} 2 \\ 1 \end{smallmatrix} \text{ and } {}_RR = \begin{smallmatrix} & 1 & \\ 1 & & 2 \end{smallmatrix} \bigoplus 2.$$

Thus R is a locally distributive right QF-2 (right serial) ring (Section 7). Then R is a left quasi-Harada ring (by Proposition 7.7). R has the representation

$$R \cong \begin{bmatrix} A & 0 \\ \operatorname{rad}(A) & A/\operatorname{rad}(A) \end{bmatrix} = \begin{bmatrix} A & \operatorname{rad}(A) \\ \operatorname{rad}(A) & A \end{bmatrix} \Big/ \begin{bmatrix} 0 & \operatorname{rad}(A) \\ 0 & \operatorname{rad}(A) \end{bmatrix},$$

where $A = e_1 R e_1$. The ring $\begin{bmatrix} A & \operatorname{rad}(A) \\ \operatorname{rad}(A) & A \end{bmatrix}$ is a certain subring of the QF ring $\begin{bmatrix} A & A \\ A & A \end{bmatrix}$, which is Morita equivalent to A. We can regard that R is constructed from the QF ring $A = e_1 R e_1$.

Left Harada rings have an almost self-duality. Left quasi-Harada rings have a right and left Morita duality. With an additional condition, the author [14] proved the following.

Theorem 6.4. *Every left quasi-Harada ring with finite ideal lattice has an almost self-duality. In particular, every left quasi-Harada ring of finite representation type has an almost self-duality.*

In the next section, we shall state some applications of results of quasi-Harada rings to locally distributive rings and Azumaya's exact rings.

7. Exact rings, locally distributive rings and Azumaya's conjecture

In this section, we treat Azumaya's exact rings and locally distributive rings. For details of the definitions and results about these rings, refer to [29, Sections 13–15].

Azumaya [6] introduced a notion of exact rings and studied self-duality of exact rings.

Definition 7.1. An artinian ring R is *exact* if R has a composition series of *two-sided ideals* $R = I_0 > I_1 > I_2 > \cdots > I_n = 0$ such that each composition factor I_i/I_{i+1} is a balanced (R,R)-bimodule (i.e., every left or right endomorphism is given by the opposite side multiplication of an element of R).

Most typical examples of exact rings are commutative artinian rings and serial rings, which have a self-duality. Azumaya proved that exact rings have a Morita duality and conjectured the self-duality of exact rings:

Conjecture 7.2 (Azumaya's conjecture). *Every exact ring has a self-duality.*

Azumaya's conjecture seems very hard. It is difficult to solve even the case of the following locally distributive rings.

Definition 7.3. An artinian ring R is *locally distributive* if eR_R and ${}_RRe$ are distributive (the lattices of submodules are distributive) for each primitive idempotent e of R.

The ring of Example 6.3 is an example of locally distributive rings.

Remark 7.4. (1) Serial rings are locally distributive and locally distributive rings are exact.

(2) For a bimodule ${}_RU_R$ over a locally distributive ring R, ${}_RU_R$ defines a weakly symmetric self-duality if and only if ${}_RU_R$ defines a good self-duality. So, over locally distributive rings (particularly serial rings), we do not distinguish between weakly symmetric self-duality and good self-duality.

As a consequence of Azumaya's conjecture, we have

Conjecture 7.5 (Subconjecture of Azumaya's conjecture). *Every locally distributive ring has a self-duality.*

This is still remained open too. Several partial results are obtained. As a result of Belzner ([29, Theorem 15.19 and Corollary 15.20]),

Theorem 7.6. *For a locally distributive ring R, if eRe is a division ring for each primitive idempotent e of R, then R has a self-duality.*

The author also obtained several results about self-duality of locally distributive rings. In the rest of this section, all results can be found in [14]. The following fact allows us to apply the study of quasi-Harada rings to certain locally distributive rings.

Proposition 7.7. *A locally distributive ring R is a left quasi-Harada ring if and only if R is right QF-2.*

The lattices of ideals of locally distributive rings are finite. Thus, as a corollary of Theorem 6.4, we have

Theorem 7.8. *Every locally distributive right QF-2 ring has an almost self-duality.*

As we stated in Section 4, the author proved the following result, which improves the self-duality of serial rings (Theorem 4.3). The proof is based on the weakly symmetric self-duality of serial QF rings (Theorem 4.2).

Theorem 7.9. *Every locally distributive right serial ring has a good self-duality.*

As another partial result about Subconjecture (Conjecture 7.5), we have

Theorem 7.10. *Let R be a locally distributive right QF-2 ring. If the number of simple right R-modules that appear in $\mathrm{soc}(R_R)$ is less than or equal to 2, then R has a good self-duality.*

We can reduce Azumaya's conjecture and Subconjecture to other problems. By a similar argument of the proof of Theorem 5.4 about Harada rings, we have the following two results.

Theorem 7.11. *The following are equivalent:*

(1) Every locally distributive right QF-2 ring has a self-duality.
(2) Every locally distributive right QF-2 ring has a weakly weakly symmetric self-duality.
(3) Every locally distributive QF ring has a weakly symmetric self-duality (Nakayama automorphism).

Theorem 7.12. *The following are equivalent:*

(1) Every locally distributive (resp. exact) ring has a self-duality.
(2) Every locally distributive (resp. exact) ring has a weakly symmetric self-duality.

8. Problems

Finally we present several problems about Subconjecture.

Let R be an indecomposable basic locally distributive ring with composition length m. It is natural to verify the existence of self-duality for R when m is small. If $m \leqq 4$, then R is a one-sided or two-sided serial ring and has a self-duality by Theorem 7.9. If $m = 5$, then R is a one-sided or two-sided serial ring, which has a self-duality, or is isomorphic to the following ring.

Problem 8.1. *Let A and B be serial local rings of composition length 2 and let M be an (A, B)-bimodule that is simple on both side. Then the upper*

triangular matrix ring

$$R = \begin{bmatrix} A & M \\ 0 & B \end{bmatrix}$$

is a locally distributive ring. Does R have a self-duality?

The existence of self-duality for such a *small* locally distributive ring is not verified yet (see Müller [19]). Thus Subconjecture seems very hard to solve. Before Subconjecture, we must answer

Problem 8.2. *Does every locally distributive right QF-2 ring have a self-duality?*

By Theorem 7.11, this is equivalent to

Problem 8.3. *Does every locally distributive QF ring have a weakly symmetric self-duality (Nakayama automorphism)?*

As we mentioned, every serial QF ring has a weakly symmetric self-duality (Nakayama automorphism) (Theorem 4.2). Indeed let R be an indecomposable basic serial QF ring with Kupisch series $e_1R, e_2R, \ldots, e_nR$. If R is weakly symmetric QF, then the identity map is a Nakayama automorphism. If R is not weakly symmetric QF, then there exists a ring automorphism σ of R such that $\sigma(e_i) = e_{i+1}$ (where $e_{n+1} = e_1$) for each e_i. Then σ^k is a Nakayama automorphism for some $k > 0$ (Haack [10]).

However Nakayama permutations of locally distributive QF rings are complicated. Indeed there exist indecomposable locally distributive QF rings that realize given any permutation as a Nakayama permutation [14]. So we have the question:

Problem 8.4. *Do locally distributive QF rings with some additional restrictions on Nakayama permutation (e.g., cyclic Nakayama permutation) have a weakly symmetric self-duality (Nakayama automorphism)?*

References

1. I. K. Amdal and F. Ringdal. Catégories unisérielles. *C. R. Acad. Sci. Paris Sér. A-B*, 267:A85–A87 and A247–A249, 1968.
2. F. W. Anderson and K. R. Fuller. *Rings and categories of modules.* Springer-Verlag, New York, 1992.
3. P. N. Ánh. Morita duality, linear compactness and AB5*: a survey. In *Abelian groups and modules (Padova, 1994)*, volume 343 of *Math. Appl.*, pages 17–28. Kluwer Acad. Publ., Dordrecht, 1995.

4. P. N. Ánh. Selfdualities of serial rings, revisited. *Bull. London Math. Soc.*, 38(3):411–420, 2006.
5. G. Azumaya. A duality theory for injective modules. (Theory of quasi-Frobenius modules). *Amer. J. Math.*, 81:249–278, 1959.
6. G. Azumaya. Exact and serial rings. *J. Algebra*, 85(2):477–489, 1983.
7. Y. Baba and K. Iwase. On quasi-Harada rings. *J. Algebra*, 185(2):544–570, 1996.
8. F. Dischinger and W. Müller. Einreihig zerlegbare artinsche Ringe sind selbstdual. *Arch. Math. (Basel)*, 43(2):132–136, 1984.
9. P. Dowbor, C. M. Ringel, and D. Simson. Hereditary Artinian rings of finite representation type. In *Representation theory, II (Proc. Second Internat. Conf., Carleton Univ., Ottawa, Ont., 1979)*, pages 232–241. Springer, Berlin, 1980.
10. J. K. Haack. Self-duality and serial rings. *J. Algebra*, 59(2):345–363, 1979.
11. J. Kado and K. Oshiro. Self-duality and Harada rings. *J. Algebra*, 211(2):384–408, 1999.
12. K. Koike. Examples of QF rings without Nakayama automorphism and H-rings without self-duality. *J. Algebra*, 241(2):731–744, 2001.
13. K. Koike. Almost self-duality and Harada rings. *J. Algebra*, 254(2):336–361, 2002.
14. K. Koike. Self-duality of quasi-Harada rings and locally distributive rings. *J. Algebra*, 302(2):613–645, 2006.
15. J. Kraemer. *Characterizations of the existence of (quasi-) self-duality for complete tensor rings.* Verlag Reinhard Fischer, Munich, 1987.
16. T. Y. Lam. *Lectures on modules and rings.* Springer-Verlag, New York, 1999.
17. K. Morita. Duality for modules and its applications to the theory of rings with minimum condition. *Sci. Rep. Tokyo Kyoiku Daigaku Sect. A*, 6:83–142, 1958.
18. B. J. Müller. Morita duality—a survey. In *Abelian groups and modules (Udine, 1984)*, volume 287 of *CISM Courses and Lectures*, pages 395–414. Springer, Vienna, 1984.
19. W. Müller. On self-duality of ring extensions over quasi-Frobenius rings. *Comm. Algebra*, 21(8):2687–2695, 1993.
20. K. Oshiro. Lifting modules, extending modules and their applications to generalized uniserial rings. *Hokkaido Math. J.*, 13(3):339–346, 1984.
21. K. Oshiro. Lifting modules, extending modules and their applications to QF-rings. *Hokkaido Math. J.*, 13(3):310–338, 1984.
22. K. Oshiro. On Harada rings. I. *Math. J. Okayama Univ.*, 31:179–188, 1989.
23. K. Oshiro. On Harada rings. II. *Math. J. Okayama Univ.*, 31:179–188, 1989.
24. K. Oshiro. On Harada rings. III. *Math. J. Okayama Univ.*, 32:111–118, 1990.
25. A. H. Schofield. *Representation of rings over skew fields.* Cambridge University Press, Cambridge, 1985.
26. D. Simson. Dualities and pure semisimple rings. In *Abelian groups, module theory, and topology (Padua, 1997)*, pages 381–388. Dekker, New York, 1998.
27. H. Tachikawa. Duality theorem of character modules for rings with minimum condition. *Math. Z.*, 68:479–487, 1958.

28. J. Waschbüsch. Self-duality of serial rings. *Comm. Algebra*, 14(4):581–589, 1986.
29. W. Xue. *Rings with Morita duality.* Springer-Verlag, Berlin, 1992.

Remarks on Divisible and Torsionfree Modules

Lixin Mao
Institute of Mathematics, Nanjing Institute of Technology
Nanjing 211167, China
E-mail: maolx2@hotmail.com

Nanqing Ding
Department of Mathematics, Nanjing University, Nanjing 210093, China
E-mail: nqding@nju.edu.cn

In this paper, some special rings such as von Neumann regular rings and PP rings are characterized in terms of endomorphisms of divisible and torsionfree modules. In addition, commutative P-coherent rings and von Neumann regular rings are characterized by properties of homomorphism modules of certain special modules.

1. Introduction

Recall that a left R-module M is called *divisible* (or *P-injective*) if $\mathrm{Ext}^1(R/Ra, M) = 0$ for all $a \in R$. A right R-module N is called *torsionfree* if $\mathrm{Tor}_1(N, R/Ra) = 0$ for all $a \in R$. The definitions of divisible and torsionfree modules coincide with the classical ones in case R is a commutative domain. The class of all divisible left R-modules is closed under extensions, direct sums, direct products and direct summands, and the class of all torsionfree right R-modules is closed under extensions, direct sums and direct summands. It is clear that a right R-module N is torsionfree if and only if the character module $N^+ = \mathrm{Hom}_{\mathbb{Z}}(N, \mathbb{Q}/\mathbb{Z})$ is divisible by the standard isomorphism $\mathrm{Ext}^1(R/Ra, N^+) \cong \mathrm{Tor}_1(N, R/Ra)^+$ for every $a \in R$. Divisible and torsionfree modules have been studied by many authors (see, for example, [2, 7, 8, 10, 13, 16]).

In this short note, we first characterize some special rings such as von Neumann regular rings and PP rings in terms of endomorphisms of divisible and torsionfree modules. Then we characterize some special commutative rings by properties of homomorphism modules of certain special modules. For instance, we prove that a commutative ring R is P-coherent if and only if Hom(A, B) is torsionfree for all injective R-modules A and B if and only if Hom(A, B) is torsionfree for all projective R-modules A and B. Finally, we study divisible and torsionfree modules under change of rings.

Throughout this article, all rings are associative with identity and all modules are unitary. We write M_R $({}_RM)$ to indicate a right (left) R-module. End$M$ stands for the endomorphism ring of a module M. Let M and N be two R-modules. $\mathrm{Hom}(M, N)$ (resp. $\mathrm{Ext}^n(M, N)$) means $\mathrm{Hom}_R(M, N)$ (resp. $\mathrm{Ext}^n_R(M, N)$), and similarly $M \otimes N$ (resp. $\mathrm{Tor}_n(M, N)$) denotes $M \otimes_R N$ (resp. $\mathrm{Tor}^R_n(M, N)$) for an integer $n \geq 1$ unless otherwise specified. We freely use the terminology and notations of [6, 12, 15].

2. Endomorphisms of divisible and torsionfree modules

Let $\mathcal{C}$ be a class of R-modules and M an R-module. Recall that a homomorphism $\phi : C \to M$ is a *$\mathcal{C}$-precover of M* [5] if $C \in \mathcal{C}$ and the abelian group homomorphism $\mathrm{Hom}(C', \phi) : \mathrm{Hom}(C', C) \to \mathrm{Hom}(C', M)$ is surjective for every $C' \in \mathcal{C}$. A $\mathcal{C}$-precover $\phi : C \to M$ is called a *$\mathcal{C}$-cover of M* if every endomorphism $g : C \to C$ such that $\phi g = \phi$ is an isomorphism. A $\mathcal{C}$-cover $\phi : C \to M$ is said to have *the unique mapping property* [3] if for any homomorphism f: $C' \to M$ with $C' \in \mathcal{C}$, there is a unique homomorphism $g : C' \to C$ such that $\phi g = f$.

Theorem 2.1. *The following are equivalent for a ring R:*

(1) R is a von Neumann regular ring.
(2) If M is a divisible left R-module and $\alpha \in EndM$, then $\ker(\alpha)$ *is divisible.*
(3) If N is a torsionfree right R-module and $\alpha \in EndN$, then coker(α) is torsionfree.
(4) Every right R-module has a torsionfree cover with the unique mapping property.

Proof. (1) $\Rightarrow$ (2) , (1) $\Rightarrow$ (3) and (1) $\Rightarrow$ (4) are clear.

(2) $\Rightarrow$ (1). For any left R-module A, there are monomorphisms $f : A \to E^0$ and $g : \mathrm{coker}(f) \to E^1$ with E^0 and E^1 injective. Define $\alpha : E^0 \oplus E^1 \to E^0 \oplus E^1$ via $(x, y) \mapsto (0, g(\overline{x}))$. Then $A \oplus E^1 \cong \ker(\alpha)$ is divisible by (2), and so A is divisible. Thus (1) follows.

(3) $\Rightarrow$ (1). For any right R-module B, there are epimorphisms $f : P_0 \to B$ and $g : P_1 \to \ker(f)$ with P_0 and P_1 projective. Define $\alpha : P_0 \oplus P_1 \to P_0 \oplus P_1$ via $(a, b) \mapsto (g(b), 0)$. Then $\mathrm{im}(\alpha) = \ker(f) \oplus 0$. Thus $(P_0/\ker(f)) \oplus P_1 \cong (P_0 \oplus P_1)/\mathrm{im}(\alpha)$ is torsionfree by (3), and hence

$B \cong P_0/\ker(f)$ is torsionfree. So R is von Neumann regular by [9, Corollary 2.6] or [2, Theorem 2.2].

(4) $\Rightarrow$ (1). Let M be a right R-module. Then M has a torsionfree cover with the unique mapping property $f : T \to M$ by (4). On the other hand, there exists an epimorphism $g : P \to \ker(f)$ with P projective. Let $\iota : \ker(f) \to T$ be the inclusion. Since $f(\iota g) = 0$, we have $\iota g = 0$. So $\ker(f) = 0$. Thus $M \cong T$ is torsionfree, as desired. ■

Recall that a ring R is called *left PP* if every principal left ideal of R is projective. R is called a *left P-coherent ring* [10] if every principal left ideal of R is finitely presented.

Theorem 2.2. *The following are equivalent for a ring R:*

(1) R is a left PP ring.

(2) If M is a divisible left R-module and $\alpha \in EndM$, then im(α) is divisible.

(3) R is a left P-coherent ring, and if N is a torsionfree right R-module and $\alpha \in EndN$, then im(α) is torsionfree.

Proof. (1) $\Rightarrow$ (2) follows from [16, Theorem 2].

(1) $\Rightarrow$ (3) comes from [9, Theorem 5.1] by noting that a module is $(1,1)$-flat if and only if it is torsionfree and a ring is left $(1,1)$-coherent if and only if it is left P-coherent.

(2) $\Rightarrow$ (1). Let A be any submodule of a divisible left R-module B. There is a monomorphism $f : B/A \to E$ with E injective. Define $\alpha : E \oplus B \to E \oplus B$ via $(x, y) \mapsto (f(\overline{y}), 0)$. Then $B/A \cong \text{im}(\alpha)$ is divisible by (2). So R is a left PP ring by [16, Theorem 2].

(3) $\Rightarrow$ (1). Let G be any submodule of a torsionfree right R-module H. There is an epimorphism $g : P \to G$ with P projective. Define $\alpha : P \oplus H \to P \oplus H$ via $(a, b) \mapsto (0, g(a))$. Then $G \cong \text{im}(\alpha)$ is torsionfree by (3). So R is a left PP ring by [9, Theorem 5.1]. ■

Theorem 2.3. *The following are equivalent for a ring R:*

(1) For any left R-module exact sequence $A \to B \to C \to 0$ with A and B divisible, C is divisible.

(2) R is a left P-coherent ring and if $0 \to N \to M \to Q$ is an exact sequence of right R-modules with M and Q torsionfree, then N is torsionfree.

(3) If M is a divisible left R-module and $\alpha \in EndM$, then coker(α) is divisible.

(4) *R is a left P-coherent ring and if N is a torsionfree right R-module and $\alpha \in EndN$, then* $\ker(\alpha)$ *is torsionfree.*

(5) *Every left R-module has a divisible cover with the unique mapping property.*

Proof. (1) $\Rightarrow$ (2). R is left P-coherent by [10, Lemma 4.9]. Now let $0 \to N \to M \to Q$ be an exact sequence of right R-modules with M and Q torsionfree. Then we get an exact sequence $Q^+ \to M^+ \to N^+ \to 0$. Since Q^+ and M^+ are divisible, N^+ is divisible by (1). So N is torsionfree.

(2) $\Rightarrow$ (1). Let $A \to B \to C \to 0$ be an exact sequence of left R-modules with A and B divisible. Then we get an exact sequence $0 \to C^+ \to B^+ \to A^+$. Note that A^+ and B^+ are torsionfree by [10, Theorem 2.7]. Thus C^+ is torsionfree by (2), and so C is divisible.

(1) $\Rightarrow$ (3) and (2) $\Rightarrow$ (4) are trivial.

(3) $\Rightarrow$ (1). Let $A \xrightarrow{f} B \xrightarrow{g} C \to 0$ be an exact sequence of left R-modules with A and B divisible. Define $\alpha : A \oplus B \to A \oplus B$ via $(x, y) \mapsto (0, f(x))$. Then $(A \oplus B)/\mathrm{im}(\alpha) \cong A \oplus (B/\mathrm{im}(f))$ is divisible by (3). So $C \cong B/\ker(g) = B/\mathrm{im}(f)$ is divisible.

(4) $\Rightarrow$ (2). Let $0 \to N \xrightarrow{\varphi} M \xrightarrow{\psi} Q$ be an exact sequence of right R-modules with M and Q torsionfree. Define $\alpha : M \oplus Q \to M \oplus Q$ via $(a, b) \mapsto (0, \psi(a))$. Then $\ker(\alpha) = \ker(\psi) \oplus Q$ is torsionfree by (4). So $N \cong \mathrm{im}(\varphi) = \ker(\psi)$ is torsionfree.

(1) $\Rightarrow$ (5). Note that R is a left P-coherent ring by [10, Lemma 4.9]. Let M be a left R-module. Then M has a divisible cover $f : F \to M$ by [10, Theorem 2.10]. It is enough to show that, for any divisible left R-module G and any homomorphism $g : G \to F$ such that $fg = 0$, we have $g = 0$. In fact, there exists $\beta : F/\mathrm{im}(g) \to M$ such that $\beta\pi = f$ since $\mathrm{im}(g) \subseteq \ker(f)$, where $\pi : F \to F/\mathrm{im}(g)$ is the natural map. Note that $F/\mathrm{im}(g)$ is divisible by (1). Thus there exists $\alpha : F/\mathrm{im}(g) \to F$ such that $\beta = f\alpha$, and so $f\alpha\pi = f$. Hence $\alpha\pi$ is an isomorphism since f is a cover. Therefore π is monic, and so $g = 0$.

(5) $\Rightarrow$ (1). Suppose that $A \xrightarrow{f} B \xrightarrow{g} C \to 0$ is an exact sequence of left R-modules with A and B divisible. Let $\theta : H \to C$ be a divisible cover with the unique mapping property. Then there exists $\delta : B \to H$ such that $g = \theta\delta$. Thus $\theta\delta f = gf = 0 = \theta 0$, and hence $\delta f = 0$, which implies that $\ker(g) = \mathrm{im}(f) \subseteq \ker(\delta)$. Therefore there exists $\gamma : C \to H$ such that $\gamma g = \delta$, and so $\theta\gamma g = g$. Thus $\theta\gamma = 1_C$ since g is epic. It follows that C is isomorphic to a direct summand of H, and hence C is divisible. ■

3. Homomorphism modules of special modules

Let R be a commutative ring. It is well known that R is a coherent ring if and only if $\mathrm{Hom}(A, B)$ is flat for all injective R-modules A and B (see [11]). By [4, Corollary 3.22], R is an *IF ring* (the ring for which every injective R-module is flat) if and only if $\mathrm{Hom}(A, B)$ is injective for all injective R-modules A and B. We continue this style of studying rings by properties of homomorphism modules of certain special R-modules.

Theorem 3.1. *The following are equivalent for a commutative ring R.*

(1) R is a P-coherent ring.
(2) $Hom(A, B)$ is torsionfree for all divisible R-modules A and all injective R-modules B.
(3) $Hom(A, B)$ is torsionfree for all injective R-modules A and B.
(4) $Hom(A, B)$ is torsionfree for all projective R-modules A and all torsionfree R-modules B.
(5) $Hom(A, B)$ is torsionfree for all projective R-modules A and B.

In this case, $A \otimes B$ is divisible for all divisible R-modules A and all flat R-modules B.

Proof. (1) $\Rightarrow$ (2). Let $a \in R$. Then Ra is finitely presented by (1). For any divisible R-module A, the exact sequence $0 \to Ra \to R \to R/Ra \to 0$ induces the exactness of $0 \to \mathrm{Hom}(R/Ra, A) \to \mathrm{Hom}(R, A) \to \mathrm{Hom}(Ra, A) \to 0$. For any injective R-module B, we get an exact sequence

$$0 \to \mathrm{Hom}(\mathrm{Hom}(Ra, A), B) \to \mathrm{Hom}(\mathrm{Hom}(R, A), B) \to \mathrm{Hom}(\mathrm{Hom}(R/Ra, A), B) \to 0$$

which gives the exactness of the sequence

$$0 \to \mathrm{Hom}(A, B) \otimes Ra \to \mathrm{Hom}(A, B) \otimes R \to \mathrm{Hom}(A, B) \otimes (R/Ra) \to 0$$

by [12, Lemma 3.60]. So $\mathrm{Hom}(A, B)$ is torsionfree.

(2) $\Rightarrow$ (3) and (4) $\Rightarrow$ (5) are trivial.

(3) $\Rightarrow$ (1). We will show that any direct product ΠR of R is torsionfree. Indeed, since R is a pure submodule of R^{++}, ΠR is a pure submodule of ΠR^{++} by [1, Lemma 1 (2)]. Note that

$$\Pi R^{++} \cong \Pi \mathrm{Hom}(R^{+} \otimes R, \mathbb{Q}/\mathbb{Z}) \cong \Pi \mathrm{Hom}(R^{+}, R^{+}) \cong \mathrm{Hom}(R^{+}, \Pi R^{+}).$$

Thus ΠR^{++} is torsionfree by (3) since R^{+} and ΠR^{+} are injective. Therefore ΠR is torsionfree, and so R is P-coherent by [10, Theorem 2.7].

(1) $\Rightarrow$ (4). For any projective R-module A, there is a projective R-module Q such that $A \oplus Q \cong \oplus R$. Then, for any torsionfree R-module B,

we have $\mathrm{Hom}(A,B) \oplus \mathrm{Hom}(Q,B) \cong \mathrm{Hom}(\oplus R, B) \cong \Pi B$. Thus $\mathrm{Hom}(A,B)$ is torsionfree since ΠB is torsionfree by (1) and [10, Theorem 2.7].

(5) $\Rightarrow$ (1). Since $\Pi R \cong \mathrm{Hom}(\oplus R, R)$ is torsionfree by (5), R is P-coherent by [10, Theorem 2.7].

In this case, let B be a flat R-module, then $B = \varinjlim P_i$ with each P_i projective. So, for any divisible R-module A, $A \otimes B = A \otimes \varinjlim P_i = \varinjlim (A \otimes P_i)$ is divisible by [10, Theorem 2.7]. ■

Theorem 3.2. *The following are equivalent for a commutative P-coherent ring R:*

(1) R is divisible as an R-module.
(2) Every injective R-module is torsionfree.
(3) Every flat R-module is divisible.
(4) $Hom(A,B)$ is divisible for all injective R-modules A and B.
(5) $Hom(A,B)$ is divisible for all projective R-modules A and B.
(6) $Hom(A,B)$ is divisible for all projective R-modules A and all flat R-modules B.
(7) $Hom(A,B)$ is torsionfree for all flat R-modules A and all injective R-modules B.
(8) $A \otimes B$ is torsionfree for all flat R-modules A and all injective R-modules B.

Proof. (1) $\Leftrightarrow$ (2) follows from [10, Proposition 4.2]. (2) $\Rightarrow$ (7) comes from [12, Theorem 3.44]. (7) $\Rightarrow$ (2), (8) $\Rightarrow$ (2) and (6) $\Rightarrow$ (5) $\Rightarrow$ (1) are clear.

(2) $\Rightarrow$ (4). For any $a \in R$ and injective R-modules A and B, there is the isomorphism $\mathrm{Ext}^1(R/Ra, \mathrm{Hom}(A,B)) \cong \mathrm{Hom}(\mathrm{Tor}_1(A, R/Ra), B)$. So $\mathrm{Hom}(A,B)$ is divisible since A is torsionfree by (2).

(4) $\Rightarrow$ (8). Let A be a flat R-module and B an injective R-module. Then $(A \otimes B)^+ \cong \mathrm{Hom}(B, A^+)$ is divisible by (4). So $A \otimes B$ is torsionfree.

(2) $\Rightarrow$ (3). Let M be a flat R-module. Then M^+ is injective and hence torsionfree by (2). Thus M^{++} is divisible, and so M is divisible by [10, Lemma 2.6] (for M is pure in M^{++}).

(3) $\Rightarrow$ (6). Let A be a projective R-module and B a flat R-module. Then there is a projective R-module C such that $A \oplus C \cong \oplus R$. So we have

$$\mathrm{Hom}(A,B) \oplus \mathrm{Hom}(C,B) \cong \mathrm{Hom}(\oplus R, B) \cong \Pi B.$$

Thus $\mathrm{Hom}(A,B)$ is divisible since ΠB is divisible by (3). ■

Combining Theorems 3.1 with 3.2, we have

Corollary 3.1. *The following are equivalent for a commutative ring R.*

(1) *R is a P-coherent ring and R is divisible as an R-module.*
(2) *Hom(A, B) is torsionfree and divisible for all injective R-modules A and B.*
(3) *Hom(A, B) is torsionfree and divisible for all projective R-modules A and B.*

Recall that R is called a *left strongly P-coherent ring* [10] if every principal left ideal I of R is cyclically presented, i.e., $I \cong R/Rr$ for some $r \in R$.

Theorem 3.3. *Let R be a commutative strongly P-coherent ring. The following are equivalent:*

(1) *R is a von Neumann regular ring.*
(2) *Hom(A, B) is divisible for all divisible R-modules A and all R-modules B.*
(3) *Hom(A, B) is divisible for all R-modules A and all divisible R-modules B.*
(4) *Hom(A, B) is divisible for all R-modules A and all torsionfree R-modules B.*
(5) *Hom(A, R) is divisible for all cyclically presented R-modules A.*
(6) *Hom(A, B) is torsionfree for all R-modules A and all divisible R-modules B.*
(7) *A⊗B is torsionfree for all R-modules A and all divisible R-modules B.*

Proof. (1) $\Rightarrow$ (2) and (1) $\Rightarrow$ (4) $\Rightarrow$ (5) are obvious.

(2) $\Rightarrow$ (7). Let A be any R-module and B any divisible R-module. Since $(A \otimes B)^+ \cong \text{Hom}(B, A^+)$ is divisible by (2), $A \otimes B$ is torsionfree.

(7) $\Rightarrow$ (5). Let A be any cyclically presented R-module. Then $\text{Hom}(A, R)^+ \cong A \otimes R^+$ is torsionfree by [12, Lemma 3.60] and (7). So $\text{Hom}(A, R)$ is divisible by [10, Theorem 2.7].

(5) $\Rightarrow$ (1). By (5), R is divisible as an R-module. Let A be any cyclically presented R-module, i.e., there is an exact sequence $0 \to Ra \to R \to A \to 0$ for some $a \in R$. Note that Ra is cyclically presented by hypothesis. So there is an exact sequence $0 \to Rb \xrightarrow{i} R \xrightarrow{f} Ra \to 0$ with $b \in R$, which yields the exact sequence $0 \to \text{Hom}(Ra, R) \xrightarrow{f^*} R \to B \to 0$. Define $p : R \to Rb$ by $p(r) = rb$ for $r \in R$. Then p induces a monomorphism $p^* : \text{Hom}(Rb, R) \to R$. It follows that $B \cong R/\text{im}(f^*) \cong \text{im}(i^*) \cong p^*(\text{im}(i^*))$ is cyclically presented. Consider the following commutative diagram with

exact rows:

$$\begin{array}{ccccccc} 0 \longrightarrow & Rb & \longrightarrow & R & \longrightarrow & Ra & \longrightarrow 0 \\ & \phi \vdots & & \downarrow & & \delta \downarrow & \\ 0 \longrightarrow & \mathrm{Hom}(B,R) & \longrightarrow & R & \longrightarrow & \mathrm{Hom}(\mathrm{Hom}(Ra,R),R). & \end{array}$$

Since δ is monic, ϕ is an isomorphism by the Five Lemma. Thus $Rb \cong \mathrm{Hom}(B,R)$ is divisible by (5), and hence Ra is divisible by [10, Lemma 4.10]. It follows that A is torsionfree by [14, Proposition 2], and so A is projective by [13, p.2047, 5(a)]. Thus R is a von Neumann regular ring.

(3) $\Leftrightarrow$ (1) $\Leftrightarrow$ (6) are easy by letting $B = R^+$. ■

Theorem 3.4. *Let R be a commutative strongly P-coherent ring. The following are equivalent:*

(1) Every R-module has a divisible cover with the unique mapping property.

(2) $Hom(A, B)$ is torsionfree for all divisible R-modules A and all R-modules B.

(3) $Hom(A, B)$ is torsionfree for all R-modules A and all torsionfree R-modules B.

(4) $Hom(A, R)$ is torsionfree for all cyclically presented R-modules A.

(5) $A \otimes B$ is divisible for all R-modules A and all divisible R-modules B.

Proof. (1) $\Rightarrow$ (3). Let A be any R-module. Then there is an exact sequence $P_1 \to P_0 \to A \to 0$ with P_0 and P_1 projective. So, for any torsionfree R-module B, we get an exact sequence $0 \to \mathrm{Hom}(A,B) \to \mathrm{Hom}(P_0,B) \to \mathrm{Hom}(P_1,B)$. By Theorem 3.1, each $\mathrm{Hom}(P_i,B)$ is torsionfree. So $\mathrm{Hom}(A,B)$ is torsionfree by Theorem 2.3.

(3) $\Rightarrow$ (4) is trivial.

(4) $\Rightarrow$ (1). Let A be any cyclically presented R-module. By the proof of (5) $\Rightarrow$ (1) in Theorem 3.3, there are exact sequences $0 \to Ra \to R \to A \to 0$, $0 \to Rb \to R \to Ra \to 0$ and $0 \to \mathrm{Hom}(Ra,R) \to R \to B \to 0$ with B cyclically presented and $Rb \cong \mathrm{Hom}(B,R)$, where $a, b \in R$. Thus Rb is torsionfree by (4), and hence it is projective by [13, p.2047, 5(a)]. Therefore A has projective dimension ≤ 2, and so every R-module has a divisible cover with the unique mapping property by [10, Theorem 4.15 and Corollary 4.16].

(1) $\Rightarrow$ (2). Let B be any R-module. Then there is an exact sequence $0 \to B \to E^0 \to E^1$ with E^0 and E^1 injective. So, for any divisible R-module A,

we get an exact sequence $0 \to \mathrm{Hom}(A, B) \to \mathrm{Hom}(A, E^0) \to \mathrm{Hom}(A, E^1)$. By Theorem 3.1, each $\mathrm{Hom}(A, E^i)$ is torsionfree. Thus $\mathrm{Hom}(A, B)$ is torsionfree by Theorem 2.3.

(2) $\Rightarrow$ (5). Let A be any R-module and B any divisible R-module. Since $(A \otimes B)^+ \cong \mathrm{Hom}(B, A^+)$ is torsionfree by (2), $A \otimes B$ is divisible by [10, Theorem 2.7].

(5) $\Rightarrow$ (4). Let A be any cyclically presented R-module. Then $\mathrm{Hom}(A, R)^+ \cong A \otimes R^+$ is divisible by [12, Lemma 3.60] and (5). So $\mathrm{Hom}(A, R)$ is torsionfree. ■

4. Divisible and torsionfree modules under change of rings

Theorem 4.1. *Let $\varphi : R \to S$ be a ring homomorphism with S flat as a right R-module.*

(1) If ${}_SM$ is a divisible left S-module, then ${}_RM$ is a divisible left R-module.

(2) If N_S is a torsionfree right S-module, then N_R is a torsionfree right R-module.

Proof. If Q is a cyclically presented left R-module, then there is an exact sequence $R \to R \to Q \to 0$ of left R-modules. Thus we have the left S-module exact sequence ${}_SS \otimes_R R \to {}_SS \otimes_R R \to {}_SS \otimes_R Q \to 0$, and so ${}_SS \otimes_R Q$ is a cyclically presented left S-module.

(1) Let ${}_SM$ be a divisible left S-module. Since S_R is flat,

$$\mathrm{Ext}^1_R(Q, M) \cong \mathrm{Ext}^1_S({}_SS \otimes_R Q, M) = 0$$

by [12, Theorem 11.65]. So ${}_RM$ is a divisible left R-module.

(2) If N_S is a torsionfree right S-module, then

$$\mathrm{Tor}^R_1(N, Q) \cong \mathrm{Tor}^S_1(N,\ {}_SS \otimes_R Q) = 0$$

by [12, Theorem 11.64] since S_R is flat. Thus N_R is torsionfree. ■

Corollary 4.1. *Let S be a multiplicative subset of a commutative ring R.*

(1) If M is a divisible $S^{-1}R$-module, then M is a divisible R-module.

(2) If N is a torsionfree $S^{-1}R$-module, then N is a torsionfree R-module.

Proof. The result holds by Theorem 4.1 since $S^{-1}R$ is a flat R-module. ■

Acknowledgments

This research was partially supported by SRFDP (No.20050284015), NSFC (No.10771096), Science Research Fund of Nanjing Institute of Technology (No.KXJ07061), Jiangsu 333 Project, and Jiangsu Qinglan Project.

References

1. T.J. Cheatham, D.R. Stone, Flat and projective character modules. *Proc. Amer. Math. Soc.* 1981, **81**(2): 175-177.
2. J. Dauns, L. Fuchs, Torsion-freeness in rings with zero-divisors. *J. Algebra Appl.* 2004, **3**: 221-237.
3. N.Q. Ding, On envelopes with the unique mapping property. *Comm. Algebra* 1996, **24**(4): 1459-1470.
4. N.Q. Ding, J.L. Chen, The flat dimensions of injective modules. *Manuscripta Math.* 1993, **78**: 165-177.
5. E.E. Enochs, Injective and flat covers, envelopes and resolvents. *Israel J. Math.* 1981, **39**: 189-209.
6. E.E. Enochs, O.M.G. Jenda, Relative Homological Algebra; GEM 30, Walter de Gruyter: Berlin-New York, 2000.
7. R. Göbel, J. Trlifaj, Approximations and Endomorphism Algebras of Modules; GEM 41, Walter de Gruyter, Berlin-New York, 2006.
8. A. Hattori, A foundation of torsion theory for modules over general rings. *Nagoya Math. J.* 1960, **17**: 147-158.
9. L.X. Mao, N.Q. Ding, On relative injective modules and relative coherent rings. *Comm. Algebra* 2006, **34**(7): 2531-2545.
10. L.X. Mao, N.Q. Ding, On divisible and torsionfree modules. *Comm. Algebra* 2008, **36**(2): 708-731.
11. E. Matlis, Commutative coherent rings. *Canad. J. Math.* 1982, **34**(6): 1240-1244.
12. J.J. Rotman, An Introduction to Homological Algebra; Academic Press: New York, 1979.
13. A. Shamsuddin, n-injective and n-flat modules. *Comm. Algebra* 2001, **29**(5): 2039-2050.
14. R.B. Warfield, Jr., Purity and algebraic compactness for modules. *Pacific J. Math.* 1969, **28**: 699-719.
15. J. Xu, Flat Covers of Modules, Lecture Notes in Math. 1634; Springer-Verlag: Berlin-Heidelberg-New York, 1996.
16. W.M. Xue, On *PP* rings. *Kobe J. Math.* 1990, **7**: 77-80.

On the Classification of Decomposable Quantum Ruled Surfaces

Izuru Mori

Department of Mathematics, Faculty of Science, Shizuoka University, Shizuoka 422-8529, JAPAN

In this paper, we will study quantum ruled surfaces defined by Van den Bergh with classification problem in mind. In particular, we will prove that "decomposable" quantum ruled surfaces over smooth projective curves X can be classified in terms of geometric triples $(X, \sigma, \mathcal{L})$ where $\sigma \in \operatorname{Aut} X$ is an automorphism of X and $\mathcal{L} \in \operatorname{Pic} X$ is an invertible $\mathcal{O}_X$-module, as in the classification of noncommutative projective curves and that of quantum projective planes.

1. Motivation

One of the active projects in algebraic geometry has been to classify low dimensional projective schemes. We follow this tradition. Since (the homogeneous coordinate rings of) noncommutative projective curves were classified by Artin and Stafford [2] in terms of geometric triples $(X, \sigma, \mathcal{L})$ (up to finite dimensional pieces) where X is a scheme, $\sigma \in \operatorname{Aut} X$ is an automorphism of X, and $\mathcal{L} \in \operatorname{Pic} X$ is an invertible $\mathcal{O}_X$-module, one of the major projects in noncommutative algebraic geometry is to classify noncommutative projective surfaces (see [13]). The classical conjecture due to Artin [1], which is still open, says that every noncommutative projective surface is birationally equivalent to one of the following: (1) a quantum projective plane, which is a noncommutative analogue of the projective plane, (2) a quantum ruled surface, which is a noncommutative analogue of a ruled surface, and (3) a surface finite over its center. Since (the homogeneous coordinate rings of) quantum projective planes were classified by Artin, Tate and Van den Bergh [3] in terms of geometric triples $(X, \sigma, \mathcal{L})$ again, the next natural goal is to classify quantum ruled surfaces. In this paper, we will study quantum ruled surfaces defined by Van den Bergh [16] with classification problem in mind. In particular, we will prove that "decomposable" quantum ruled surfaces over smooth projective curves X can also be classified in terms of geometric triples $(X, \sigma, \mathcal{L})$.

2. $\mathbb{Z}$-Algebras over a Quasi-scheme

First, we recall some of the basic notations and terminologies used in noncommutative algebraic geometry.

Definition 2.1. [12,15] A quasi-scheme X is a Grothendieck category $\operatorname{Mod} X$. We say that two quasi-schemes X and Y are isomorphic, denoted by $X \cong Y$, if the categories $\operatorname{Mod} X$ and $\operatorname{Mod} Y$ are equivalent. We say that a quasi-scheme X is noetherian if the category $\operatorname{Mod} X$ is locally noetherian. In this case, we denote by $\operatorname{mod} X$ the full subcategory of $\operatorname{Mod} X$ consisting of noetherian objects.

If X is a reasonably nice scheme (e.g. quasi-compact and quasi-separated), then the category of quasi-coherent sheaves on X is a Grothendieck category. We always view X as a quasi-scheme by $\operatorname{Mod} X$ where $\operatorname{Mod} X$ is the category of quasi-coherent sheaves on X. If X is noetherian, then $\operatorname{mod} X$ is the category of coherent sheaves on X.

In order to extend the notion of blowing up to noncommutative settings, Van den Bergh introduced a notion of bimodule over quasi-schemes.

Definition 2.2. [15] Let X and Y be quasi-schemes. An X-Y bimodule M is an adjoint pair of functors with the following suggestive notations:

$$\begin{aligned} - \otimes_X M \quad &: \operatorname{Mod} X \to \operatorname{Mod} Y \\ \mathcal{H}om_Y(M, -) &: \operatorname{Mod} Y \to \operatorname{Mod} X. \end{aligned}$$

We denote by $\operatorname{BiMod}(X, Y)$ the category of X-Y bimodules viewed as the full subcategory of the opposite category of the category of left exact functors $\operatorname{Mod} Y \to \operatorname{Mod} X$.

Let X be a quasi-scheme. By an X-bimodule, we mean an X-X bimodule. The identity functor $\operatorname{Mod} X \to \operatorname{Mod} X$ can be viewed as an X-bimodule. We denote it by $o_X : \operatorname{Mod} X \to \operatorname{Mod} X$, which plays a role of the structure sheaf on X. If X, Y, Z are quasi-schemes, and $M \in \operatorname{BiMod}(X, Y), N \in \operatorname{BiMod}(Y, Z)$ are bimodules, then we define the bimodule $M \otimes_Y N \in \operatorname{BiMod}(X, Z)$ as the composition of functors

$$\begin{aligned} - \otimes_X (M \otimes_Y N) &:= (- \otimes_X M) \otimes_Y N &&: \operatorname{Mod} X \to \operatorname{Mod} Z \\ \mathcal{H}om_Z(M \otimes_Y N, -) &:= \mathcal{H}om_Y(M, \mathcal{H}om_Z(N, -)) &&: \operatorname{Mod} Z \to \operatorname{Mod} X. \end{aligned}$$

Now we define a notion of $\mathbb{Z}$-algebra over a sequence of quasi-schemes.

Definition 2.3. Let $\tilde{X} = \{X_i\}_{i \in \mathbb{Z}}$ be a sequence of quasi-schemes. A $\mathbb{Z}$-algebra over $\tilde{X}$ is a set of objects $A = \{A_{ij}\}_{i,j \in \mathbb{Z}}$ where $A_{ij} \in$

$\mathrm{BiMod}(X_i, X_j)$ with morphisms $\eta_i : o_{X_i} \to A_{ii}$ in $\mathrm{BiMod}(X_i, X_i)$ (units) and $\mu_{ijk} : A_{ij} \otimes_{X_j} A_{jk} \to A_{ik}$ in $\mathrm{BiMod}(X_i, X_k)$ (multiplications) such that

$$\mu_{iii} \circ (\eta_i \otimes \mathrm{Id}_{ii}) \cong \mu_{iii} \circ (\mathrm{Id}_{ii} \otimes \eta_i) \cong \mathrm{Id}_{ii},$$
$$\mu_{ijl} \circ (\mathrm{Id}_{ij} \otimes \mu_{jkl}) \cong \mu_{ikl} \circ (\mu_{ijk} \otimes \mathrm{Id}_{kl})$$

for all $i, j, k, l \in \mathbb{Z}$ where $\mathrm{Id}_{ij} : A_{ij} \to A_{ij}$ are the identity morphisms.

An A-module is a sequence of objects $M = \{M_i\}_{i \in \mathbb{Z}}$ where $M_i \in \mathrm{Mod}\, X_i$ with morphisms $h_{ij}^M : M_i \otimes_{X_i} A_{ij} \to M_j$ in $\mathrm{Mod}\, X_j$ (actions) such that

$$h_{ii}^M \circ (\mathrm{Id}_i \otimes \eta_i) \cong \mathrm{Id}_i,$$
$$h_{ik}^M \circ (\mathrm{Id}_i \otimes \mu_{ijk}) \cong h_{jk}^M \circ (h_{ij}^M \otimes \mathrm{Id}_{jk})$$

for all $i, j, k \in \mathbb{Z}$ where $\mathrm{Id}_i : M_i \to M_i$ are the identity morphisms.

We denote by $\mathrm{GrMod}\, A$ the category of A-modules. A morphism $f : M \to N$ in $\mathrm{GrMod}\, A$ is a sequence of morphisms $f = \{f_i\}_{i \in \mathbb{Z}}$ where $f_i \in \mathrm{Hom}_{X_i}(M_i, N_i)$ such that

$$h_{ij}^N \circ (f_i \otimes \mathrm{Id}_{ij}) \cong f_j \circ h_{ij}^M$$

for all $i, j \in \mathbb{Z}$.

Let A be a $\mathbb{Z}$-algebra over a sequence of noetherian quasi-schemes $\tilde{X}$. We say that A is noetherian if the category $\mathrm{GrMod}\, A$ is locally noetherian. In this case, we say that an A-module $M \in \mathrm{GrMod}\, A$ is right bounded if $M_i = 0$ for all $i \gg 0$. We denote by $\mathrm{Tors}\, A$ the full subcategory of $\mathrm{GrMod}\, A$ consisting of direct limits of right bounded modules, and $\mathrm{Tails}\, A = \mathrm{GrMod}\, A / \mathrm{Tors}\, A$ the quotient category. The quotient functor is denoted by $\pi : \mathrm{GrMod}\, A \to \mathrm{Tails}\, A$, and its right adjoint is denoted by $\omega : \mathrm{Tails}\, A \to \mathrm{GrMod}\, A$ (see [5,15]). For each $i \in \mathbb{Z}$, we define a functor $- \otimes_{X_i} e_i A : \mathrm{Mod}\, X_i \to \mathrm{GrMod}\, A$ by $V \otimes_{X_i} e_i A := \{V \otimes_{X_i} A_{ij}\}_{j \in \mathbb{Z}}$.

Definition 2.4. Let A be a noetherian $\mathbb{Z}$-algebra over a sequence of noetherian quasi-schemes $\tilde{X}$. The noncommutative projective scheme associated to A is a quasi-scheme $\mathrm{Proj}\, A$ defined by $\mathrm{Mod}(\mathrm{Proj}\, A) = \mathrm{Tails}\, A$.

We tacitly assume that $\tilde{X}$ is a sequence of noetherian quasi-schemes and A is a noetherian $\mathbb{Z}$-algebra over $\tilde{X}$ whenever we write $\mathrm{Tails}\, A$ or $\mathrm{Proj}\, A$.

Let X be a quasi-scheme. If A is a $\mathbb{Z}$-algebra over $\tilde{X} = \{X_i\}$ where $X_i = X$ for all $i \in \mathbb{Z}$, then we simply call A a $\mathbb{Z}$-algebra over X. If A is a noetherian $\mathbb{Z}$-algebra over a noetherian quasi-scheme X, then the structure map $f : \mathrm{Proj}\, A \to X$ is defined by the adjoint pair of functors

$$\begin{aligned} f^*(-) &:= \pi(- \otimes_X e_0 A) : \mathrm{Mod}\, X \to \mathrm{Tails}\, A \\ f_*(-) &:= (\omega(-))_0 \qquad\quad : \mathrm{Tails}\, A \to \mathrm{Mod}\, X. \end{aligned}$$

Definition 2.5. Let A and A' be $\mathbb{Z}$-algebras over a quasi-scheme X. We write $\operatorname{GrMod} A \cong_X \operatorname{GrMod} A'$ if there exists an equivalence functor $F : \operatorname{GrMod} A \to \operatorname{GrMod} A'$ such that $(-)_0 \cong (F(-))_0 : \operatorname{GrMod} A \to \operatorname{Mod} X$.

If X, A, A' are all noetherian, then we say that $\operatorname{Proj} A$ and $\operatorname{Proj} A'$ are isomorphic over X, denoted by $\operatorname{Proj} A \cong_X \operatorname{Proj} A'$, if there exists an equivalence functor $F : \operatorname{Tails} A \to \operatorname{Tails} A'$ such that $f'_* \circ F \cong f_* : \operatorname{Tails} A \to \operatorname{Mod} X$ where $f : \operatorname{Proj} A \to X$ and $f' : \operatorname{Proj} A' \to X$ are the structure maps.

There is a way of twisting $\mathbb{Z}$-algebras, which induces an equivalence of module categories.

Definition 2.6. Let A be a $\mathbb{Z}$-algebra over a sequence of quasi-schemes $\tilde{X} = \{X_i\}$ and $\tilde{L} = \{L_i\}$ a sequence of autoequivalences $L_i : \operatorname{Mod} X_i \to \operatorname{Mod} X_i$. A twist of A by $\tilde{L}$ is a $\mathbb{Z}$-algebra $A \otimes_{\tilde{X}} \tilde{L}$ over $\tilde{X}$ defined by

$$(A \otimes_{\tilde{X}} \tilde{L})_{ij} = L_i^{-1} \otimes_{X_i} A_{ij} \otimes_{X_j} L_j$$

with units

$$o_{X_i} \cong L_i^{-1} \otimes_{X_i} o_{X_i} \otimes_{X_i} L_i \xrightarrow{L_i^{-1} \otimes \eta_i \otimes L_i} L_i^{-1} \otimes_{X_i} A_{ii} \otimes_{X_i} L_i = (A \otimes_{\tilde{X}} \tilde{L})_{ii},$$

and multiplications

$$\begin{aligned}
& (A \otimes_{\tilde{X}} \tilde{L})_{ij} \otimes_{X_j} (A \otimes_{\tilde{X}} \tilde{L})_{jk} \\
= \;& (L_i^{-1} \otimes_{X_i} A_{ij} \otimes_{X_j} L_j) \otimes_{X_j} (L_j^{-1} \otimes_{X_j} A_{jk} \otimes_{X_k} L_k) \\
\cong \;& L_i^{-1} \otimes_{X_i} (A_{ij} \otimes_{X_j} A_{jk}) \otimes_{X_k} L_k \\
\longrightarrow \;& L_i^{-1} \otimes_{X_i} A_{ik} \otimes_{X_k} L_k \qquad \text{(induced by } \mu_{ijk}) \\
= \;& (A \otimes_{\tilde{X}} \tilde{L})_{ik}.
\end{aligned}$$

A twist of an A-module M by $\tilde{L}$ is an $A \otimes_{\tilde{X}} \tilde{L}$-module $M \otimes_{\tilde{X}} \tilde{L}$ defined by

$$(M \otimes_{\tilde{X}} \tilde{L})_i = M_i \otimes_{X_i} L_i$$

with actions

$$\begin{aligned}
& (M \otimes_{\tilde{X}} \tilde{L})_i \otimes_{X_i} (A \otimes_{\tilde{X}} \tilde{L})_{ij} \\
= \;& (M_i \otimes_{X_i} L_i) \otimes_{X_i} (L_i^{-1} \otimes_{X_i} A_{ij} \otimes_{X_j} L_j) \\
\cong \;& (M_i \otimes_{X_i} A_{ij}) \otimes_{X_j} L_j \\
\longrightarrow \;& M_j \otimes_{X_j} L_j \qquad \text{(induced by } h_{ij}^M) \\
= \;& (M \otimes_{\tilde{X}} \tilde{L})_j.
\end{aligned}$$

Lemma 2.1. *[16] If A is a (noetherian) $\mathbb{Z}$-algebra over a sequence of (noetherian) quasi-schemes $\tilde{X} = \{X_i\}$ and $\tilde{L} = \{L_i\}$ is a sequence of autoequivalences $L_i : \operatorname{Mod} X_i \to \operatorname{Mod} X_i$, then the functor $- \otimes_{\tilde{X}} \tilde{L} : \operatorname{GrMod} A \to$*

$\mathrm{GrMod}(A \otimes_{\tilde{X}} \tilde{L})$ *is an equivalence of categories, which induces an isomorphism* $\mathrm{Proj}\, A \to \mathrm{Proj}(A \otimes_{\tilde{X}} \tilde{L})$.

In particular, if A *is a (noetherian)* $\mathbb{Z}$*-algebra over a (noetherian) quasi-scheme* X *and* $L_0 \cong o_X$*, then* $\mathrm{GrMod}\, A \cong_X \mathrm{GrMod}(A \otimes_X \tilde{L})$ *and* $\mathrm{Proj}\, A \cong_X \mathrm{Proj}(A \otimes_X \tilde{L})$.

3. Sheaf Bimodules

For the rest of this paper, we fix a field k. Unless otherwise specified, we assume that all schemes will be separated and of finite type over a field k, all maps between schemes are over k, and all categories and functors are k-linear.

There is another notion of bimodule over schemes, called a sheaf bimodule. In this section, we review the definition and basic properties of sheaf bimodules (see [8]).

Let X and Y be schemes. We denote by pr_1 and pr_2 the canonical projections $pr_1 : X \times Y \to X$ and $pr_2 : X \times Y \to Y$.

Definition 3.1. [14] A coherent $\mathcal{O}_X$-$\mathcal{O}_Y$ bimodule is a coherent sheaf $\mathcal{M}$ on $X \times Y$ such that the restrictions $pr_1|_{\mathrm{Supp}\,\mathcal{M}} : \mathrm{Supp}\,\mathcal{M} \to X$ and $pr_2|_{\mathrm{Supp}\,\mathcal{M}} : \mathrm{Supp}\,\mathcal{M} \to Y$ are both finite. We denote by $\mathrm{Bimod}(\mathcal{O}_X, \mathcal{O}_Y)$ the full subcategory of $\mathrm{mod}(X \times Y)$ consisting of coherent $\mathcal{O}_X$-$\mathcal{O}_Y$ bimodules.

If X, Y, Z are schemes, and $\mathcal{M} \in \mathrm{Bimod}(\mathcal{O}_X, \mathcal{O}_Y), \mathcal{N} \in \mathrm{Bimod}(\mathcal{O}_Y, \mathcal{O}_Z)$ are bimodules, then we define the bimodule$\mathcal{M} \otimes_{\mathcal{O}_Y} \mathcal{N} \in \mathrm{Bimod}(\mathcal{O}_X, \mathcal{O}_Z)$ by

$$\mathcal{M} \otimes_{\mathcal{O}_Y} \mathcal{N} := pr_{13*}(pr_{12}^* \mathcal{M} \otimes_{\mathcal{O}_{X \times Y \times Z}} pr_{23}^* \mathcal{N}) \tag{1}$$

where

$$pr_{13} : X \times Y \times Z \to X \times Z$$
$$pr_{12} : X \times Y \times Z \to X \times Y$$
$$pr_{23} : X \times Y \times Z \to Y \times Z$$

are the canonical projections. It is possible to tensor a module and a bimodule in the appropriate order. For example, if $\mathcal{F} \in \mathrm{Mod}\, X$ and $\mathcal{M} \in \mathrm{Bimod}(\mathcal{O}_X, \mathcal{O}_Y)$, then $\mathcal{F} \otimes_{\mathcal{O}_X} \mathcal{M} \in \mathrm{Mod}\, Y$ is defined by

$$\mathcal{F} \otimes_{\mathcal{O}_X} \mathcal{M} := pr_{2*}(pr_1^* \mathcal{F} \otimes_{\mathcal{O}_{X \times Y}} \mathcal{M}). \tag{2}$$

In particular,

$$\mathcal{O}_X \otimes_{\mathcal{O}_X} \mathcal{M} := pr_{2*}(pr_1^* \mathcal{O}_X \otimes_{\mathcal{O}_{X \times Y}} \mathcal{M}) \cong pr_{2*}(\mathcal{O}_{X \times Y} \otimes_{\mathcal{O}_{X \times Y}} \mathcal{M}) \cong pr_{2*}\mathcal{M}$$

in $\operatorname{Mod} Y$. By symmetry, $\mathcal{M} \otimes_{\mathcal{O}_Y} \mathcal{O}_Y \cong pr_{1*}\mathcal{M}$ in $\operatorname{Mod} X$.

If $\mathcal{M} \in \operatorname{Bimod}(\mathcal{O}_X, \mathcal{O}_Y)$, then the functor $-\otimes_{\mathcal{O}_X} \mathcal{M} : \operatorname{Mod} X \to \operatorname{Mod} Y$ defined by (2) has a right adjoint, so $\mathcal{M} \in \operatorname{Bimod}(\mathcal{O}_X, \mathcal{O}_Y)$ can be viewed as an X-Y bimodule in the earlier sense by $- \otimes_X \mathcal{M} := - \otimes_{\mathcal{O}_X} \mathcal{M} : \operatorname{Mod} X \to \operatorname{Mod} Y$. This defines a fully faithful functor $\operatorname{Bimod}(\mathcal{O}_X, \mathcal{O}_Y) \to \operatorname{BiMod}(X, Y)$ compatible with tensor products [9, Lemma 5.1] [16, Lemma 3.1.1]. It follows that a bimodule $\mathcal{M} \in \operatorname{Bimod}(\mathcal{O}_X, \mathcal{O}_Y)$ is uniquely determined by the functor $- \otimes_{\mathcal{O}_X} \mathcal{M} : \operatorname{Mod} X \to \operatorname{Mod} Y$.

Definition 3.2. [14] Let X and Y be schemes. A bimodule $\mathcal{E} \in \operatorname{Bimod}(\mathcal{O}_X, \mathcal{O}_Y)$ is called locally free (of rank r) if $pr_{1*}\mathcal{E}$ and $pr_{2*}\mathcal{E}$ are locally free on X and Y respectively (of the same rank r).

If X and Y are smooth schemes, and $\mathcal{E} \in \operatorname{Bimod}(\mathcal{O}_X, \mathcal{O}_Y)$ is a locally free bimodule of rank r, then there exist locally free bimodules $\mathcal{E}^*, {}^*\mathcal{E} \in \operatorname{Bimod}(\mathcal{O}_Y, \mathcal{O}_X)$ of rank r such that $-\otimes_{\mathcal{O}_Y} \mathcal{E}^* : \operatorname{Mod} Y \to \operatorname{Mod} X$ is a right adjoint to $- \otimes_{\mathcal{O}_X} \mathcal{E} : \operatorname{Mod} X \to \operatorname{Mod} Y$, and $- \otimes_{\mathcal{O}_Y} {}^*\mathcal{E} : \operatorname{Mod} Y \to \operatorname{Mod} X$ is a left adjoint to $- \otimes_{\mathcal{O}_X} \mathcal{E} : \operatorname{Mod} X \to \operatorname{Mod} Y$ [16, Section 3]. Inductively, we define

$$\mathcal{E}^{i*} := \begin{cases} (\mathcal{E}^{(i-1)*})^* & i \geq 1, \\ \mathcal{E} & i = 0, \\ {}^*(\mathcal{E}^{(i+1)*}) & i \leq -1. \end{cases}$$

By a coherent $\mathcal{O}_X$-bimodule, we mean a coherent $\mathcal{O}_X$-$\mathcal{O}_X$ bimodule. Let Δ_X be the diagonal in $X \times X$. The coherent $\mathcal{O}_X$-bimodule $\mathcal{O}_{\Delta_X}$ is called the trivial $\mathcal{O}_X$-bimodule, which corresponds to the identity functor o_X. A bimodule $\mathcal{L} \in \operatorname{Bimod}(\mathcal{O}_X, \mathcal{O}_Y)$ is called invertible if there exists a bimodule $\mathcal{L}^{-1} \in \operatorname{Bimod}(\mathcal{O}_Y, \mathcal{O}_X)$ such that $\mathcal{L} \otimes_{\mathcal{O}_Y} \mathcal{L}^{-1} \cong \mathcal{O}_{\Delta_X}$ and $\mathcal{L}^{-1} \otimes_{\mathcal{O}_X} \mathcal{L} \cong \mathcal{O}_{\Delta_Y}$. If $\mathcal{L} \in \operatorname{Bimod}(\mathcal{O}_X, \mathcal{O}_Y)$ is an invertible bimodule, then $\mathcal{L}$ is a locally free bimodule of rank 1 and $\mathcal{L}^* \cong {}^*\mathcal{L} \cong \mathcal{L}^{-1} \in \operatorname{Bimod}(\mathcal{O}_Y, \mathcal{O}_X)$ is an invertible bimodule.

If $\mathcal{F} \in \operatorname{mod} X$ is a coherent $\mathcal{O}_X$-module, and $\sigma \in \operatorname{Aut} X$ is an automorphism of X, then we define an $\mathcal{O}_X$-bimodule $\mathcal{F}_\sigma$ by $\mathcal{F}_\sigma := pr_1^*\mathcal{F} \otimes_{\mathcal{O}_{X\times X}} \mathcal{O}_{\Delta_\sigma}$ where $\Delta_\sigma = \{(p, \sigma(p)) \mid p \in X\} \subset X \times X$ is the graph of X under σ. By [4, Proposition 2.15], a coherent $\mathcal{O}_X$-bimodule is invertible if and only if it is isomorphic to $\mathcal{L}_\sigma$ for some automorphism $\sigma \in \operatorname{Aut} X$ and some invertible $\mathcal{O}_X$-module $\mathcal{L} \in \operatorname{Pic} X$.

We use the following facts later.

Lemma 3.1. *Let X be a scheme, $\mathcal{F} \in \operatorname{mod} X$ a coherent $\mathcal{O}_X$-module,*

$\mathcal{L} \in \operatorname{mod} X$ *a locally free* $\mathcal{O}_X$*-module of finite rank, and* $\sigma \in \operatorname{Aut} X$ *an automorphism of* X*. Then*

(1) $\mathcal{F} \otimes_{\mathcal{O}_X} \mathcal{L}_\sigma \cong \sigma_*(\mathcal{F} \otimes_{\mathcal{O}_X} \mathcal{L})$.
(2) $\mathcal{L}_\sigma \otimes_{\mathcal{O}_X} \mathcal{F} \cong \mathcal{L} \otimes_{\mathcal{O}_X} \sigma^* \mathcal{F}$.

Lemma 3.2. *[4, Lemma 2.14] Let* X *be a scheme. For automorphisms* $\sigma, \tau \in \operatorname{Aut} X$*, and invertible* $\mathcal{O}_X$*-modules* $\mathcal{L}, \mathcal{M} \in \operatorname{Pic} X$*,*

$$\mathcal{L}_\sigma \otimes_{\mathcal{O}_X} \mathcal{M}_\tau \cong (\mathcal{L} \otimes_{\mathcal{O}_X} \sigma^* \mathcal{M})_{\tau\sigma}$$

in $\operatorname{Bimod}(\mathcal{O}_X, \mathcal{O}_X)$.

In particular, $(\mathcal{L}_\sigma)^{-1} \cong (\sigma_*(\mathcal{L}^{-1}))_{\sigma^{-1}}$ *in* $\operatorname{Bimod}(\mathcal{O}_X, \mathcal{O}_X)$.

4. Noncommutative Symmetric Algebras

For the rest of this paper, we assume that all schemes are smooth over k. Let $\tilde{X} = \{X_i\}$ be a sequence of schemes, $\mathcal{E} = \{\mathcal{E}_i\}$ a sequence of bimodules where $\mathcal{E}_i \in \operatorname{Bimod}(\mathcal{O}_{X_i}, \mathcal{O}_{X_{i+1}})$, and $\mathcal{Q} = \{\mathcal{Q}_i\}$ a sequence of subbimodules where $\mathcal{Q}_i \subset \mathcal{E}_i \otimes_{\mathcal{O}_{X_{i+1}}} \mathcal{E}_{i+1}$. A sheaf $\mathbb{Z}$-algebra over $\tilde{X}$ generated by $\mathcal{E}$ subject to the relation $\mathcal{Q}$ is a $\mathbb{Z}$-algebra over $\tilde{X}$ defined by

$$\begin{array}{ll} j < i: & \mathcal{A}_{ij} := 0, \\ j = i: & \mathcal{A}_{ii} := \mathcal{O}_{\Delta_{X_i}}, \\ j = i+1: & \mathcal{A}_{i,i+1} := \mathcal{E}_i, \\ j > i+1: & \mathcal{A}_{ij} := \dfrac{\mathcal{E}_i \otimes_{\mathcal{O}_{X_{i+1}}} \cdots \otimes_{\mathcal{O}_{X_{j-1}}} \mathcal{E}_{j-1}}{\sum_{k=i}^{j-2} \mathcal{E}_i \otimes_{\mathcal{O}_{X_{i+1}}} \cdots \otimes_{\mathcal{O}_{X_{k-1}}} \mathcal{E}_{k-1} \otimes_{\mathcal{O}_{X_k}} \mathcal{Q}_k \otimes_{\mathcal{O}_{X_{k+2}}} \mathcal{E}_{k+2} \otimes_{\mathcal{O}_{X_{k+3}}} \cdots \otimes_{\mathcal{O}_{X_{j-1}}} \mathcal{E}_{j-1}}, \end{array}$$

where the units are $\eta_i = \operatorname{Id} : \mathcal{O}_{\Delta_{X_i}} \to \mathcal{A}_{ii}$, and the multiplications

$$\mu_{ijk} : \mathcal{A}_{ij} \otimes_{\mathcal{O}_{X_j}} \mathcal{A}_{jk} \to \mathcal{A}_{ik}$$

are induced by the commutative diagrams

$$\begin{array}{ccc} (\mathcal{E}_i \otimes_{\mathcal{O}_{X_{i+1}}} \cdots \otimes_{\mathcal{O}_{X_{j-1}}} \mathcal{E}_{j-1}) \otimes_{\mathcal{O}_{X_j}} (\mathcal{E}_j \otimes_{\mathcal{O}_{X_{j+1}}} \cdots \otimes_{\mathcal{O}_{X_{k-1}}} \mathcal{E}_{k-1}) & \longrightarrow & \mathcal{A}_{ij} \otimes_{\mathcal{O}_{X_j}} \mathcal{A}_{jk} \\ \wr\wr & & \downarrow \mu_{ijk} \\ \mathcal{E}_i \otimes_{\mathcal{O}_{X_{i+1}}} \cdots \otimes_{\mathcal{O}_{X_{j-1}}} \mathcal{E}_{j-1} \otimes_{\mathcal{O}_{X_j}} \mathcal{E}_j \otimes_{\mathcal{O}_{X_{j+1}}} \cdots \otimes_{\mathcal{O}_{X_{k-1}}} \mathcal{E}_{k-1} & \longrightarrow & \mathcal{A}_{ik}. \end{array}$$

In the above construction, we will often choose suitable subbimodules $\mathcal{Q}_i \subset \mathcal{E}_i \otimes_{\mathcal{O}_{X_{i+1}}} \mathcal{E}_{i+1}$ defined below.

Definition 4.1. Let X, Y, Z be schemes, and $\mathcal{E} \in \operatorname{Bimod}(\mathcal{O}_X, \mathcal{O}_Y), \mathcal{F} \in \operatorname{Bimod}(\mathcal{O}_Y, \mathcal{O}_Z)$ locally free bimodules. An invertible subbimodule $\mathcal{Q} \subset \mathcal{E} \otimes_{\mathcal{O}_Y} \mathcal{F}$ is called non-degenerate if the composition

$$\mathcal{E}^* \otimes_{\mathcal{O}_X} \mathcal{Q} \to \mathcal{E}^* \otimes_{\mathcal{O}_X} \mathcal{E} \otimes_{\mathcal{O}_Y} \mathcal{F} \to \mathcal{F}$$

induced by the natural morphism $\mathcal{E}^* \otimes_{\mathcal{O}_X} \mathcal{E} \to \mathcal{O}_{\Delta_Y}$ is an isomorphism in $\mathrm{Bimod}(\mathcal{O}_Y, \mathcal{O}_Z)$.

Definition 4.2. [16] Let X be a scheme, and $\mathcal{E} \in \mathrm{Bimod}(\mathcal{O}_X, \mathcal{O}_X)$ a locally free $\mathcal{O}_X$-bimodule. The noncommutative symmetric algebra $\mathcal{S}(\mathcal{E})$ over X generated by $\mathcal{E}$ is the sheaf $\mathbb{Z}$-algebra over X generated by $\{\mathcal{E}^{i*}\}$ subject to the relation $\{\mathcal{Q}_i\}$ where $\mathcal{Q}_i \subset \mathcal{E}^{i*} \otimes_{\mathcal{O}_X} \mathcal{E}^{(i+1)*}$ are the images of the natural morphisms

$$\mathcal{O}_{\Delta_X} \to \mathcal{E}^{i*} \otimes_{\mathcal{O}_X} (\mathcal{E}^{i*})^* = \mathcal{E}^{i*} \otimes_{\mathcal{O}_X} \mathcal{E}^{(i+1)*}.$$

A quantum $\mathbb{P}^1$-bundle over X is a quasi-scheme of the form $\mathbb{P}_X(\mathcal{E}) := \mathrm{Proj}\,\mathcal{S}(\mathcal{E})$ for some locally free $\mathcal{O}_X$-bimodule $\mathcal{E}$ of rank 2. A quantum $\mathbb{P}^1$-bundle over a curve X is called a quantum ruled surface over X.

In fact, if $\mathcal{E}$ is a locally free $\mathcal{O}_X$-bimodule of rank 2, then $\mathcal{S}(\mathcal{E})$ is noetherian [16, Section 6]. Presumably, if the rank of $\mathcal{E}$ is greater than 2, then $\mathcal{S}(\mathcal{E})$ is not noetherian, so we tacitly assume that $\mathcal{E}$ is of rank 2 whenever we write $\mathbb{P}_X(\mathcal{E})$.

Lemma 4.1. *[16, Section 4] Let X be a scheme, and $\mathcal{E} \in \mathrm{Bimod}(\mathcal{O}_X, \mathcal{O}_X)$ a locally free $\mathcal{O}_X$-bimodule. If $\mathcal{A}$ is a sheaf $\mathbb{Z}$-algebra over X generated by $\{\mathcal{E}^{i*}\}$ subject to the relation $\{\mathcal{Q}_i\}$ where $\mathcal{Q}_i \subset \mathcal{E}^{i*} \otimes_{\mathcal{O}_X} \mathcal{E}^{(i+1)*}$ are non-degenerate invertible subbimodules, then $\mathrm{GrMod}\,\mathcal{A} \cong_X \mathrm{GrMod}\,\mathcal{S}(\mathcal{E})$. Moreover, if $\mathcal{E}$ is of rank 2, then $\mathrm{Proj}\,\mathcal{A} \cong_X \mathrm{Proj}\,\mathcal{S}(\mathcal{E}) = \mathbb{P}_X(\mathcal{E})$.*

From now on, we often omit subscripts under the tensor symbol $\otimes$ in order to save the space. In the proof below, we sometimes omit even the tensor symbol $\otimes$ itself.

Theorem 4.1. *Let X be a scheme, and $\mathcal{E}$ a locally free $\mathcal{O}_X$-bimodule. If $\mathcal{L}$ is an invertible $\mathcal{O}_X$-bimodule, then $\mathrm{GrMod}\,\mathcal{S}(\mathcal{E} \otimes_{\mathcal{O}_X} \mathcal{L}) \cong_X \mathrm{GrMod}\,\mathcal{S}(\mathcal{E})$. Moreover, if $\mathcal{E}$ is of rank 2, then $\mathbb{P}_X(\mathcal{E} \otimes_{\mathcal{O}_X} \mathcal{L}) \cong_X \mathbb{P}_X(\mathcal{E})$.*

Proof. Let $\mathcal{A} = \mathcal{S}(\mathcal{E})$, $\mathcal{F} = \mathcal{E} \otimes \mathcal{L}$, and $\tilde{\mathcal{L}} = \{\mathcal{L}_i\}$ where $\mathcal{L}_i = \begin{cases} \mathcal{O}_{\Delta_X} & i \equiv 0 \pmod 2 \\ \mathcal{L} & i \equiv 1 \pmod 2. \end{cases}$ Since

$$\mathcal{F}^{i*} = (\mathcal{E} \otimes \mathcal{L})^{i*} \cong \left\{ \begin{matrix} \mathcal{E}^{i*} \otimes \mathcal{L} & i \equiv 0 \pmod 2 \\ \mathcal{L}^{-1} \otimes \mathcal{E}^{i*} & i \equiv 1 \pmod 2 \end{matrix} \right\} \cong \mathcal{L}_i^{-1} \otimes \mathcal{E}^{i*} \otimes \mathcal{L}_{i+1},$$

it is easy to see that

$$\mathcal{F}^{i*} \otimes \cdots \otimes \mathcal{F}^{(j-1)*} \cong \mathcal{L}_i^{-1} \otimes \mathcal{E}^{i*} \otimes \cdots \otimes \mathcal{E}^{(j-1)*} \otimes \mathcal{L}_j.$$

Let $\mathcal{Q}_i$ be the images of the natural morphisms

$$\mathcal{O}_{\Delta_X} \to \mathcal{E}^{i*} \otimes (\mathcal{E}^{i*})^* = \mathcal{E}^{i*} \otimes \mathcal{E}^{(i+1)*},$$

and $\mathcal{R}_i$ the images of the compositions

$$\mathcal{L}_i^{-1} \otimes \mathcal{Q}_i \otimes \mathcal{L}_{i+2} \to \mathcal{L}_i^{-1} \otimes \mathcal{E}^{i*} \otimes \mathcal{E}^{(i+1)*} \otimes \mathcal{L}_{i+2} \cong \mathcal{F}^{i*} \otimes \mathcal{F}^{(i+1)*}.$$

It is easy to see that the following diagrams

$$\begin{array}{ccc} \mathcal{L}_i^{-1}\mathcal{E}^{i*}\cdots\mathcal{E}^{(k-1)*}\mathcal{Q}_k\mathcal{E}^{(k+2)*}\cdots\mathcal{E}^{(j-1)*}\mathcal{L}_j & \cong & \mathcal{F}^{i*}\cdots\mathcal{F}^{(k-1)*}\mathcal{R}_k\mathcal{F}^{(k+2)*}\cdots\mathcal{F}^{(j-1)*} \\ \cap & & \cap \\ \mathcal{L}_i^{-1}\mathcal{E}^{i*}\cdots\mathcal{E}^{(j-1)*}\mathcal{L}_j & \cong & \mathcal{F}^{i*}\cdots\mathcal{F}^{(j-1)*} \end{array}$$

commute. We will show that $\mathcal{A}\otimes\tilde{\mathcal{L}}$ is isomorphic to the sheaf $\mathbb{Z}$-algebra over X generated by $\{\mathcal{F}^{i*}\}$ subject to the relation $\{\mathcal{R}_i\}$.

$j < i$: $\quad (\mathcal{A}\otimes\tilde{\mathcal{L}})_{ij} = \mathcal{L}_i^{-1} \otimes \mathcal{A}_{ij} \otimes \mathcal{L}_j = \mathcal{L}_i^{-1} \otimes 0 \otimes \mathcal{L}_j \cong 0.$

$j = i$: $\quad (\mathcal{A}\otimes\tilde{\mathcal{L}})_{ii} = \mathcal{L}_i^{-1} \otimes \mathcal{A}_{ii} \otimes \mathcal{L}_i = \mathcal{L}_i^{-1} \otimes \mathcal{O}_{\Delta_X} \otimes \mathcal{L}_i \cong \mathcal{O}_{\Delta_X}.$

$j = i+1$:

$$\begin{aligned} (\mathcal{A}\otimes\tilde{\mathcal{L}})_{i,i+1} &= \mathcal{L}_i^{-1} \otimes \mathcal{A}_{i,i+1} \otimes \mathcal{L}_{i+1} \\ &= \mathcal{L}_i^{-1} \otimes \mathcal{E}^{i*} \otimes \mathcal{L}_{i+1} \\ &\cong \mathcal{F}^{i*}. \end{aligned}$$

$j > i+1$:

$$\begin{aligned} (\mathcal{A}\otimes\tilde{\mathcal{L}})_{ij} &= \mathcal{L}_i^{-1} \otimes \mathcal{A}_{ij} \otimes \mathcal{L}_j \\ &= \mathcal{L}_i^{-1} \otimes \frac{\mathcal{E}^{i*}\otimes\cdots\otimes\mathcal{E}^{(j-1)*}}{\sum_{k=i}^{j-2}\mathcal{E}^{i*}\otimes\cdots\otimes\mathcal{E}^{(k-1)*}\otimes\mathcal{Q}_k\otimes\mathcal{E}^{(k+2)*}\otimes\cdots\otimes\mathcal{E}^{(j-1)*}} \otimes \mathcal{L}_j \\ &\cong \frac{\mathcal{L}_i^{-1}\otimes\mathcal{E}^{i*}\otimes\cdots\otimes\mathcal{E}^{(j-1)*}\otimes\mathcal{L}_j}{\sum_{k=i}^{j-2}\mathcal{L}_i^{-1}\otimes\mathcal{E}^{i*}\otimes\cdots\otimes\mathcal{E}^{(k-1)*}\otimes\mathcal{Q}_k\otimes\mathcal{E}^{(k+2)*}\otimes\cdots\otimes\mathcal{E}^{(j-1)*}\otimes\mathcal{L}_j} \\ &\cong \frac{\mathcal{F}^{i*}\otimes\cdots\otimes\mathcal{F}^{(j-1)*}}{\sum_{k=i}^{j-2}\mathcal{F}^{i*}\otimes\cdots\otimes\mathcal{F}^{(k-1)*}\otimes\mathcal{R}_k\otimes\mathcal{F}^{(k+2)*}\otimes\cdots\otimes\mathcal{F}^{(j-1)*}}. \end{aligned}$$

Since the diagrams

$$\begin{array}{ccc}
(\mathcal{F}^{i*}\cdots\mathcal{F}^{(j-1)*})(\mathcal{F}^{j*}\cdots\mathcal{F}^{(k-1)*}) & \longrightarrow & \mathcal{F}^{i*}\cdots\mathcal{F}^{(k-1)*} \\
\wr & & \wr \\
(\mathcal{L}_i^{-1}\mathcal{E}^{i*}\cdots\mathcal{E}^{(j-1)*}\mathcal{L}_j)(\mathcal{L}_j^{-1}\mathcal{E}^{j*}\cdots\mathcal{E}^{(k-1)*}\mathcal{L}_k) & \longrightarrow & \mathcal{L}_i^{-1}\mathcal{E}^{i*}\cdots\mathcal{E}^{(k-1)*}\mathcal{L}_k \\
\downarrow & & \downarrow \\
(\mathcal{L}_i^{-1}\mathcal{A}_{ij}\mathcal{L}_j)(\mathcal{L}_j^{-1}\mathcal{A}_{jk}\mathcal{L}_k) & \xrightarrow{\mathcal{L}_i^{-1}\mu_{ijk}\mathcal{L}_k} & \mathcal{L}_i^{-1}\mathcal{A}_{ik}\mathcal{L}_k \\
\| & & \| \\
(\mathcal{A}\tilde{\mathcal{L}})_{ij}(\mathcal{A}\tilde{\mathcal{L}})_{jk} & \longrightarrow & (\mathcal{A}\tilde{\mathcal{L}})_{ik}
\end{array}$$

commute, $\mathcal{A}\otimes\tilde{\mathcal{L}}$ is isomorphic to the sheaf $\mathbb{Z}$-algebra over X generated by $\{\mathcal{F}^{i*}\}$ subject to the relation $\{\mathcal{R}_i\}$.

Since $\mathcal{Q}_i \subset \mathcal{E}^{i*}\otimes\mathcal{E}^{(i+1)*}$ are non-degenerate invertible $\mathcal{O}_X$-subbimodules, the compositions

$$(\mathcal{E}^{i*})^*\otimes\mathcal{Q}_i \to (\mathcal{E}^{i*})^*\otimes\mathcal{E}^{i*}\otimes\mathcal{E}^{(i+1)*} \to \mathcal{E}^{(i+1)*}$$

are isomorphisms. Since the diagrams

$$\begin{array}{ccccc}
(\mathcal{F}^{i*})^*\mathcal{R}_i & \to & (\mathcal{F}^{i*})^*\mathcal{F}^{i*}\mathcal{F}^{(i+1)*} & \to & \mathcal{F}^{(i+1)*} \\
\wr & & \wr & & \wr \\
(\mathcal{L}_i^{-1}\mathcal{E}^{i*}\mathcal{L}_{i+1})^*(\mathcal{L}_i^{-1}\mathcal{Q}_i\mathcal{L}_{i+2}) & \to & (\mathcal{L}_i^{-1}\mathcal{E}^{i*}\mathcal{L}_{i+1})^*(\mathcal{L}_i^{-1}\mathcal{E}^{i*}\mathcal{E}^{(i+1)*}\mathcal{L}_{i+2}) & \to & \mathcal{L}_{i+1}^{-1}\mathcal{E}^{(i+1)*}\mathcal{L}_{i+2} \\
\wr & & \wr & & \| \\
\mathcal{L}_{i+1}^{-1}(\mathcal{E}^{i*})^*\mathcal{Q}_i\mathcal{L}_{i+2} & \to & \mathcal{L}_{i+1}^{-1}(\mathcal{E}^{i*})^*\mathcal{E}^{i*}\mathcal{E}^{(i+1)*}\mathcal{L}_{i+2} & \to & \mathcal{L}_{i+1}^{-1}\mathcal{E}^{(i+1)*}\mathcal{L}_{i+2}
\end{array}$$

commute, the top horizontal compositions are isomorphisms, so $\mathcal{R}_i \subset \mathcal{F}^{i*}\otimes\mathcal{F}^{(i+1)*}$ are non-degenerate invertible $\mathcal{O}_X$-subbimodules.

Since $\mathcal{A}\otimes\tilde{\mathcal{L}}$ is isomorphic to the sheaf $\mathbb{Z}$-algebra over X generated by $\{\mathcal{F}^{i*}\}$ subject to the non-degenerate relation $\{\mathcal{R}_i\}$,

$$\begin{aligned}
&\operatorname{GrMod}(\mathcal{A}\otimes\tilde{\mathcal{L}}) \cong_X \operatorname{GrMod}\mathcal{S}(\mathcal{F}) = \operatorname{GrMod}\mathcal{S}(\mathcal{E}\otimes\mathcal{L}),\\
&\operatorname{Proj}(\mathcal{A}\otimes\tilde{\mathcal{L}}) \cong_X \operatorname{Proj}\mathcal{S}(\mathcal{F}) = \operatorname{Proj}\mathcal{S}(\mathcal{E}\otimes\mathcal{L}) = \mathbb{P}_X(\mathcal{E}\otimes\mathcal{L})
\end{aligned}$$

by Lemma 4.1. On the other hand, since $\mathcal{L}_0 = \mathcal{O}_{\Delta_X}$,

$$\begin{aligned}
&\operatorname{GrMod}(\mathcal{A}\otimes\tilde{\mathcal{L}}) \cong_X \operatorname{GrMod}\mathcal{A} = \operatorname{GrMod}\mathcal{S}(\mathcal{E}),\\
&\operatorname{Proj}(\mathcal{A}\otimes\tilde{\mathcal{L}}) \cong_X \operatorname{Proj}\mathcal{A} = \operatorname{Proj}\mathcal{S}(\mathcal{E}) = \mathbb{P}_X(\mathcal{E})
\end{aligned}$$

by Lemma 2.1, hence the result. □

In order to prove $\mathbb{P}_X(\mathcal{L}\otimes_{\mathcal{O}_X}\mathcal{E}) \cong \mathbb{P}_X(\mathcal{E})$, we need the following lemma, whose proof is a tedious but routine check.

Lemma 4.2. *If X is a scheme, and $\mathcal{E}$ is a locally free $\mathcal{O}_X$-bimodule, then the functor $F : \operatorname{GrMod}\mathcal{S}(\mathcal{E}) \to \operatorname{GrMod}\mathcal{S}(\mathcal{E}^*)$ defined by $\{F(\mathcal{M})_i\} = \{\mathcal{M}_{i+1}\}$ is an equivalence of categories, which induces an isomorphism $\mathbb{P}_X(\mathcal{E}) \to \mathbb{P}_X(\mathcal{E}^*)$ if $\mathcal{E}$ is of rank 2.*

Unfortunately, the isomorphism $\mathbb{P}_X(\mathcal{E}) \to \mathbb{P}_X(\mathcal{E}^*)$ induced by $F : \operatorname{GrMod}\mathcal{S}(\mathcal{E}) \cong \operatorname{GrMod}\mathcal{S}(\mathcal{E}^*)$ as above is not over X. Below, we give two examples when $\mathbb{P}_X(\mathcal{E}^*) \cong_X \mathbb{P}_X(\mathcal{E})$ as in the commutative case.

Example 4.1. Let X be a scheme, and $\mathcal{E}$ a locally free $\mathcal{O}_X$-bimodule. If $\mathcal{E} \otimes_{\mathcal{O}_X} \mathcal{E}$ admits a non-degenerate invertible $\mathcal{O}_X$-subbimodule $\mathcal{Q}$, then $\mathcal{E}^* \cong \mathcal{E} \otimes_{\mathcal{O}_X} \mathcal{Q}^{-1}$, so $\operatorname{GrMod}\mathcal{S}(\mathcal{E}^*) \cong_X \operatorname{GrMod}\mathcal{S}(\mathcal{E})$ and $\mathbb{P}_X(\mathcal{E}^*) \cong_X \mathbb{P}_X(\mathcal{E})$ by Theorem 4.1.

Example 4.2. Let X be a scheme and $\mathcal{E} = \mathcal{L} \oplus \mathcal{M}$ where $\mathcal{L}$ and $\mathcal{M}$ are invertible $\mathcal{O}_X$-bimodules. If $\mathcal{L} \otimes \mathcal{M} \cong \mathcal{M} \otimes \mathcal{L}$ in $\operatorname{Bimod}(\mathcal{O}_X, \mathcal{O}_X)$, then

$$\begin{aligned}\mathcal{E}^* \otimes (\mathcal{L} \otimes \mathcal{M}) = (\mathcal{L} \oplus \mathcal{M})^* \otimes (\mathcal{L} \otimes \mathcal{M}) &\cong (\mathcal{L}^{-1} \oplus \mathcal{M}^{-1}) \otimes (\mathcal{L} \otimes \mathcal{M}) \\ &\cong \mathcal{M} \oplus (\mathcal{M}^{-1} \otimes \mathcal{L} \otimes \mathcal{M}) \cong \mathcal{M} \oplus \mathcal{L} \cong \mathcal{E}.\end{aligned}$$

Since $\mathcal{L} \otimes \mathcal{M}$ is an invertible $\mathcal{O}_X$-bimodule, $\operatorname{GrMod}\mathcal{S}(\mathcal{E}^*) \cong_X \operatorname{GrMod}\mathcal{S}(\mathcal{E})$ and $\mathbb{P}_X(\mathcal{E}^*) \cong_X \mathbb{P}_X(\mathcal{E})$ by Theorem 4.1.

Theorem 4.2. *Let X be a scheme, and $\mathcal{E}$ a locally free $\mathcal{O}_X$-bimodule. If $\mathcal{L}$ is an invertible $\mathcal{O}_X$-bimodule, then $\operatorname{GrMod}\mathcal{S}(\mathcal{L} \otimes_{\mathcal{O}_X} \mathcal{E}) \cong \operatorname{GrMod}\mathcal{S}(\mathcal{E})$. Moreover, if $\mathcal{E}$ is of rank 2, then $\mathbb{P}_X(\mathcal{L} \otimes_{\mathcal{O}_X} \mathcal{E}) \cong \mathbb{P}_X(\mathcal{E})$.*

Proof. By Lemma 4.2 and Theorem 4.1,

$$\begin{aligned}\operatorname{GrMod}\mathcal{S}(\mathcal{L} \otimes_{\mathcal{O}_X} \mathcal{E}) &\cong \operatorname{GrMod}\mathcal{S}((\mathcal{L} \otimes_{\mathcal{O}_X} \mathcal{E})^*) \\ &\cong_X \operatorname{GrMod}\mathcal{S}(\mathcal{E}^* \otimes_{\mathcal{O}_X} \mathcal{L}^{-1}) \\ &\cong_X \operatorname{GrMod}\mathcal{S}(\mathcal{E}^*) \\ &\cong \operatorname{GrMod}\mathcal{S}(\mathcal{E}).\end{aligned}$$

The second formula follows similarly. □

Let $\mathcal{E}$ and $\mathcal{E}'$ be locally free $\mathcal{O}_X$-bimodules of rank 2. Combining Theorem 4.1 and Theorem 4.2, if there exist invertible $\mathcal{O}_X$-bimodules $\mathcal{L}$ and $\mathcal{M}$ such that $\mathcal{E}' \cong \mathcal{L} \otimes_{\mathcal{O}_X} \mathcal{E} \otimes_{\mathcal{O}_X} \mathcal{M}$, then $\mathbb{P}_X(\mathcal{E}') \cong_X \mathbb{P}_X(\mathcal{L} \otimes_{\mathcal{O}_X} \mathcal{E} \otimes_{\mathcal{O}_X} \mathcal{M}) \cong \mathbb{P}_X(\mathcal{E})$. It is known that the converse of this fact holds for arithmetic noncommutative $\mathbb{P}^1$'s [10, Corollary 1.3]. However, for an invertible $\mathcal{O}_X$-bimodule $\mathcal{L}$, there rarely exists an invertible $\mathcal{O}_X$-bimodule $\mathcal{L}'$ such that $\mathcal{L} \otimes_{\mathcal{O}_X} - \cong - \otimes_{\mathcal{O}_X} \mathcal{L}' : \operatorname{Bimod}(\mathcal{O}_X, \mathcal{O}_X) \to \operatorname{Bimod}(\mathcal{O}_X, \mathcal{O}_X)$, so there is no guarantee that $\mathbb{P}_X(\mathcal{L} \otimes_{\mathcal{O}_X} \mathcal{E} \otimes_{\mathcal{O}_X} \mathcal{M}) \cong_X \mathbb{P}_X(\mathcal{E})$ in general.

5. Decomposable Quantum Ruled Surfaces

We apply the results in the previous section to study "decomposable" quantum ruled surfaces.

Definition 5.1. Let X be a scheme. We say that a locally free $\mathcal{O}_X$-bimodule $\mathcal{E}$ is decomposable if there exist (nonzero) locally free $\mathcal{O}_X$-subbimodules $\mathcal{E}_1, \mathcal{E}_2 \subset \mathcal{E}$ such that $\mathcal{E} = \mathcal{E}_1 \oplus \mathcal{E}_2$.

By abuse of terminology, we say that a quantum rulod surface over a smooth projective curve X is decomposable if it is isomorphic to $\mathbb{P}_X(\mathcal{E})$ where $\mathcal{E}$ is a decomposable locally free $\mathcal{O}_X$-bimodule of rank 2. The following result says that "decomposable" quantum ruled surfaces can be classified in term of geometric triples $(X, \sigma, \mathcal{L})$ where X is a smooth projective curve, $\sigma \in \operatorname{Aut} X$ is an automorphism of X, and $\mathcal{L} \in \operatorname{Pic} X$ is an invertible $\mathcal{O}_X$-module, as in the classification of quantum projective planes [3] and that of noncommutative projective curves [2].

Theorem 5.1. *If X is a smooth projective curve and $\mathcal{E}$ is a decomposable locally free $\mathcal{O}_X$-bimodule of rank 2, then $\mathbb{P}_X(\mathcal{E}) \cong_X \mathbb{P}_X(\mathcal{O}_{\Delta_X} \oplus \mathcal{L}_\sigma)$ for some automorphism $\sigma \in \operatorname{Aut} X$ and some invertible $\mathcal{O}_X$-module $\mathcal{L} \in \operatorname{Pic} X$.*

Proof. Let $\mathcal{E} = \mathcal{L}_1 \oplus \mathcal{L}_2$ where $\mathcal{L}_1$ and $\mathcal{L}_2$ are locally free $\mathcal{O}_X$-bimodules of rank 1. Since X is a smooth curve, $\mathcal{L}_1$ and $\mathcal{L}_2$ are invertible $\mathcal{O}_X$-bimodules by [11, Theorem 2.8]. By Theorem 4.1,

$$\mathbb{P}_X(\mathcal{E}) = \mathbb{P}_X(\mathcal{L}_1 \oplus \mathcal{L}_2) \cong_X \mathbb{P}_X((\mathcal{L}_1 \oplus \mathcal{L}_2) \otimes \mathcal{L}_1^{-1}) \cong_X \mathbb{P}_X(\mathcal{O}_{\Delta_X} \oplus (\mathcal{L}_2 \otimes \mathcal{L}_1^{-1})).$$

Since $\mathcal{L}_2 \otimes \mathcal{L}_1^{-1}$ is an invertible $\mathcal{O}_X$-bimodule, it is isomorphic to $\mathcal{L}_\sigma$ for some $\sigma \in \operatorname{Aut} X$ and $\mathcal{L} \in \operatorname{Pic} X$, hence the result. $\square$

We now know that "decomposable" quantum ruled surfaces are parameterized by geometric triples $(X, \sigma, \mathcal{L})$. The natural question is to ask which geometric triples give isomorphic quantum ruled surfaces. The analogous question for quantum projective planes is not completely understood (see [7]). In this paper, we can only give some suggestions for how to approach this question.

Definition 5.2. Let X be a smooth projective curve and $\mathcal{E}$ a locally free $\mathcal{O}_X$-bimodule of rank 2. We say that $\mathcal{E}$ is normalized if $pr_{2*}\mathcal{E}$ is normalized, that is, $\mathrm{H}^0(X, pr_{2*}\mathcal{E}) \neq 0$ but $\mathrm{H}^0(X, (pr_{2*}\mathcal{E}) \otimes_{\mathcal{O}_X} \mathcal{L}) = 0$ for any invertible $\mathcal{O}_X$-module $\mathcal{L} \in \operatorname{Pic} X$ such that $\deg \mathcal{L} < 0$.

As in the commutative case, the following result is expected.

Theorem 5.2. *Any quantum ruled surface over a smooth projective curve X is isomorphic to $\mathbb{P}_X(\mathcal{E})$ over X for some normalized locally free $\mathcal{O}_X$-bimodule $\mathcal{E}$ of rank 2.*

Proof. Let $\mathbb{P}_X(\mathcal{E})$ be a quantum ruled surface over X where $\mathcal{E}$ is a locally free $\mathcal{O}_X$-bimodule of rank 2. For $\mathcal{L} \in \operatorname{Pic} X$,

$$\begin{aligned} pr_{2*}(\mathcal{E} \otimes_{\mathcal{O}_X} \mathcal{L}_{\mathrm{id}}) &\cong \mathcal{O}_X \otimes_{\mathcal{O}_X} (\mathcal{E} \otimes_{\mathcal{O}_X} \mathcal{L}_{\mathrm{id}}) \\ &\cong (\mathcal{O}_X \otimes_{\mathcal{O}_X} \mathcal{E}) \otimes_{\mathcal{O}_X} \mathcal{L}_{\mathrm{id}} \\ &\cong (pr_{2*}\mathcal{E}) \otimes_{\mathcal{O}_X} \mathcal{L}_{\mathrm{id}} \\ &\cong \mathrm{id}_*((pr_{2*}\mathcal{E}) \otimes_{\mathcal{O}_X} \mathcal{L}) \\ &\cong (pr_{2*}\mathcal{E}) \otimes_{\mathcal{O}_X} \mathcal{L} \end{aligned}$$

by Lemma 3.1. Since $pr_{2*}\mathcal{E}$ is a locally free $\mathcal{O}_X$-module of rank 2, there exists $\mathcal{L} \in \operatorname{Pic} X$ such that $(pr_{2*}\mathcal{E}) \otimes_{\mathcal{O}_X} \mathcal{L}$ is normalized. Since $\mathcal{L}_{\mathrm{id}}$ is an invertible $\mathcal{O}_X$-bimodule, $\mathbb{P}_X(\mathcal{E} \otimes_{\mathcal{O}_X} \mathcal{L}_{\mathrm{id}}) \cong_X \mathbb{P}_X(\mathcal{E})$ by Theorem 4.1 where $\mathcal{E} \otimes_{\mathcal{O}_X} \mathcal{L}_{\mathrm{id}}$ is normalized. □

Using Lemma 3.1 and Lemma 3.2, we can prove the following lemma, whose proof is omitted.

Lemma 5.1. *Let X be a smooth projective curve. If $\mathcal{L}$ is an invertible $\mathcal{O}_X$-bimodule, then* $\deg pr_{2*}(\mathcal{L}^{-1}) = -\deg pr_{2*}\mathcal{L}$.

Let X be a smooth projective curve, and $\mathcal{L} \in \operatorname{Pic} X$ an invertible $\mathcal{O}_X$-module. It is easy to see that $\mathcal{O}_X \oplus \mathcal{L}$ is a normalized locally free $\mathcal{O}_X$-module of rank 2 if and only if $\deg \mathcal{L} \leq 0$, hence the following result.

Proposition 5.1. *Let X be a smooth projective curve. For any decomposable locally free $\mathcal{O}_X$-bimodule $\mathcal{E}$ of rank 2, there exists a normalized locally free $\mathcal{O}_X$-bimodule of rank 2 of the form $\mathcal{O}_{\Delta_X} \oplus \mathcal{L}$ where $\mathcal{L}$ is an invertible $\mathcal{O}_X$-bimodule with* $\deg pr_{2*}\mathcal{L} \leq 0$ *such that* $\mathbb{P}_X(\mathcal{O}_{\Delta_X} \oplus \mathcal{L}) \cong_X \mathbb{P}_X(\mathcal{E})$.

Proof. Since

$$\begin{aligned} pr_{2*}(\mathcal{O}_{\Delta_X} \oplus \mathcal{L}) &\cong \mathcal{O}_X \otimes_{\mathcal{O}_X} (\mathcal{O}_{\Delta_X} \oplus \mathcal{L}) \\ &\cong (\mathcal{O}_X \otimes_{\mathcal{O}_X} \mathcal{O}_{\Delta_X}) \oplus (\mathcal{O}_X \otimes_{\mathcal{O}_X} \mathcal{L}) \\ &\cong \mathcal{O}_X \oplus pr_{2*}\mathcal{L}, \end{aligned}$$

$\mathcal{O}_{\Delta_X} \oplus \mathcal{L}$ is normalized if and only if $\deg pr_{2*}\mathcal{L} \leq 0$.

By Theorem 5.1, we may assume that $\mathcal{E} = \mathcal{O}_{\Delta_X} \oplus \mathcal{L}$ for some invertible $\mathcal{O}_X$-bimodule $\mathcal{L}$. If $\deg pr_{2*}\mathcal{L} \leq 0$, then we are done. Suppose that $\deg pr_{2*}\mathcal{L} > 0$. Since

$\mathbb{P}_X(\mathcal{O}_{\Delta_X} \oplus \mathcal{L}^{-1}) \cong_X \mathbb{P}_X((\mathcal{O}_{\Delta_X} \oplus \mathcal{L}^{-1}) \otimes_{\mathcal{O}_X} \mathcal{L}) \cong_X \mathbb{P}_X(\mathcal{O}_{\Delta_X} \oplus \mathcal{L}) = \mathbb{P}_X(\mathcal{E})$

by Theorem 4.1 and $\deg pr_{2*}(\mathcal{L}^{-1}) = -\deg pr_{2*}\mathcal{L} < 0$ by Lemma 5.1, the result follows. $\square$

By Proposition 5.1, for any decomposable locally free $\mathcal{O}_{\mathbb{P}^1}$-bimodule $\mathcal{E}$ of rank 2, there exists a normalized locally free $\mathcal{O}_{\mathbb{P}^1}$-bimodule of rank 2 of the form $\mathcal{O}_\Delta \oplus \mathcal{O}_{\mathbb{P}^1}(-n)_\sigma$ where $\sigma \in \operatorname{Aut}\mathbb{P}^1 = \mathrm{PGL}(2,k)$ and $n = -\deg pr_{2*}(\mathcal{O}_{\mathbb{P}^1}(-n)_\sigma) \in \mathbb{N}$ such that $\mathbb{P}(\mathcal{O}_\Delta \oplus \mathcal{O}_{\mathbb{P}^1}(-n)_\sigma) \cong_{\mathbb{P}^1} \mathbb{P}(\mathcal{E})$, so "decomposable" quantum ruled surfaces over $\mathbb{P}^1$ (quantum Hirzebruch surfaces) can be classified in terms of the pairs $(\sigma, n) \in \mathrm{PGL}(2,k) \times \mathbb{N}$.

Lemma 5.2. *For any $\sigma, \tau \in \operatorname{Aut}\mathbb{P}^d$ and $m, n \in \mathbb{Z}$,*

$$\mathcal{O}_{\mathbb{P}^d}(-m)_\sigma \otimes_{\mathbb{P}^d} \mathcal{O}_{\mathbb{P}^d}(-n)_\tau \cong \mathcal{O}_{\mathbb{P}^d}(-m-n)_{\tau\sigma}$$

in $\mathrm{Bimod}(\mathcal{O}_{\mathbb{P}^d}, \mathcal{O}_{\mathbb{P}^d})$.

Proof. For any $\sigma \in \operatorname{Aut}\mathbb{P}^d$ and $n \in \mathbb{Z}$, $\sigma^*(\mathcal{O}_{\mathbb{P}^d}(-n)) \cong \mathcal{O}_{\mathbb{P}^d}(-n)$ in $\operatorname{Mod}\mathcal{O}_{\mathbb{P}^d}$ by [6, Chapter II, Example 7.1.1], so the result follows from Lemma 3.2. $\square$

We do not know which pairs (σ, n) give isomorphic quantum Hirzebruch surfaces. At this time, we have the following result.

Theorem 5.3. *Let $\sigma, \sigma' \in \operatorname{Aut}\mathbb{P}^d$ and $n \in \mathbb{N}$. If σ and σ' are conjugate, then $\mathbb{P}(\mathcal{O}_\Delta \oplus \mathcal{O}_{\mathbb{P}^d}(-n)_\sigma) \cong \mathbb{P}(\mathcal{O}_\Delta \oplus \mathcal{O}_{\mathbb{P}^d}(-n)_{\sigma'})$.*

Proof. Let $\tau \in \operatorname{Aut}\mathbb{P}^d$ such that $\sigma' = \tau^{-1}\sigma\tau$. By Theorem 4.1, Theorem 4.2, and Lemma 5.2,

$$\begin{aligned}
&\mathbb{P}(\mathcal{O}_\Delta \oplus \mathcal{O}_{\mathbb{P}^d}(-n)_\sigma)\\
\cong&\mathbb{P}(\mathcal{O}_{\mathbb{P}^d,\tau} \otimes_{\mathbb{P}^d} (\mathcal{O}_{\mathbb{P}^d,\mathrm{id}} \oplus \mathcal{O}_{\mathbb{P}^d}(-n)_\sigma) \otimes_{\mathbb{P}^d} \mathcal{O}_{\mathbb{P}^d,\tau^{-1}})\\
\cong&\mathbb{P}((\mathcal{O}_{\mathbb{P}^d,\tau} \otimes_{\mathbb{P}^d} \mathcal{O}_{\mathbb{P}^d,\mathrm{id}} \otimes_{\mathbb{P}^d} \mathcal{O}_{\mathbb{P}^d,\tau^{-1}}) \oplus (\mathcal{O}_{\mathbb{P}^d,\tau} \otimes_{\mathbb{P}^d} \mathcal{O}_{\mathbb{P}^d}(-n)_\sigma \otimes_{\mathbb{P}^d} \mathcal{O}_{\mathbb{P}^d,\tau^{-1}}))\\
\cong&\mathbb{P}(\mathcal{O}_{\mathbb{P}^d,\tau^{-1}\mathrm{id}\,\tau} \oplus \mathcal{O}_{\mathbb{P}^d}(-n)_{\tau^{-1}\sigma\tau})\\
\cong&\mathbb{P}(\mathcal{O}_{\mathbb{P}^d,\mathrm{id}} \oplus \mathcal{O}_{\mathbb{P}^d}(-n)_{\sigma'})\\
\cong&\mathbb{P}(\mathcal{O}_\Delta \oplus \mathcal{O}_{\mathbb{P}^d}(-n)_{\sigma'}).
\end{aligned}$$

$\square$

It is known that a sort of the converse to the above result holds for arithmetic noncommutative $\mathbb{P}^1$'s [10, Theorem 1.2].

References

1. M. Artin, Some Problems on Three-dimensional Graded Domains, *Representation theory and algebraic geometry*, LMS Lecture Note Series **238** Cambridge Univ. Press (1997), 1-19.
2. M. Artin and J.T. Stafford, Noncommutative Graded Domains with Quadratic Growth, *Invent. Math.* **122** (1995), 231-276.
3. M. Artin, J. Tate and M. Van den Bergh, Some Algebras Associated to Automorphisms of Elliptic Curves, *The Grothendieck Festschrift* Vol. **1**, Birkhauser (1990), 33-85.
4. M. Artin and M. Van den Bergh, Twisted Homogeneous Coordinate Rings, *J. Algebra* **133** (1990), 249-271.
5. M. Artin and J.J. Zhang, Noncommutative Projective Schemes, *Adv. Math.* **109** (1994), 228-287.
6. R. Hartshorne, *Algebraic Geometry*, GTM **52**, Springer-Verlag, 1977.
7. I. Mori, Noncommutative Projective Schemes and Point Schemes, *Algebras, rings, and their representations*, World Sci. Publ. (2006), 215-239.
8. I. Mori, Intersection Theory over Quantum Ruled Surfaces, *J. Pure Appl. Algebra* **211** (2007), 25-41.
9. I. Mori and S.P. Smith, The Grothendieck Group of a Quantum Projective Space Bundle, *K-theory* **37** (2006), 263-289.
10. A. Nyman, Arithmetic Noncommutative $\mathbb{P}^1$'s, preprint.
11. D. Patrick, Noncommutative Ruled Surfaces, Ph. D. Thesis, MIT, 1997.
12. A.L. Rosenberg, *Non-commutative Algebraic Geometry and Representations of Quantized Algebras*, Math. and its Appl., Vol. **330**, Kluwer Acad. Publ., Dordrecht, 1995.
13. J.T. Stafford and M. Van den Bergh, Noncommutative Curves and Noncommutative Surfaces, *Bull. Amer. Math. Soc.* **38** (2001), 171-216.
14. M. Van den Bergh, A Translation Principle for the Four-dimensional Sklyanin Algebras, *J. Algebra,* **184** (1996) 435-490.
15. M. Van den Bergh, Blowing up of Non-commutative Smooth Surfaces, *Mem. Amer. Math. Soc.*, **154** (2001).
16. M. Van den Bergh, Non-commutative $\mathbb{P}^1$-bundles over Commutative Schemes, preprint.

Rings and Modules with n-acc

Patrick F. Smith

Dedicated to the memory of my friend and colleague Harry Dowson (1939-2008)

Rings and modules with the property that they satisfy the ascending chain condition on right ideals or submodules with a bounded number of generators are investigated. Although much of the paper is a survey of known results, there are a number of new theorems and examples. Several questions are raised.

1. Pontrjagin's Theorem

All rings have identity elements and all modules are unital right modules, unless stated otherwise. Let R be a ring and let M be an R-module. Given a positive integer n, the module M satisfies *n-acc* provided every ascending chain of n-generated submodules terminates. Moreover, the module M satisfies *pan-acc* in case M satisfies n-acc for every positive integer n. The ring R is said to satisfy *right n-acc*, for a given positive integer n, provided the right R-module R satisfies n-acc. Similarly, the ring R satisfies *right pan-acc* if R_R satisfies pan-acc. These chain conditions are discussed in many places (see, for example,[1-5,7,9-12,13-15,17,20-22,23] and[25]).

Before proceeding to investigate these chain conditions in rings and modules we want to point out that the corresponding descending chain conditions are of less interest because of the following result.

Proposition 1.1. *The following statements are equivalent for a module M.*

(i) M satisfies the descending chain condition on finitely generated submodules.

(ii) M satisfies the descending chain condition on n-generated submodules, for every positive integer n.

(iii) M satisfies the descending chain condition on n-generated submodules, for some positive integer n.

(iv) M satisfies the descending chain condition on cyclic submodules.

Proof. (i) $\Rightarrow$ (ii) $\Rightarrow$ (iii) $\Rightarrow$ (iv) Clear.

(iv) ⇒ (i) By.[6] □

Note the following simple fact. Let R be any ring, M any R-module and n any positive integer. Let $N_1 \subseteq N_2 \subseteq \ldots$ be any ascending chain of n-generated submodules of M. Let $N = \cup_{i=1}^{\infty} N_i$. Clearly N is a countably generated submodule of M. It follows that M *satisfies* n*-acc if and only if every countably generated submodule of* M *satisfies* n*-acc.*

Let R be a ring and let M be any R-module. Then the *singular submodule* $Z(M)$ of M is defined to be the set of elements m in M such that $mE = 0$ for some essential right ideal E of R. The *second singular submodule* of M is the submodule $Z_2(M)$ of M containing $Z(M)$ such that $Z_2(M)/Z(M)$ is the singular submodule of the module $M/Z(M)$. In the Goldie torsion theory, a module M is torsion if $M = Z_2(M)$ and is torsion-free if it is nonsingular, i.e. $Z(M) = 0$ (see[26] for more details). The ring R is called *right nonsingular* provided R_R is a nonsingular module.

Let R be a ring. A non-zero R-module M has *finite uniform dimension* provided it does not contain an infinite direct sum of non-zero submodules and in this case there exists a positive integer n, called the *uniform dimension* or *Goldie dimension* or *Goldie rank* of M and denoted by $\mathrm{u}(M)$, such that every direct sum of non-zero submodules of M has at most n submodules (see[19] for more details). The ring R has *finite right uniform dimension* provided R_R has finite uniform dimension. A submodule K of a general module M is called *closed* (in M) if K has no proper essential extension in M. For example, every direct summand of M is closed in M. Also every submodule K of M such that M/K is nonsingular is closed in M. Note the following result.

Lemma 1.1. *Let R be any ring and let M, M_1, …, M_k be R-modules each with finite uniform dimension, for some positive integer k. Then*

(i) For every submodule N of M, N has finite uniform dimension and $u(N) \leq u(M)$. Moreover, $u(N) = u(M)$ if and only if N is an essential submodule of M.

(ii) For every closed submodule K of M, M/K has finite uniform dimension and $u(M) = u(K) + u(M/K)$.

(iii) The module $M_1 \oplus \cdots \oplus M_k$ has finite uniform dimension and $u(M_1 \oplus \cdots \oplus M_k) = u(M_1) + \cdots + u(M_k)$.

Proof. By [11, Section 5]. □

It is well known and very easy to prove that, for any ring R, an R-module

M satisfies the ascending chain condition on finitely generated submodules if and only if M is Noetherian. It is also clear that every Noetherian R-module satisfies pan-acc. The converse is false and we begin with a theorem which is due to Baumslag and Baumslag [5, Theorem 8] and Renault [22, Corollaire 2.3] and which in some sense is typical of a number of results of a similar type.

Theorem 1.1. *Let R be a right Noetherian right nonsingular ring and let M be any free right R-module. Then M satisfies pan-acc.*

Proof. Let n be any positive integer and let $N_1 \subseteq N_2 \subseteq \dots$ be any ascending chain of n-generated submodules of M. For each positive integer k, the nonsingular n-generated submodule N_k has uniform dimension at most nm, where $m = \mathrm{u}(R_R)$, by Lemma 1.1. Again using Lemma 1.1, there exists a positive integer t such that N_t is essential in N_i for all $i \geq t$. There exist submodules M_1 and M_2 of M such that M_1 is finitely generated, $N_t \subseteq M_1$ and $M = M_1 \oplus M_2$. Let $i \geq t$. Note that N_i/N_t is a singular module so that $(N_i + M_1)/M_1 \cong N_i/(N_i \cap M_1)$ is singular because $N_t \subseteq N_i \cap M_1$. Because $M/M_1 \cong M_2$, the module M/M_1 is nonsingular and hence $N_i \subseteq M_1$. Thus $N_t \subseteq N_{t+1} \subseteq \dots$ is an ascending chain of submodules of the Noetherian R-module M_1. There exists a positive integer $t' \geq t$ such that $N_{t'} = N_{t'+1} = \dots$. The result follows. □

Theorem 1.1 can be strengthened as follows.

Theorem 1.2. *(See [2, Theorem 1.5].) Let R be a ring with finite right uniform dimension and let n be a positive integer. Let M_i $(i \in I)$ be any collection of nonsingular right R-modules. Then the direct product $\prod_{I \in I} M_i$ satisfies n-acc if and only if $\oplus_{j \in J} M_j$ satisfies n-acc for every finite subset J of I.*

Note that Theorem 1.2 shows in particular that if R is a ring with finite right uniform dimension then every direct product of nonsingular Noetherian right R-modules satisfies pan-acc.

Corollary 1.1. *(See [3, Theorem 1.7].) Let R be a right Noetherian ring. Then every right R-module M contains a submodule N such that N satisfies pan-acc and M/N is Goldie torsion.*

We wish to mention two other results for right Noetherian rings. The first is taken from [23, Theorem 4.2].

Theorem 1.3. *Let R be a right Noetherian ring and let n be any positive integer. Then every finite direct sum of right R-modules each satisfying n-acc also satisfies n-acc.*

Theorem 1.4. *Let R be a right Noetherian ring, let n be any positive integer and let M_i $(i \in I)$ be nonsingular right R-modules. Then the direct product $\prod_{i \in I} M_i$ satisfies n-acc if and only if M_i satisfies n-acc for all $i \in I$.*

Proof. By Theorems 1.2 and 1.3. □

Recall that, for any ring R, an R-module M is called *torsionless* provided M embeds in the direct product of a collection of copies of R_R (see [19, 3.4.2]), equivalently, for each non-zero element m in M there exists a homomorphism $\varphi : M \to R$ such that $\varphi(m) \neq 0$. Theorem 1.3 has the following immediate consequence which generalizes Theorem 1.1.

Corollary 1.2. *Let R be a right Noetherian right nonsingular ring. Then every torsionless right R-module satisfies pan-acc.*

We now give some examples of modules which do not satisfy 1-acc. The first result is elementary.

Lemma 1.2. *Let P_i $(i \in I)$ be an infinite collection of maximal ideals of a commutative ring R and let M denote the R-module $\oplus_{i \in I}(R/P_i)$. Then M does not satisfy 1-acc.*

Of course, the reason why Lemma 1.2 is true is that

$$(R/P_1) \oplus \cdots \oplus (R/P_k) \cong R/(P_1 \cap \cdots \cap P_k)$$

for every positive integer k and distinct maximal ideals P_i $(1 \leq i \leq k)$. Next we have the following result of Renault [22, Lemme 1.1].

Lemma 1.3. *The following statements are equivalent for a ring R with Jacobson radical J.*

() R/J is a semiprime Artinian ring (i.e. R is semilocal).

(i) Every semisimple R module satisfies pan-acc.

(ii) Every semisimple R module satisfies 1-acc.

Let R be a ring and M an R-module. Then the *annihilator* of M in R will be denoted by $ann_R(M)$ so that $ann_R(M) = \{r \in R : Mr = 0\}$. We have the following related fact.

Lemma 1.4. *Let R be a ring and let U be a simple R-module with annihilator P in R such that the ring R/P is not (right) Artinian. Then the R-module $U^{(\mathbb{N})}$ does not satisfy 1-acc.*

Let R be a commutative domain. An R-module M is called *divisible* in case $M = cM$ for every non-zero element c of R. Recall that the domain R is Dedekind if and only if every divisible R-module is injective (see, for example, [24, Theorem 4.25]). For a general domain R, an R-module M is called *reduced* if 0 is the only divisible submodule of M. The next result is taken from [23, Lemma 2.2].

Lemma 1.5. *Let R be commutative domain which is not a field and let M be a non-zero divisible R-module. Then M does not satisfy* 1-*acc.*

The starting point of the study of rings and modules satisfying n-acc for some positive integer n or satisfying pan-acc is the following theorem of Pontrjagin.[21] He proves it for the ring $\mathbb{Z}$ but it is true equally for any (commutative) principal ideal domain (PID).

Theorem 1.5. *Let R be any PID. Then a torsion-free R-module M satisfies pan-acc if and only if every countably generated submodule of M is free.*

Proof. The sufficiency follows by Theorem 1.1. Conversely suppose that M satisfies pan-acc. Let n be any positive integer and let N be any n-generated submodule of M. Because M satisfies n-acc, there exists an n-generated submodule K of M containing N and maximal with respect to these properties. Note that the free module K has rank n, otherwise K can be generated by $(n-1)$ elements and so is properly contained in an n-generated submodule of M. Let $m \in M$ such that $cm \in K$ for some $0 \neq c \in R$. Then $(Rm + K)/K$ is a torsion R-module so that $Rm + K$ also has rank n. Thus $Rm + K$ is n-generated and $Rm + K = K$ by the choice of K. It follows that $m \in K$. Thus M/K is torsion-free.

Let $H = Rh_1 + Rh_2 + \dots$ be any countably generated submodule of M. By the above argument there exists a finitely generated submodule H_1 of M such that $h_1 \in H_1$ and M/H_1 is torsion-free. Similarly there exists a finitely generated submodule H_2 of M such that $Rh_2 + H_1 \subseteq H_2$ and M/H_2 is torsion-free. Repeating this process there exist an ascending chain $H_1 \subseteq H_2 \subseteq \dots$ of finitely generated submodules of M such that $h_i \in H_i$ and M/H_i is torsion-free for all $i \geq 1$. For each $i \geq 1$, the module H_{i+1}/H_i

is finitely generated torsion-free and hence free so there exists a finitely generated free submodule G_i of H_{i+1} such that $H_{i+1} = H_i \oplus G_i$. Now it is easy to check that $H = H_1 \oplus G_1 \oplus G_2 \oplus \ldots$ and so H is free. □

Adapting the above argument Nicolas [20, Théorème 2.2] proves the following result.

Lemma 1.6. *Let R be any Dedekind domain. Then a torsion-free R-module M satisfies pan-acc if and only if every countably generated submodule of M is projective.*

Recall that a commutative domain R is called a *Prüfer doman* provided every finitely generated ideal is projective. Heinzer and Lantz[16] prove that a Prüfer doman which satisfies 2-acc is Noetherian and we shall give their proof next. First we need some preliminary facts. Given ideals A and B in a commutative ring R we denote by $(A{:}B)$ the set of elements r in R such that $rB \subseteq A$.

Lemma 1.7. *Let a and b be non-zero elements of a Prüfer domain R. Then the ideal $(Ra{:}Rb)$ is 2-generated.*

Proof. Let Q denote the field of fractions of R and let $X = \{q \in Q : q(Ra+Rb) \subseteq R\}$. The ideal $Ra+Rb$ is invertible and hence $R = X(Ra : Rb)$. There exist elements q_1 and q_2 in X such that $1 = q_1a + q_2b$. Note that $q_1b \in R$ and $q_2b \in R$ so that $q_1a, q_2a \in (Ra : Rb)$. Let $r \in (Ra : Rb)$. Then $rb = sa$ for some $s \in R$. Moreover,

$$r = rq_1a + rq_2b = rq_1a + sq_2a \in Rq_1a + Rq_2a.$$

It follows that $(Ra : Rb) = Rq_1a + Rq_2a$. □

Corollary 1.3. *Let q be any element in Q. Then $\{r \in R : qr \in R\}$ is a 2-gen- erated ideal of R*

Proof. If $q = 0$ then there is nothing to prove. Suppose that $q \neq 0$. Then $q = b/a$ for some non-zero $a, b \in R$. Clearly $\{r \in R : qr \in R\} = (Ra : Rb)$, which is 2-generated by the lemma. □

Theorem 1.6. *(See [16, Proposition 4.1].) Every Prüfer domain satisfying 2-acc is Noetherian.*

Proof. Let P be any prime ideal of R. Let A be a 2-generated ideal of R maximal with respect to the property $A \subseteq P$. Suppose that $A \neq P$. Let $x \in P \backslash A$. Then $B = A + Rx$ is a finitely generated ideal of R and hence B is invertible. There exist a positive integer n, elements $b_i \in B$ and elements $q_i \in \{q \in Q : qB \subseteq R\}$, for all $1 \leq i \leq n$, such that

$$1 = q_1 b_1 + \cdots + q_n b_n.$$

Let $C_i = \{r \in R : q_i R \subseteq R\}$ for all $1 \leq i \leq n$. Then $B \subseteq C_1 \cap \cdots \cap C_n$. Let $r \in C_1 \cap \cdots \cap C_n$. Then

$$r = (q_1 r) b_1 + \cdots + (q_n r) b_n \in Rb_1 + \cdots + Rb_n \subseteq B.$$

Thus $B = C_1 \cap \cdots \cap C_n$ and hence $C_1 \ldots C_n \subseteq B \subseteq P$. Because P is a prime ideal, $C_i \subseteq P$ for some $1 \leq i \leq n$. By Corollary 1.3, C_i is a 2-generated ideal of R. But A is properly contained in B and $B \subseteq C_i$, and this contradicts the choice of A. Thus $A = P$. We have proved that every prime ideal of R is 2-generated. By Cohen's Theorem (see, for example, [18, Theorem 8]), R is a Noetherian ring. □

Corollary 1.4. *Let R be a Prüfer domain. Then a non-zero torsion-free R-module M satisfies pan-acc if and only if R is a Noetherian ring and every countably generated submodule of M is projective.*

Proof. Suppose that M satisfies pan-acc. Let $0 \neq m \in M$. Then $R \cong Rm$ so that R satisfies 2-acc and hence R is Noetherian by the theorem. The result now follows by Lemma 1.6. □

In view of Theorem 1.6 we have the following question.

Question 1
Let R be a prime ring which is either right hereditary or right Goldie right semihereditary such that R satisfies right pan-acc. Is R right Noetherian?

In Question 1 we are interested only in prime rings because we can easily give an example to show that right hereditary rings which satisfy right pan-acc need not be right Noetherian.

Example 1.1. Let K be a field, let S be the polynomial ring $K[x]$ in an indeterminate x over K and let M be any free S-module of infinite rank. Let R denote the formal "matrix ring"

$$\begin{bmatrix} K & M \\ 0 & S \end{bmatrix}$$

consisting of all "matrices" of the form

$$\begin{bmatrix} k & m \\ 0 & s \end{bmatrix}$$

where $k \in K, m \in M$ and $s \in S$. Then R is a right hereditary ring which satisfies right pan-acc but R is not right Noetherian.

Proof. The ring R is right hereditary by [8, Lemma 3.8]. Clearly R is not right Noetherian because M_S is not finitely generated. Let n be any positive integer and let $L_1 \subseteq L_2 \subseteq \ldots$ be any ascending chain of n-generated right ideals of R. Let A denote the ideal of R consisting of all matrices in R with (1,1)-entry 0. Suppose that there exists $i \geq 1$ such that $L_i \nsubseteq A$. It is easy to show that in this case L_i contains the ideal

$$\begin{bmatrix} K & M \\ 0 & 0 \end{bmatrix}$$

and hence L_i must be an ideal of the form

$$\begin{bmatrix} K & M \\ 0 & B_i \end{bmatrix}$$

for some ideal B_i of S. Since T is a PID (and hence Noetherian) it follows that in this case $L_k = L_{k+1} = \ldots$ for some integer $k \geq i$.

Now suppose that $L_i \subseteq A$ for all $i \geq 1$. By Theorem 1.1, the S-module $M \oplus S$ satisfies n-acc and hence $L_s = L_{s+1} = \ldots$ for some positive integer s. Thus R satisfies right pan-acc. □

Baumslag and Baumslag [5, Theorem $3^\sharp$] characterize which $\mathbb{Z}$-modules satisfy n-acc, for a given positive integer n. Essentially the problem of characterizing such Abelian groups falls into the torsion case and the torsion-free case. In fact we have the following more general situation.

Lemma 1.8. *Let R be a Prüfer domain and let n be a positive integer. Then an R-module M with torsion submodule T satisfies n-acc if and only if*

(i) T satisfies n-acc, and

(ii) every countably generated submodule N of M such that $N \cap T = 0$ satisfies n-acc.

Proof. The necessity is clear. Conversely, suppose that M satisfies (i) and (ii). Let K be any n-generated submodule of M. Then $K/(K \cap T) \cong (K+T)/T$ which is a finitely generated torsion-free R-module and hence is projective. It follows that $K \cap T$ is a direct summand of K so that $K \cap T$ is also n-generated.

Let $L_1 \subseteq L_2 \subseteq L_3 \subseteq \dots$ be any ascending chain of n-generated submodules of M. By the above remark, $L_1 \cap T \subseteq L_2 \cap T \subseteq \dots$ is an ascending chain of n-generated submodules of T. There exists a positive integer k such that $L_k \cap T = L_{k+1} \cap T = \dots$. Let $L = \bigcup_{i \geq 1} L_i$. Then $L \cap T = L_k \cap T$. It follows that $L \cap T$ is an n-generated torsion R-module and so $c(L \cap T) = 0$ for some $0 \neq c \in R$. Let $y \in T \cap cL$. Then $y = cx$ for some $x \in L$ and $ay = 0$ for some $0 \neq a \in R$. This implies that $(ac)x = 0$ so that $x \in L \cap T$ and hence $cx = 0$, i.e. $y = 0$. Thus cL is a countably generated submodule of M such that $T \cap cL = 0$. Moreover, $cL_k \subseteq cL_{k+1} \subseteq \dots$ is an ascending chain of n-generated submodules of cL. By (ii) there exists a positive integer $t \geq k$ such that $cL_t = cL_{t+1} = \dots$. Let $i \geq t$. Let $u \in L_{i+1}$. Then there exists $v \in L_i$ such that $cu = cv$ so that $u - v \in L_{i+1} \cap T \subseteq L_i$. Thus $u \in L_i$. It follows that $L_t = L_{t+1} = \dots$, as required. □

The following theorem generalizes [5, Theorem $3^\sharp$]. If R is a ring, A an ideal of R and M an R-module then we set $ann_M(A) = \{m \in M : mA = 0\}$.

Theorem 1.7. *Let R be a Dedekind domain and let n be any positive integer. Then an R-module M with torsion submodule T satisfies n-acc if and only if*

(i) T is reduced and $ann_M(P) \neq 0$ for only a finite number of maximal ideals P of R, and

(ii) every countably generated torsion-free submodule of M satisfies n-acc.

Corollary 1.5. *Let R be a Dedekind domain. Then an R-module M with torsion submodule T satisfies pan-acc if and only if*

(i) T is reduced and $ann_M(P) \neq 0$ for only a finite number of maximal ideals P of R, and

(ii) every countably generated torsion-free submodule of M is projective.

The corollary clearly follows by Theorem 1.7 and Lemma 1.6. To prove Theorem 1.7 one merely follows the proof of the corresponding result of Baumslag and Baumslag (see[23] for details). The corollary has the following immediate consequence (see Lemmas 1.2 and 1.5).

Corollary 1.6. *Let R be a Dedekind domain. Then a torsion R-module M satisfies pan-acc if and only if M satisfies 1-acc.*

Heinzer and Lantz [16, p. 272] point out that Fuchs [14, p. 125] has shown that, for each positive integer n, there exist a torsion-free $\mathbb{Z}$-module A_n such that A_n satisfies n-acc but not $(n+1)$-acc. For each positive integer n, if R_n is the commutative ring which is the trivial extension of A_n by $\mathbb{Z}$ then the ring R_n satisfies n-acc but not $(n+1)$-acc. None of these rings R_n $(n \geq 1)$ is a domain and this leaves the following question of Heinzer and Lantz [16, Section 4].

Question 2

For every positive integer n does there exist a commutative domain R_n such that R_n satisfies n-acc but not $(n+1)$-acc? In particular, given a positive integer n, if A_n is the above group and K is a field does the group algebra $K[A_n]$ satisfy n-acc but not $(n+1)$-acc?

Grams[15] gives an example of a Prüfer domain R which satisfies 1-acc but not 2-acc (see Theorem 1.6). Moreover Heinzer and Lantz [16, Example 4.2] give an example of a regular UFD which satisfies 2-acc but not 3-acc. For other examples see [16, Section 4]. This brings us to the following question.

Question 3

Let R be a commutative Noetherian domain. Which R-modules satisfy pan-acc?

By Corollary 1.2 we know that torsionless modules over commutative Noetherian domains satisfy pan-acc. Are there other classes of modules with the same property? Provided the Noetherian domain is close (in some sense) to a Dedekind domain then a lot can be said. The following result is taken from [23, Theorem 3.4].

Theorem 1.8. *Let S be a subring of a Dedekind domain S' such that S is also a Dedekind domain and S' is a finitely generated S-module. Let A be any non-zero ideal of S' and let R be any subring of S' such that $S + A \subseteq R$. Then an R-module M with torsion submodule T satisfies pan-acc if and only if*

(i) T is reduced and $ann_M(P) \neq 0$ for at most a finite number of maximal ideals P of R, and

(ii) every countably generated torsion-free submodule of M can be embedded in a free R-module.

In particular, a torsion R-module satisfies pan-acc if and only if it satisfies 1-acc.

The situation described in the theorem arises naturally. Let S be *any* Dedekind domain and let K denote the field of fractions of S. Let K' be any finite separable extension of K and let S' denote the integral closure of S in K'. By [27, p. 264 Theorem 7], S' is a finitely generated S-module and, by [27, p. 281 Theorem 19], S' is a Dedekind domain. Let A be any non-zero ideal of S'. Then any subring R of S' such that $S+A \subseteq R$ satisfies the hypotheses of Theorem 1.8. For example, if $S = \mathbb{Z}$, $K = \mathbb{Q}$ and K' is the field $\mathbb{Q}[\sqrt{d}]$, for any square-free integer d in $\mathbb{Z}$, then $S' = \mathbb{Z}[\sqrt{d}]$ or $\mathbb{Z}[(1+\sqrt{d})/2]$ (see, for example, [10, p. 86 Theorem 1.3]). It follows that the domain $R = \mathbb{Z}[\sqrt{d}]$ satisfies the hypotheses of Theorem 1.8 for every square-free rational integer d. Note that the ring $\mathbb{Z}[\sqrt{d}]$ is not a Dedekind domain if $d \equiv 1 \pmod 4$ (see [10, p. 86 Theorem 1.3]).

Heinzer and Lantz [16, Corollary 3.5] prove that if R is a commutative Noetherian ring and X any family of indeterminates then the polynomial ring $R[X]$ satisfies pan-acc. They prove this result using localization techniques. This raises the following question.

Question 4
Let R be a right and left Noetherian ring and let X be any family of indeterminates. Does the polynomial ring $R[X]$ satisfy right (and left) pan-acc?

2. Renault's Theorem

In,[22] Renault investigates when free modules over right Noetherian rings satisfy pan-acc. We begin with an example due to Renault[22] which depends heavily on Lemma 1.4.

Example 2.1. Let K be a field and let S be a right Noetherian simple K-algebra. Let U be a simple right S-module and let R denote the formal "matrix ring"

$$\begin{bmatrix} K & U \\ 0 & S \end{bmatrix}$$

consisting of all "matrices" of the form

$$\begin{bmatrix} k & u \\ 0 & s \end{bmatrix}$$

where $k \in K, u \in U$ and $s \in S$. Then R is a right (but not left) Noetherian ring such that the free right R-module $R^{(\mathbb{N})}$ does not satisfy 1-acc.

Now we ask:

Question 5

Let R be a right Noetherian ring such that every free right R-module satisfies 1-acc. Does every free right R-module satisfy pan-acc?

Let R be a ring. An R-module M will be said to *satisfy the direct sum condition* provided every countably generated submodule is contained in a direct sum of finitely generated submodules of M. Clearly every free module and every semisimple module satisfies the direct sum condition. More generally, every direct sum of finitely generated R-modules satisfies the direct sum condition. Note also that if M_i is an R-module satisfying the direct sum condition, for all i in some index set I, then the R-module $\bigoplus_{i \in I} M_i$ also satisfies the direct sum condition. For, let N be any countably generated submodule of the module $M = \bigoplus_{i \in I} M_i$. For each $i \in I$, let $\pi_i : M \to M_i$ denote the canonical projection. Because, for each $i \in I$, $\pi_i(N)$ is a countably generated submodule of M_i, there exists a submodule K_i of M_i such that K_i is a direct sum of finitely generated submodules and $\pi_i(N) \subseteq K_i$. Let $K = \bigoplus_{i \in I} K_i$. Then N is contained in the submodule K of M and K is a direct sum of finitely generated submodules.

The next result is taken from [25, Theorem 8]. It generalizes Theorem 1.1.

Theorem 2.1. *Let R be a right Noetherian ring, let M be a right R-module which satisfies the direct sum condition and let n be a positive integer. Then M satisfies n-acc if and only if $Z_2(M)$ satisfies n-acc.*

Note that Theorem 2.1 shows in particular that, for any right Noetherian ring R, every nonsingular R-module which satisfies the direct sum condition satisfies pan-acc. For other results on nonsingular modules which satisfy pan-acc see.[2]

Given a non-empty subset X of a ring R, $\mathbf{r}(X)$ will denote the right annihilator of X in R. Renault [22, Théorème 3.2] proves the next result.

Theorem 2.2. *Let R be a right Noetherian ring such that for each prime ideal P there exists a finite subset X of R such that $P = \mathbf{r}(X)$. Then every free right R-module satisfies pan-acc.*

By using [25, Lemmas 10 and 14], the conclusion of Theorem 2.2 can be strengthened to read : *every torsionless right R-module which satisfies the direct sum condition satisfies pan-acc.*

Renault [22, Corollaire 3.3] points out that the hypotheses of Theorem 2.2 are satisfied by right and left Noetherian rings and by right FBN rings. Recall that a ring R is *right FBN* if R is right Noetherian and, for each prime ideal P of R, every essential right ideal of the ring R/P contains a non-zero two-sided ideal of R/P. In addition, if a right Noetherian ring satisfies the descending chain condition on right annihilators then the hypotheses of Theorem 2.2 are met and hence every free right R-module satisfies pan-acc (see [3, Theorem 1] for a related result).

This raises the following question.

Question 6

Let R be a right Noetherian ring such that for each prime ideal P there exists a finite subset X of R such that $P = \mathbf{r}(X)$. Does every torsionless right R-module satisfy pan-acc?

There is some positive evidence relating to Question 6. For example, note the following result.

Theorem 2.3. *(See [25, Corollary 17].) Let R be a right and left Noetherian ring. Then every torsionless R-module satisfies pan-acc.*

Note that Frohn[13] proved Theorem 2.3 in the case of commutative rings by a different method of proof. In view of the above remarks we ask the following:

Question 6*

Let R be a right Noetherian ring which is either right FBN or satisfies the descending chain condition on right annihilators. Does every torsionless right R-module satisfy pan-acc?

For right FBN rings we have the following result.

Theorem 2.4. *(See [25, Theorem 13].) Let R be a right FBN ring and let M be a right R-module which satisfies the direct sum condition. Then M satisfies pan-acc if and only if for each positive integer n and each ascending chain $L_1 \subseteq L_2 \subseteq \dots$ of n-generated submodules L_i $(i \geq 1)$ of M there exists a positive integer k such that $ann(L_k) = ann(L_{k+1}) = \dots$.*

3. Jonah's Theorem

Rings R with the property that every right module satisfies pan-acc are right perfect as the following result shows.

Theorem 3.1. *(See[17] or [22, Proposition 1.2].) The following statements are equivalent for a ring R.*

(i) R is right perfect.

(ii) Every right R-module satisfies 1-acc.

(iii) Every right R-module satisfies pan-acc.

Corollary 3.1. *(See [22, Proposition 1.3].) Let R be a commutative domain such that R/I is a perfect ring for every non-zero ideal I of R. Then every free R-module satisfies pan-acc.*

This allows Renault [22, Corollaire 1.5] to give an example of a non-Noetherian commutative domain R such that every free R-module satisfies pan-acc. We give another example below. The proof of Corollary 3.1 depends on the following surprisingly useful lemma.

Lemma 3.1. *Let A be an ideal of a ring R such that R/A is a right perfect ring and let $N_1 \subseteq N_2 \subseteq \dots$ be an ascending chain of n-generated submodules of a right R-module M such that $N_1A = N_2A = \dots$. Then $N_k = N_{k+1} = \dots$ for some positive integer k.*

Proof. Let $N = \cup_{i \geq 1} N_i$. Then N is a submodule of M and for every x in N there exists $i \geq 1$ such that $x \in N_i$ and hence $xA \subseteq N_iA = N_1A \subseteq N_1$. Thus N/N_1 is a right (R/A)-module. By Theorem 3.1, the ascending chain $N_1/N_1 \subseteq N_2/N_1 \subseteq \dots$ of n-generated submodules of N/N_1 terminates and hence $N_k = N_{k+1} = \dots$ for some positive integer k. □

We now give some further applications of Theorem 3.1. The first is based on [1, Theoreme 1].

Theorem 3.2. *Let R be a subring of a ring S such that there exists an ideal A of R which is also a left ideal of S with the ring R/A right perfect and let M be a right S-module which satisfies n-acc, for some positive integer n. Then the right R-module M satisfies n-acc.*

Proof. Let $L_1 \subseteq L_2 \subseteq \dots$ be any ascending chain of n-generated R-submodules of M. Then $L_1S \subseteq L_2S \subseteq \dots$ is an ascending chain of n-generated S-submodules of M. By hypothesis, there exists $k \geq 1$ such that $L_kS = L_{k+1}S = \dots$ and hence $L_kA = L_{k+1}A = \dots$. By Lemma 3.1 there exists $t \geq k$ such that $L_t = L_{t+1} = \dots$. □

Corollary 3.2. *Let A, R and S be as in the theorem. Let L be a right R-module such that L is an R-submodule of an S-module M such that M_S satisfies n-acc, for some positive integer n. Then L_R satisfies n-acc.*

Proof. Apply Theorem 3.2 to the module M_S to obtain that M_R, and hence L_R, satisfies n-acc. □

Corollary 3.3. *Let S be a ring such that every free (respectively, torsionless) right S-module satisfies n-acc, for some positive integer n, and let R be a subring of S such that there exists an ideal A of R which is also a left ideal of S with the ring R/A right perfect. Then every free (respectively, torsionless) right R-module satisfies n-acc.*

Proof. Let L be a free R-module. Then $L \cong R_R^{(I)} \subseteq S_S^{(I)} = M_S$, for some index set I. Now apply Corollary 3.2. The torsionless case is similar. □

Theorem 3.3. *Let R be a subring of a ring S such that there exists an ideal A of R which is also a finitely generated right ideal of S with the ring R/A right perfect. Let M be a right S-module which satisfies pan-acc. Then the right R-module M satisfies pan-acc.*

Proof. Let $L_1 \subseteq L_2 \subseteq \dots$ be any ascending chain of n-generated submodules of the R-module M. Let $K = L_i$ for some $i \geq 1$. Then there exist $x_i \in K \, (1 \leq i \leq n)$ such that $K = x_1R + \cdots + x_nR$ and hence $KA = x_1RA + \cdots + x_nRA = x_1A + \cdots + x_nA$. Suppose that $A = a_1S + \cdots + a_kS$ for some positive integer k Then $KA = \sum_{i=1}^{n}\sum_{j=1}^{k} x_ia_jS$ and we obtain that L_iA is an nk-generated submodule of M_S for every $i \geq 1$. Hence the chain $L_1A \subseteq L_2A \subseteq \dots$ terminates. By Lemma 3.1, there exists a positive integer t such that $L_t = L_{t+1} = \dots$. □

There is an analogue of Corollary 3.2 which we will not state. Moreover, Corollary 3.3 has the following analogue. The proof is similar and is omitted.

Corollary 3.4. *Let S be a ring such that every free (respectively, torsionless) right S-module satisfies pan-acc, for some positive integer n, and let R be a subring of S such that there exists an ideal A of R which is also a finitely generated right ideal of S with the ring R/A right perfect. Then every free (respectively, torsionless) right R-module satisfies pan-acc.*

Corollary 3.5. *Let R be a subring of a right and left Noetherian ring S such that there exists an ideal A of R which is also a left or right ideal of S with the ring R/A right perfect. Then every torsionless right R-module satisfies pan-acc.*

Proof. By Theorem 2.3 and Corollaries 3.3 and 3.4. □

To illustrate this last corollary, let F be a subfield of a field K such that the field extension K/F is infinite. Let R denote the subring $F + xK[x]$ of the commutative PID $S = K[x]$. Let A denote the ideal $xK[x]$ of the ring S and note that A is contained in R and $R/A \cong F$. By Corollary 3.5 every torsionless R-module satisfies pan-acc. Note that, because K/F is infinite, R is a non-Noetherian commutative domain.

Theorem 3.3 has the following partial converse.

Theorem 3.4. *Let R be a subring of a ring S such that the right R-module S is finitely generated and such that there exists an ideal A of S contained in R with A finitely generated as a right ideal of R and the ring R/A right perfect. Let M be a right R-module which satisfies pan-acc and which is an R-submodule of a right S-module X. Then the right S-module MS satisfies pan-acc.*

Proof. Let n be any positive integer. Let $L_1 \subseteq L_2 \subseteq \dots$ be any ascending chain of n-generated S-submodules of MS. Because $S = s_1R + \cdots + s_tR$ for some positive integer t and elements $s_i \in S\,(1 \leq i \leq t)$, L_i is a tn-generated R-module for each $i \geq 1$. On the other hand, $L_1A \subseteq L_2A \subseteq \dots$ is an ascending chain of R-submodules of $MSA = MA \subseteq M_R$. The R-module L_iA is tnk-generated for every $i \geq 1$, where $A = a_1R + \cdots + a_kR$ for some positive integer k and elements $a_i \in A\,(1 \leq i \leq k)$. By hypothesis, we can suppose without loss of generality that $L_1A = L_2A = \dots$. By Lemma 3.1, $L_s = L_{s+1} = \dots$ for some positive integer s. □

The next result is taken from [4, Proposition 3.10] but the proof is quite short so we give it.

Theorem 3.5. *Let A and B be ideals of a ring R such that $AB = 0$, the ring R/B is right perfect and every (finitely generated) free right (R/A)-module satisfies n-acc, for some fixed positive integer n. Then every (finitely generated) free right R-module satisfies n-acc.*

Proof. Let F be any (finitely generated) free right R-module and let $N_1 \subseteq N_2 \subseteq \dots$ be any ascending chain of n-generated submodules of F. Then $(N_1 + FA)/FA \subseteq (N_2 + FA)/FA \subseteq \dots$ is an ascending chain of n-generated submodule of the (finitely generated) free right R-module F/FA. By hypothesis, there exists a positive integer k such that $N_k + FA = N_{k+1} + FA = \dots$. Now $AB = 0$ gives $N_k B = N_{k+1} B = \dots$. By Lemma 3.1, there exists a positive integer t such that $N_t = N_{t+1} = \dots$. Thus F satisfies n-acc. □

Consider again Example 2.1 but now suppose in addition that the ring S is left as well as right Noetherian. Note that Theorem 3.5 shows that in this case the ring R has a free right module which does not satisfy 1-acc but has the additional property that every free *left* R-module satisfies pan-acc. To see why this is the case, let A denote the ideal in R consisting of all matrices whose (2,2) entry is 0 and let B denote the ideal in R whose (1,1) entry is 0. Clearly $BA = 0$, $R/B \cong K$ so that R/B is a field and $R/A \cong T$ so that R/A is a simple Noetherian ring. Every free left (R/A)-module satisfies pan-acc by Theorem 2.3. Now apply Theorem 3.5.

4. Cohn's Problem

In [9, Exercises 0.1, Question 1], Cohn asks the following question.

Question 7
Let R be a ring which satisfies right n-acc, for some positive integer n. Does every finitely generated free right R-module satisfy n-acc?

Question 7 can be generalised as follows.

Question 7*
Let R be a ring and let M_1 and M_2 be R-modules which each satisfy n-acc, for some positive integer n? Does the R-module $M_1 \oplus M_2$ also satisfy n-acc?

We have already noted above that if R is a right Noetherian ring then every finite direct sum of right R-modules each satisfying n-acc, for some fixed positive integer n, also satisfies n-acc (see Theorem 1.3). In addition we have the following simple result for Prüfer domains.

Lemma 4.1. *Let R be a Prüfer domain and let M_1 and M_2 be R-modules which both satisfy n-acc, for some positive integer n. Suppose further that M_2 is torsion-free. Then the module $M_1 \oplus M_2$ satisfies n-acc.*

Proof. Let M denote the R-module $M_1 \oplus M_2$ and let $\pi : M \to M_2$ denote the canonical projection. Let $L_1 \subseteq L_2 \subseteq \ldots$ be any ascending chain of n-generated submodules of M. For each $i \geq 1$, $\pi(L_i)$ is a finitely generated torsion-free R-module so that $\pi(L_i)$ is projective and hence $L_i \cap M_1$ is a direct summand of L_i. Thus $L_i \cap M_1$ is n-generated for all $i \geq 1$. It follows that we have an ascending chain $\pi(L_1) \subseteq \pi(L_2) \subseteq \ldots$ of n-generated submodules of M_2 which must terminate and an ascending chain $L_1 \cap M_1 \subseteq L_2 \cap M_1 \subseteq \ldots$ of n-generated submodules of M_1 which must also terminate. It is now easy to prove that $L_k = L_{k+1} = \ldots$ for some positive integer k.

□

The following result is due to Frohn [12, Lemma].

Lemma 4.2. *Let R be a commutative ring and let M_1 and M_2 be R-modules which each satisfy 1-acc. Then the R-module $M_1 \oplus M_2$ also satisfies 1-acc.*

Proof. Let $R(x_1, y_1) \subseteq R(x_2, y_2) \subseteq \ldots$ be any ascending chain of cyclic submodules of $M = M_1 \oplus M_2$ for some $x_i \in M_1, y_i \in M_2\,(i \geq 1)$. Then $Rx_1 \subseteq Rx_2 \subseteq \ldots$ and $Ry_1 \subseteq Ry_2 \subseteq \ldots$ are ascending chains of cyclic submodules in M_1, M_2, respectively. Thus we can suppose without loss of generality that $Rx_1 = Rx_2 = \ldots$ and $Ry_1 = Ry_2 = \ldots$. There exist elements a, b and c in R such that $x_2 = ax_1$, $y_2 = by_1$ and $(x_1, y_1) = c(x_2, y_2)$. Note that $x_1 = cx_2 = acx_1$ so that $(1 - ac)x_1 = 0$. Similarly, $(1 - bc)y_1 = 0$. Let $d = a + b - abc \in R$. It is easy to check that $(x_2, y_2) = d(x_1, y_1)$ and hence $R(x_1, y_1) = R(x_2, y_2)$. Thus $R(x_1, y_1) = R(x_2, y_2) = \ldots$. □

Using this last lemma Frohn [12, Theorem] proves the following result.

Theorem 4.1. *Let R be a commutative ring which satisfies 1-acc. Then every free R-module satisfies 1-acc.*

Now note the following result.

Theorem 4.2. *(See [4, Theorem 2.1].) Let R be a right and left Öre domain and let n be a positive integer such that the free right R-module $R_R^{(n)}$ satisfies n-acc. Then every free right R-module satisfies n-acc.*

There is a companion result for commutative domains (see [4, Theorem 2.6]).

Theorem 4.3. *Let R be a commutative domain and let n be a positive integer such that the free R-module $R_R^{(n-1)}$ satisfies n-acc. Then every free R-module satisfies n-acc.*

Corollary 4.1. *Let R be a commutative domain which satisfies 2-acc. Then every free R-module satisfies 2-acc.*

We end with the following question.

Question 8
Let R be a right Noetherian ring such that every free right R-module satisfies pan-acc) (respectively, n-acc, for some positive integer n). Does every torsionless right R-module satisfy pan-acc (respectively, n-acc)?

Acknowledgement

This paper was given at the Fifth China-Japan-Korea International Symposium on Ring Theory in Tokyo, September 2007. The author would like to thank the organizers for their financial support.

References

1. M. E. Antunes Simões and A.-M. Nicolas, Exemples d'anneaux n-acc, Comm. Algebra 12, 1653-1665 (1984).
2. M. E. Antunes Simões and P. F. Smith, Direct products satisfying the ascending chain condition for submodules with a bounded number of generators, Comm. Algebra 23, 3525-3540 (1995).
3. M. E. Antunes Simões and P. F. Smith, On the ascending chain condition for submodules with a bounded number of generators, Comm. Algebra 24, 1713-1721 (1996).
4. M. E. Antunes Simões and P. F. Smith, Rings whose free modules satisfy the ascending chain condition on submodules with a bounded number of generators, J. Pure Appl. Algebra 123, 51-66 (1998).
5. B. Baumslag and G. Baumslag, On ascending chain conditions, Proc. London Math. Soc. (3) 22, 681-704 (1971).
6. J.-E. Björk, Rings satisfying a minimum condition on principal ideals, J. Reine Angew. Math. 236, 112-119 (1969).
7. F. Bonang, Noetherian rings whose subidealizer subrings have pan-a.c.c., Comm. Algebra 17, 1137-1146 (1989).
8. A. W. Chatters and P. F. Smith, A note on hereditary rings, J. Algebra 44, 181-190 (1977).
9. P. M. Cohn, *Free Rings and their Relations* (Academic Press, London 1971).
10. P. M. Cohn, *Algebraic Numbers and Algebraic Functions* (Chapman and Hall, London 1991).
11. N. V. Dung, D. V. Huynh, P. F. Smith and R. Wisbauer, *Extending Modules* (Longman, Harlow 1994).
12. D. Frohn, A counterexample concerning ACCP in power series rings, Comm. Algebra 30, 2961-2966 (2002).
13. D. Frohn, Modules with n-acc and the acc on certain types of annihilators, J. Algebra 256, 467-483 (2002).
14. L. Fuchs, *Infinite Abelian Groups Vol II* (Academic Press, New York 1973).
15. A. Grams, Atomic rings and the ascending chain condition for principal ideals, Proc. Cambridge Phil. Soc. 75 (1974), 321-329.
16. W. Heinzer and D. Lantz, Commutative rings with ACC on n-generated ideals, J. Algebra 80, 261-278 (1983).
17. D. Jonah, Rings with the minimum condition for principal right ideals have the maximum condition for principal left ideals, Math. Z. 113, 106-112 (1970).

18. I. Kaplansky, *Commutative Rings* (Allyn and Bacon, Boston 1970).
19. J. C. McConnell and J. C. Robson, *Noncommutative Noetherian Rings* (John Wiley and Sons, Chichester 1987).
20. A.-M. Nicolas, Sur les modules tels que toute suite croissante de sous-modules engendrés par n générateurs soit stationnaire, J. Algebra 60, 249-260 (1979).
21. L. S. Pontrjagin, *Topological Groups* (Gordon and Breach, New York 1966).
22. G. Renault, Sur des conditions de chaînes ascendantes dans des modules libres, J. Algebra 47, 268-275 (1977).
23. E. Sanchez Campos and P. F. Smith, Certain chain conditions in modules over Dedekind domains and related rings, to appear in the Proceedings of the Porto Conference 2006 to commemorate Robert Wisbauer's 65th Birthday.
24. D. W. Sharpe and P. Vamos, *Injective Modules*, Cambridge Tracts in Mathematics and Mathematical Physics, No. 62 (Cambridge University Press, London 1972).
25. P. F. Smith, Modules satisfying the ascending chain condition on submodules with a bounded number of generators, Int. Electron. J. Algebra 2 (2007), 71-82.
26. B. Stenström, *Rings of Quotients* (Springer-Verlag, Berlin 1975).
27. O. Zariski and P. Samuel, *Commutative Algebra Vol I* (Van Nostrand, Princeton 1958).

PART C

General Lectures

On Generalized (α, β)-derivations in Rings and Modules

Mohammad Ashraf, Shakir Ali and Nadeem-ur-Rehman

Department of Mathematics, Aligarh Muslim University,
Aligarh 202002
India

E-mail:mashraf80@hotmail.com, shakir50@rediffmail.com, rehman100@gmail.com

Let R, S be rings and α, β be homomorphisms of S into R. Suppose that M is an R-bimodule. An additive mapping $F : S \longrightarrow M$ is called a generalized (α, β) - derivation on S if there exists an (α, β)-derivation $d : S \longrightarrow M$ such that $F(xy) = F(x)\alpha(y) + \beta(x)d(y)$ holds for all $x, y \in S$. In the present paper, we extend some well known results concerning generalized derivations of certain classes of rings to generalized (α, β)-derivations.

Keywords: Bi-modules, derivation; (α, β)-derivation and generalized (α, β)-derivation.
2000 Mathematics Subject Classification: $16W25, 16N60, 16U80$.

1. Introduction

Throughout the present paper R, S will denote associative rings, and $M \neq (0)$ an R-bimodule. For each $x, y \in S$, the symbol $[x, y]$ will represent the commutator $xy - yx$ and the symbol $x \circ y$ stands for the anti-commutator $xy + yx$. Also, we set $[x, y]_{\alpha,\beta} = x\alpha(y) - \beta(y)x$, where α and β are homomorphisms of S into R. We shall do a great deal of calculations with commutators and anti-commutators, routinely using the following identities:

$$[xy, z]_{\alpha,\beta} = x[y, z]_{\alpha,\beta} + [x, \beta(z)]y = x[y, \alpha(z)] + [x, z]_{\alpha,\beta}y,$$

$$[x, yz]_{\alpha,\beta} = \beta(y)[x, z]_{\alpha,\beta} + [x, y]_{\alpha,\beta}\alpha(z),$$

$$(x \circ (yz))_{\alpha,\beta} = (x \circ y)_{\alpha,\beta}\alpha(z) - \beta(y)[x, z]_{\alpha,\beta} = \beta(y)(x \circ z)_{\alpha,\beta} + [x, y]_{\alpha,\beta}\alpha(z),$$

$$\text{and} \quad ((xy) \circ z)_{\alpha,\beta} = x(y \circ z)_{\alpha,\beta} - [x, \beta(z)]y = (x \circ z)_{\alpha,\beta}y + x[y, \alpha(z)].$$

Recall that a ring R is prime if for any $x, y \in R$, $xRy = (0)$ implies that either $x = 0$ or $y = 0$. A module M is said to be n-torsion free whenever $na = 0$; with an integer n and $a \in M$ implies that $a = 0$. An additive mapping $d : R \longrightarrow M$ is called a derivation if $d(xy) = d(x)y + xd(y)$ holds for all $x, y \in R$. An additive mapping $F : R \longrightarrow M$ is called a generalized derivation if there exists a derivation $d : R \longrightarrow M$ such that $F(xy) = F(x)y + xd(y)$, holds for all $x, y \in R$. It is easy to check that the notion of a generalized derivation covers the notions of a derivation as well as of a left multiplier *i.e.*, an additive mapping f on R satisfying $f(xy) = f(x)y$, for all $x, y \in R$.

Let R, S and M be as above and α, β be the homomorphisms of S into R. Following [7], an additive mapping $d : S \longrightarrow M$ is called an (α, β)-derivation if $d(xy) = d(x)\alpha(y) + \beta(x)d(y)$ holds for all $x, y \in S$. An additive mapping $F : S \longrightarrow M$ is called a generalized (α, β)-derivation if there exists an (α, β)-derivation $d : S \longrightarrow M$ such that $F(xy) = F(x)\alpha(y) + \beta(x)d(y)$ holds for all $x, y \in S$. Note that for I_S the identity map on S, an (I_S, I_S)-genaralized derivation is called simply a generalized derivation. In the remaining part of the paper, a generalized (α, β)-derivation $F : S \longrightarrow M$ associated with a nonzero (α, β)-derivation $d : S \longrightarrow M$ will be denoted as (F, d).

The purpose of this paper is to extend some well known results in the setting of generalized (α, β)-derivation of certain classes of rings. In fact, our theorems generalize the results proved in [2], [3], [5], [8] and [10].

2. Main Results

The main result of the present paper is as follows:

Theorem 2.1. *Let R, S be rings and M be an R-bimodule such that $xRm = (0)$ with $x \in R, m \in M$ implies that either $x = 0$ or $m = 0$. Suppose that α, β are homomorphisms of S into R such that β is one-one and onto. If (F, d) is a generalized (α, β)-derivation of S into M such that $[F(x), x]_{\alpha,\beta} = 0$, for all $x \in S$, then R is commutative.*

Proof. We have

$$[F(x), x]_{\alpha,\beta} = 0, \quad \text{for all } x \in S. \tag{1}$$

On linearizing (1) and using (1), we find that

$$[F(x), y]_{\alpha,\beta} + [F(y), x]_{\alpha,\beta} = 0, \quad \text{for all } x, y \in S. \tag{2}$$

Replace y by yx in (2), to get

$$[F(x), yx]_{\alpha,\beta} + [F(y)\alpha(x) + \beta(y)d(x), x]_{\alpha,\beta} = 0, \quad \text{for all } x, y \in S.$$

This implies that

$$[F(x), y]_{\alpha,\beta}\alpha(x) + \beta(y)[F(x), x]_{\alpha,\beta} + [F(y), x]_{\alpha,\beta}\alpha(x) +$$

$$F(y)[\alpha(x), \alpha(x)] + \beta(y)[d(x), x]_{\alpha,\beta} + [\beta(y), \beta(x)]d(x) = 0. \qquad (3)$$

On combining (1), (2) and (3), we obtain

$$\beta(y)[d(x), x]_{\alpha,\beta} + [\beta(y), \beta(x)]d(x) = 0, \quad \text{for all } x, y \in S. \qquad (4)$$

Replacing y by zy in (4) and using (4), we get

$$[\beta(z), \beta(x)]\beta(y)d(x) = 0 \text{ for all } x, y, z \in S. \qquad (5)$$

Since β is onto, the above equation yields that $[\beta(z), \beta(x)]Rd(x) = (0)$, for all $x, z \in S$. Thus, for each $x \in S$, we have either $[\beta(z), \beta(x)] = 0$ or $d(x) = 0$. Now, let $H = \{x \in S \mid d(x) = 0\}$ and $K = \{x \in S \mid [\beta(z), \beta(x)] = 0, \text{ for all } z \in S\}$. Then, H and K are additive subgroups of S whose union is S. But a group can not be union of two of its proper subgroups and hence $S = H$ or $S = K$. If $S = H$, then $d(x) = 0$ for all $x \in S$, a contradiction. On the other hand if $S = K$, then $[\beta(z), \beta(x)] = 0$ for all $x, z \in S$. Since β is onto, the last expression implies that R is commutative. This completes the proof of the theorem.

□

Remark 2.1. If M is an R-bimodule with the property that $xRm = (0)$ with $x \in R$ and $m \in M$ implies that either $x = 0$ or $m = 0$, then by using similar arguments as used in Remark 5 of [7] we find that the ring R is prime. In fact if $xRy = (0)$ for some $x, y \in R$, then for any nonzero $m \in M$ we have $xR(yRm) = (0)$ and hence either $x = 0$ or $y = 0$. Therefore, in view of the above result we get the following corollary.

Corollary 2.1. *([2, Theorem 1]) Let R be a 2-torsion free prime ring and α, β be automorphisms of R. Suppose there exists an (α, β)-derivation $d : R \longrightarrow R$ such that $[d(x), x]_{\alpha,\beta} = 0$, for all $x \in R$, then either $d = 0$ or R is commutative.*

Theorem 2.2. *Let R, S be rings and M be an R-bimodule such that $xRm = (0)$ with $x \in R, m \in M$ implies that either $x = 0$ or $m = 0$. Suppose that α, β are homomorphisms of S into R such that β is one-one and onto. If (F, d) is a generalized (α, β)-derivation of S into M such that $F([x, y]) = 0$, for all $x, y \in S$, then R is commutative.*

Proof. By the hypothesis, we have

$$F([x, y]) = 0, \quad \text{for all } x, y \in S. \tag{6}$$

Replacing y by yx in (6) using (6), we find that

$$0 = F([x, yx]) = F([x, y]x) = \beta([x, y])d(x), \text{ for all } x, y \in S. \tag{7}$$

Now, taking y as zy in equation (7), we get $[\beta(x), \beta(z)]\beta(y)d(x) = 0$, for all $x, y, z \in S$. This implies that $[\beta(x), \beta(z)])Rd(x) = (0)$, for all $x, z \in R$. Now, using the similar arguments as we have used in the end of the proof of Theorem 2.1, we get the required result.

□

Theorem 2.3. *Let R, S be rings and M be an R-bimodule such that $xRm = (0)$ with $x \in R, m \in M$ implies that either $x = 0$ or $m = 0$. Suppose that α, β are homomorphisms of S into R such that β is one-one and onto. If (F, d) is a generalized (α, β)-derivation of S into M such that $F([x, y]) = [x, y]_{\alpha,\beta}$ for all $x, y \in S$, then R is commutative.*

Proof. Since F is a generalized (α, β)-derivation, we have

$$F([x, y]) = [x, y]_{\alpha,\beta} \text{ for all } x, y \in S. \tag{8}$$

Replace y by zy in (8), to get

$$F(z[x, y] + [x, z]y) = [x, zy]_{\alpha,\beta}, \text{ for all } x, y, z \in S.$$

This gives that

$$F(z)\alpha([x, y]) + \beta(z)d([x, y]) + F([x, z])\alpha(y) + \beta([x, z])d(y)$$

$$= [x, z]_{\alpha,\beta}\alpha(y) + \beta(z)[x, y]_{\alpha,\beta}, \text{ for all } x, y, z \in S. \tag{9}$$

Application of (8), yields that

$$F(z)\alpha([x,y]) + \beta(z)d([x,y]) + \beta([x,z])d(y) = \beta(z)[x,y]_{\alpha,\beta}. \tag{10}$$

Again, replacing y by yx in (10), we get

$$F(z)\alpha([x,y]x) + \beta(z)d([x,y]x) + \beta([x,z])(d(y)\alpha(x) + \beta(y)d(x))$$

$$= \beta(z)([x,y]_{\alpha,\beta}\alpha(x) + \beta(y)[x,x]_{\alpha,\beta}) \text{ for all } x,y,z \in S.$$

This yields that

$$\{F(z)\alpha([x,y])+\beta(z)d([x,y])+\beta([x,z])d(y)\}\alpha(x)+\beta(z)\beta([x,y])d(x)+\beta([x,z])\beta(y)d(x)$$

$$= \beta(z)([x,y]_{\alpha,\beta}\alpha(x) + \beta(z)\beta(y)[x,x]_{\alpha,\beta}, \quad \text{for all } x,y,z \in S. \tag{11}$$

Now using (10) in (11), we obtain

$$\beta(z)\beta([x,y])d(x) + \beta([x,z])\beta(y)d(x) = \beta(z)\beta(y)[x,x]_{\alpha,\beta}. \tag{12}$$

Put $z = z_1 z$ in (12), to get

$$\beta(z_1)\beta(z)\beta([x,y])d(x) + \beta(z_1[x,y] + [x,z_1]z)\beta(y)d(x)$$
$$= \beta(z_1)\beta(z)\beta(y)[x,x]_{\alpha,\beta}. \tag{13}$$

On combining (12) and (13), we find that

$$\beta([x,z_1])\beta(z)\beta(y)d(x) = 0, \text{ for all } x,y,z,z_1 \in S. \tag{14}$$

This implies that, $[\beta(x),\beta(z_1)])R\beta(y)d(x) = (0)$, for all $x, z_1 \in S$. Thus, our hypothesis forces that for each $x \in S$, either $\beta(y)d(x) = 0$ or $[\beta(x),\beta(z_1)] = 0$. Now, we put $U = \{x \in S \mid \beta(y)d(x) = 0, \text{ for all } y \in S\}$ and $V = \{x \in S \mid [\beta(x),\beta(z_1)] = 0, \text{ for all } z_1 \in S\}$. By the discussion given, S is the union of U and V. Using the analogous arguments as above one can prove that either $U = S$ or $V = S$. If $V = S$, then $[\beta(x),\beta(z_1)] = 0$ for all $x, z_1 \in S$ and hence R is commutative. On the other hand if $U = S$, then $\beta(y)d(x) = 0$ for all $x, y \in S$. Therefore, we find that $[\beta(y),\beta(z)]Rd(x) = (0)$ for all $x,y,z \in S$. Using the similar techniques as we have used above, we find that $d(x) = 0$, for all $x \in S$ or $[\beta(y),\beta(z)] = 0$ for all $y,z \in S$. But $d(x) = 0$, for all $x \in S$, a contradiction. Therefore, we have only the remaining possibility that $[\beta(y),\beta(z)] = 0$ for all $y,z \in S$. Hence, R is commutative. This completes the proof of our theorem.

□

Using similar arguments as above with necessary variations we can prove the following:

Theorem 2.4. *Let R, S be rings and M be an R-bimodule such that $xRm = (0)$ with $x \in R, m \in M$ implies that either $x = 0$ or $m = 0$. Suppose that α, β are homomorphisms of S into R such that β is one-one and onto. If (F, d) is a generalized (α, β)-derivation of S into M such that $F([x, y]) + [x, y]_{\alpha,\beta} = 0$, for all $x, y \in S$, then R is commutative.*

In view of the Remark 2.1 and Thorems 2.3 & 2.4, we obtain the following result:

Corollary 2.2. *Let R be prime ring. Suppose that α, β are endomorphisms of R such that β is one-one and onto. If (F, d) is a generalized (α, β)-derivation of R such that $F([x, y]) - [x, y]_{\alpha,\beta} = 0$ (resp. $F([x, y]) + [x, y]_{\alpha,\beta} = 0$), for all $x, y \in R$, then R is commutative.*

Theorem 2.5. *Let R, S be rings and M be an R-bimodule such that $xRm = (0)$ with $x \in R, m \in M$ implies that either $x = 0$ or $m = 0$. Suppose that α, β are homomorphisms of S into R such that β is one-one and onto. If (F, d) is a generalized (α, β)-derivation of S into M such that $F([x, y]) = \alpha([x, y])$, for all $x, y \in S$, then R is commutative.*

Proof. Notice that $F([x, y]) = \alpha([x, y])$, for all $x, y \in S$. This can be rewritten as

$$F(x)\alpha(y) + \beta(x)d(y) - F(y)\alpha(x) - \beta(y)d(x) = \alpha([x, y]) \tag{15}$$

Replacing y by yx in (15), we get

$$\begin{aligned} &F(x)\alpha(y)\alpha(x) + \beta(x)d(y)\alpha(x) + \beta(x)\beta(y)d(x) - F(y)\alpha(x)\alpha(x) \\ &-\beta(y)d(x)\alpha(x) - \beta(y)\beta(x)d(x) = \alpha([x, y])\alpha(x), for\ all\ x, y \in S. \end{aligned} \tag{16}$$

From (15) and (16), we find that $\beta([x, y])d(x) = 0$, for all $x, y \in S$. Taking y as zy in the last expression, we get $[\beta(x), \beta(z)]\beta(y)d(x) = 0$, for all $x, y, z \in S$ and hence $[\beta(x), \beta(z)]Rd(x) = (0)$, for all $x, z \in S$. Thus, the result follows from the proof given in the last paragraph of Theorem 2.1.

□

Remark 2.2. Using the same techniques as we have used in Theorem 2.5 with necessary variations, one can obtain the similar result for the property; $F([x, y]) + \alpha([x, y]) = 0$, for all $x, y \in S$.

Now, we shall study the behaviour of the ring satisfying the properties; $F(x \circ y) = 0$ and $F(x \circ y) = \pm(x \circ y)_{\alpha,\beta}$.

Theorem 2.6. *Let R, S be rings and M be an R-bimodule such that $xRm = (0)$ with $x \in R, m \in M$ implies that either $x = 0$ or $m = 0$. Suppose that α, β are homomorphisms of S into R such that β is one-one and onto. If (F, d) is a generalized (α, β)-derivation of S into M such that $F(x \circ y) = 0$, for all $x, y \in S$, then R is commutative.*

Proof. We have

$$F(x \circ y) = 0, \quad \text{for all } x, y \in S. \tag{17}$$

Replacing y by yx in (17) and using (17), we obtain

$$0 = F(x \circ yx) = F((x \circ y)x) = \beta(x \circ y)d(x), \text{ for all } x, y \in S. \tag{18}$$

Further, replace y by zy in (18) and use (18), to get $\beta([x, z]y)d(x) = 0$, for all $x, y, z \in S$. This forces that $[\beta(x), \beta(z)]Rd(x) = (0)$, for all $x, z \in S$. Hencefourth, using same approach as used in the last paragraph of the proof of Thorem 2.1, we get the required result.

□

Theorem 2.7. *Let R, S be rings and M be an R-bimodule such that $xRm = (0)$ with $x \in R, m \in M$ implies that either $x = 0$ or $m = 0$. Suppose that α, β are homomorphisms of S into R such that β is one-one and onto. If (F, d) is a generalized (α, β)-derivation of S into M such that $F(x \circ y) = (x \circ y)_{\alpha,\beta}$ for all $x, y \in S$, then R is commutative.*

Proof. For any $x, y \in S$, we have $F(x \circ y) = (x \circ y)_{\alpha,\beta}$. This can be rewritten as

$$F(x)\alpha(y) + \beta(x)d(y) + F(y)\alpha(x) + \beta(y)d(x) = (x \circ y)_{\alpha,\beta}. \tag{19}$$

Replace y by yx in (19), to get

$$F(x)\alpha(y)\alpha(x)+\beta(x)d(y)\alpha(x)+\beta(x)\beta(y)d(x)+F(y)\alpha(x)\alpha(x)+\beta(y)d(x)\alpha(x)+$$

$$\beta(y)\beta(x)d(x) = (x \circ y)_{\alpha,\beta}\alpha(x) - \beta(y)[x, x]_{\alpha,\beta}, \quad \text{for all } x, y \in S. \tag{20}$$

Combining (19) and (20), we get $\beta(x \circ y)d(x) + \beta(y)[x, x]_{\alpha,\beta} = 0$, for all $x, y \in S$. Again replacing y by zy, we find that $\beta([x, z])\beta(y)d(x) = 0$, for all $x, y, z \in S$. This implies that $[\beta(x), \beta(z)]Rd(x) = (0)$, for all $x, z \in S$. Hence, proof is follows from the last paragraph of Theorem 2.1.

□

In view of Theorem 2.7, we can prove the following:

Theorem 2.8. *Let R, S be rings and M be an R-bimodule such that $xRm = (0)$ with $x \in R, m \in M$ implies that either $x = 0$ or $m = 0$. Suppose that α, β are homomorphisms of S into R such that β is one-one and onto. If (F, d) is a generalized (α, β)-derivation of S into M such that $F(x \circ y) + (x \circ y)_{\alpha,\beta} = 0$ for all $x, y \in S$, then R is commutative.*

Acknowledgments

The authors would like to thank the organizers of the Fifth China-Japan-Korea ISORT held at Tokyo, Japan, for providing warm hospitality during the symposium. The support received from DST, NBHM and INSA to attend this meeting is gratefully acknowledged.

References

1. Ashraf, M. and Wafa, S. M. Al-Shammakh, *On generalized (θ, ϕ)-derivations in rings*, Internat. J. Math. Game Theory & Algebra 12(2002), 295-300.
2. Ashraf, M. and Rehman, N., *On (σ, τ)-derivations in prime rings*, Arch. Math(Brno). 38(2002), 259-264.
3. Ashraf, M. and Rehman, N., *On commutativity of rings with derivations*, Results Math. 42(2002), 3-8.
4. Ashraf, M. and Rehman, N., *On derivations and commutativity in prime rings*, East West J. Math. 3(1)(2001), 87-91.
5. Bell, H. E. and Daif, M.N., *On derivations and commutativity in prime rings*, Acta Math. Hungar. 66(4)(1995), 337-343.
6. Bresar, M., *On the distance of the compositions of two derivations to the generalized derivations*, Glasgow Math. J. 33(1)(1991), 89-93.
7. Bresar, M. and Vukman, J., *Jordan (θ, ϕ)-derivations*, Glasnik Mat. 6(1991), 13-17.
8. Daif, M. N. and Bell, H.E., *Remarks on derivations on semiprime rings*, Internat. J. Math. Math. Sci. 15 (1)(1992), 205-206.
9. Hvala, B., *Generalized derivations in rings*, Comm. Algebra 26(1998), 1149-1166.
10. Rehman, N., *On commutativity of rings with generalized derivations*, Math. J. Okayama Univ. 44(2002), 43-49.

ON COLOCAL PAIRS

YOSHITOMO BABA
Department of Mathematics
Osaka Kyoiku University
Kashiwara, Osaka 582-8582 JAPAN
E-mail: ybaba@cc.osaka-kyoiku.ac.jp

In [10, Theorem 3.1] K. R. Fuller characterized indecomposable injective projective modules over artinian rings using i-pairs. In [3] the author generalized this theorem to indecomposable projective quasi-injective modules and indecomposable quasi-projective injective modules over artiniain rings. In [2] the author and K. Oshiro studied the above Fuller's theorem minutely. And M. Morimoto and T. Sumioka generzlized these results to modules in [17]. Further in [13] M. Hoshino and T. Sumioka extended the results in [3] to perfect rings and consider the condition "colocal pairs". Furthermore in [7] the auther studied the results in [3] from the point of view of [2] and [13] and gave results on colocal pairs. The purpose of this note is to report about this development.

Keywords: Ring; module; quasi-projective, quasi-injective

1. ON FULLER'S THEOREM

Throughout this paper, we let R be a semiperfect ring. By M_R (resp. ${}_RM$) we stress that M is a unitary right (resp. left) R-module. For an R-module M, we denote the injective hull, the Jacobson radical, the socle, the top $M/J(M)$, and the composition length of M by $E(M)$, $J(M)$, $S(M)$, $T(M)$, and $|M|$, respectively. Further we denote the right (resp. left) annihilator of T in S by $r_S(T)$ (resp. $l_S(T)$).

Definition 1.1. Let M, N be R-modules. We say that M is N-*injective* if, for any submodule X of N and any R-homomorphism $\varphi : X \to M$, there exists $\tilde{\varphi} : N \to M$ with $\tilde{\varphi}|_X = \varphi$. And we say that M is N-*simple-injective* if, for any submodule X of N and any R-homomorphism $\varphi : X \to M$ with Imφ simple, there exists $\tilde{\varphi} : N \to M$ with $\tilde{\varphi}|_X = \varphi$.

Definition 1.2. Let e, f be primitive idempotents in R and let g be an idempotent in R. We say that R satisfies $\alpha_r[e, g, f]$ if $r_{gRf}l_{eRg}(X) = X$ for any right fRf-submodule X of gRf with $r_{gRf}(eRg) \subseteq X$. And we say that R satisfies $\alpha_l[e,g,f]$ if $l_{eRg}r_{gRf}(Y) = Y$ for any left eRe-submodule Y of eRg with $l_{eRg}(gRf) \subseteq Y$. Further we say that (eR, Rf) is an *injective pair* (abbreviated *i-pair*) if $S(eR_R) \cong T(fR_R)$ and $S({}_RRf) \cong T({}_RRe)$.

The following theorem is given by K. R. Fuller in [10]. By this theorem, indecomposable projective injective right R-modules over right artinian rings are characterized using i-pairs.

Theorem 1.1 ([10, Theorem 3.1).] *Let R be a right artinian ring and let e be an idempotent in R. Then the following are equivalent.*

(a) *eR_R is injective.*
(b) *For each e_i in a basic set of idempotents for e, there is a primitive idempotent f_i in R such that (e_iR, Rf_i) is an i-pair.*
(c) *There exists an idempotent f in R such that*
  (i) *$l_{eR}(Rf) = 0 = r_{Rf}(eR)$;*
  (ii) *eRf defines a duality between the category of finitely generated left eRe-modules and the category of finitely generated right fRf-modules.*

Moreover, if eR_R is injective, then the ${}_RRf_i$ of (b) and the ${}_RRf$ of (c) are also injective.

In [2] Theorem 2 is minutely studied by the author and K. Oshiro over semiprimary rings as follows.

Theorem 1.2 ([2, Theorems 1,2,3 and Corollary 1).] *Let R be a semiprimary ring.*

(I) *Let e be a primitive idempotent in R. Then the following are equivalent.*
  (a) *eR_R is injective.*
  (b) (i) *There exists a primitive idempotent f in R with (eR, Rf) an i-pair.*
    (ii) *R satisfies $\alpha_r[e,1,f]$.*

(II) *Let e, f be primitive idempotents in R with (eR, Rf) an i-pair.*
  (1) *If ACC holds on right annihilator ideals, then $\alpha_r[e,1,f]$ holds.*
  (2) *The following are equivalent.*
    (a) $|{}_{eRe}eR| < \infty$.
    (b) $|Rf_{fRf}| < \infty$.

(c) *Both eR_R and $_RRf$ are injective.*

(III) *Let e, f be primitive idempotents in R with (eR, Rf) an i-pair and $|_{eRe}eR| < \infty$. Then the folloiwg hold.*

(1) *$_{eRe}eRf$ and eRf_{fRf} are injective.*

(2) *There exists a duality between the category of finitely generated left eRe-modules and the category of finitely generated right fRf-modules*

Theorem 3 is more simply provod in [17] by M. Morimoto and T. Sumioka and is further considered over perfect rings in [13] by M. Hoshino and T. Sumioka as follows.

Theorem 1.3 ([13, Theorems 3.6, 3.7). *] Let R be a left perfect ring.*

(I) *Let e be primitive idempotents in R. The following are equivalent.*

(a) *eR_R is R-simple-injective.*

(b) *There exists a primitive idempotent f in R such that*

(i) *(eR, Rf) is an i-pair.*

(ii) *R satisfies $\alpha_r[e, 1, f]$.*

(II) *Let e, f be primitive idempotents in R with (eR, Rf) an i-pair. Then the following are equivalent.*

(a) $|_{eRe}eR| < \infty$.

(b) $|Rf_{fRf}| < \infty$.

(c) *Both eR_R and $_RRf$ are injective.*

2. COLOCAL PAIRS OF MODULES

The results in §1 is generalized to module theory in [17] by M. Morimoto and T. Sumioka as follows.

Definition 2.1. We say that a module M is *colocal* if $S(M)$ is simple and essential in M.

Definition 2.2. Let S, T be rings and $_SM$, N_T, $_SU_T$ be a left S-module, a right T-module and a left S-right T-bimodule, respectively. We say that a map $\varphi : M \times N \to U$ is *S-T-bilinear map* if it satisfies the following three conditions for any $x, x_1, x_2 \in M$, $y, y_1, y_2 \in N$, $s \in S$ and $t \in T$.

(i) $\varphi(x_1 + x_2, y) = \varphi(x_1, y) + \varphi(x_2, y)$.

(ii) $\varphi(x, y_1 + y_2) = \varphi(x, y_1) + \varphi(x, y_2)$.

(iii) $\varphi(sx, yt) = s\varphi(x, y)t$.

And we say that (M, N) is a *colocal pair* (*with respect to* φ) if M, N have the φ and both ${}_SU$ and U_T are colocal.

The following are a generalization of Thoerem 1.2 (I),(II).

Theorem 2.1 ([17, Theorems 2.7, 2.9). *] Let R be a semiprimary ring, let f be a primitive idempotent in R and let M be an indecomposable right R-module.*

(I) *Then the following are equivalent.*

  (*a*) *M is injective with $S(M_R) \cong T(fR_R)$.*

  (*b*) (*i*) *$({}_{\mathrm{End}(M_R)}M,\ Rf_{fRf})$ is colocal pair with respect to $\varphi : M \times Rf \to Mf,\ (m, rf) \mapsto mrf$.*

   (*ii*) *${}_{\mathrm{End}(M_R)}Mf_{fRf}$ is faithful both as a left* End(M_R)-*module and as a right fRf-module.*

   (*iii*) *$r_{Rf}l_M(X) = X$ for any submodule X of Rf_{fRf}.*

(II) *Suppose that*

  (1) *$({}_{\mathrm{End}(M_R)}M,\ Rf_{fRf})$ is a colocal pair with respect to $\varphi : M \times Rf \to Mf,\ (m, rf) \mapsto mrf$,*

  (2) *${}_{\mathrm{End}(M_R)}Mf$ is faithful both as a left* End(M_R)*-module and as a right fRf-module.*

 Then the following are equivalent.

  (*a*) *$|Mf_{fRf}| < \infty$.*

  (*b*) (*i*) *$l_M r_{Rf}(X) = X$ for any submodule X of ${}_{\mathrm{End}(M_R)}M$.*

   (*ii*) *$r_{Rf}l_M(Y) = Y$ for any submodule Y of Rf_{fRf}.*

  (*c*) (*i*) *M_R is injective.*

   (*ii*) *$l_M r_{Rf}(X) = X$ for any submodule X of ${}_{\mathrm{End}(M_R)}M$.*

Further in [14] M. Hoshino and T. Sumioka generalized Theorem 2.1 and gave several sufficient conditions for colocal modules over a left or right perfect rings to be injective.

Theorem 2.2 ([14, Theorems 4.2, 4.4). *]*

(I) *Let R be a semiperfect ring and let L be a colocal right R-module with finite Loewy length. Let f be a primitive idempotent in R with $S(L_R) \cong T(fR_R)$. Then the following are equivalent.*

  (*a*) *L_R is injective.*

  (*b*) (*i*) *${}_{\mathrm{End}(L_R)}Lf_{fRf}$ is colocal both as a left* End(L_R)*-module and as a right fRf-module.*

   (*ii*) *$M = r_{Rf}l_L(M)$ for any submodule M of Rf_{fRf}.*

(II) *Let R be a left or right perfect ring, let L be a colocal right R-module. Let f be a primitive idempotent in R with $S(L_R) \cong T(fR_R)$. Suppose that $|Rf/r_{Rf}(L)_{fRf}| < \infty$. Then the following are equivalent.*

(a) *L_R is injective.*

(b) (i) *${}_{\mathrm{End}(L_R)}Lf_{fRf}$ is a colocal both as a left $\mathrm{End}(L_R)$-module and right fRf-module.*

(ii) *$r_{Rf}(L) = 0$.*

(c) (i) *${}_{\mathrm{End}(L_R)}Lf_{fRf}$ is a colocal both as a left $\mathrm{End}(L_R)$-module and right fRf-module.*

(ii) *$M = r_{Rf}l_L(M)$ for any submodule M of Rf_{fRf}.*

3. QUASI-PROJECTIVE MODULES AND QUASI-INJECTIVE MODULES

On the other hand, in [3] the author generalized Theorem 2 to indecomposable projective quasi-injective modules and indecomposable quasi-projective injective modules over artiniain rings as follows.

Theorem 3.1 (). *3, Theorem 1]] Let R be a semiprimary ring and let e, f be primitive idempotents in R. Suppose that DCC holds on $\{ r_{Rf}(I) \mid {}_{eRe}I \subseteq eR\}$. Then the following are equivalent.*

(a) *eR_R is quasi-injective with $S(eR_R) \cong T(fR_R)$.*

(b) *$E(T({}_RRe))$ is quasi-projective of the form ${}_RRf/r_{Rf}(eR)$.*

(c) *$S(eR_R) \cong T(fR_R)$ and $S({}_{eRe}eRf)$ is simple.*

(d) (i) *$l_{eR}(Rf) = 0$;*

(ii) *eRf defines a Morita duality between the category of finitely generated left eRe-mdoules and the category of finitely generated right $fRf/r_{Rf}(eR)$-modules.*

Theorem 3.1 is very useful result and using this theorem "Quasi-Harada rings" are defined in [4] by K. Iwase and the author. The class of Quasi-Harada rings contains the classes of Nakayama rings, QF-rings and Harada rings. And these are studied, for instance, [5] ,]6] ,]7] and [16] .

We note that Theorem 3.1 is reproved by M. Morimoto and T. Sumioka in [17] using colocal pairs. Further they generalized Theorem 3.1 to modules as follows.

Theorem 3.2 ([17, Theorem 2.8). *] Let R be a semiprimary ring, let f be a primitive idempotent in R and let M be a right R-module. Suppose* *that*

(*i*) $({}_{\mathrm{End}(M_R)}M,\ Rf_{fRf})$ *is a colocal pair,*
(*ii*) ${}_{\mathrm{End}(M_R)}Mf$ *is faithful,*
(*iii*) $|Mf_{fRf}| < \infty$.

Then M_R *is quasi-injective with* $S(M_R) \cong T(fR_R)$.

Further, if Mf_{fRf} *is also faithful, then* M_R *is injective.*

4. COLOCAL PAIRS

We define colocal pairs which are special cases of colocal pairs defined in §2.

Definition 4.1. Let e, f be primitive idempotents in R. We say that (eR, Rf) is a *colocal pair* (abbreviated *c-pair*) if both eRf_{fRf} and ${}_{eRe}eRf$ are colocal.

In [14] M. Hoshino and T. Sumioka extended Theorem 3.1 to left perfect rings as follows.

Theorem 4.1 ([14, Theorem 6.2). *] Let* R *be a left perfect ring, let* e, f *be primitive idemptents in* R *with* $|Rf/r_{Rf}(eR)_{fRf}| < \infty$. *Then the following are equivalent.*

(*a*) eR_R *is quasi-injective with* $S(eR_R) \cong T(fR_R)$.
(*b*) $E(T({}_RRe))$ *is quasi-projective with a projective cover* Rf.
(*c*) (*i*) (eR, Rf) *is a c-pair.*
  (*ii*) $l_{eR}(Rf) = 0$.
(*d*) (*i*) ${}_{eRe}eRf$ *is colocal.*
  (*ii*) $S(eR_R) \cong T(fR_R)$.

In [8] the author characterize $\alpha_r[e, g, f]$ using a quasi-projective right R-module $eR/l_{eR}(Rf)_R$ in case that (eR, Rf) is a c-pair.

Proposition 4.1 ([8, Proposition 1.3). *] Let* e, f *be primitive idempotents in* R *and let* g *be an idempotent in* R. *Suppose that* (eR, Rf) *is a c-pair.*

(1) *Consider the following two conditions:*

  (*a*) R *satisfies* $\alpha_r[e, g, f]$.
  (*b*) $eR/l_{eR}(Rf)_R$ *is* $gR/r_{gR}(eRg)$*-simple-injective.*

  Then (*a*) $\Rightarrow$ (*b*) *holds. And, if* fRf *is a right or left perfect ring, then the converse also holds.*

(2) *The following are equivalent.*

(c) $eR/l_{eR}(Rf)_R$ *is* $gR/l_{gR}(Rf)$*-simple-injective.*

(d) (i) *The condition* (b) *holds.*

(ii) $r_{gRf}(eRg) = 0$.

Definition 4.2. Let M be an R-module. We say that M is *simple-quasi-injective* if M is M-simple-injective.

In [8] the author gave an equivalent condition for a quasi-projective module $eR/l_{eR}(Rf)$ to be simple-quasi-injective, which is a generalization of Theorem 1.3 (I).

Theorem 4.2 ([8, Theorem 1.5).] *Let R be a left perfect ring and let e, f be primitive idempotents in R with $eRf \neq 0$. The following are equivalent.*

(a) $eR/l_{eR}(Rf)_R$ *is simple-quasi-injective.*

(b) (i) (eR, Rf) *is a c-pair.*

(ii) *R satisfies* $\alpha_r[e, e, f]$.

As a corollary we have the following interesting result.

Corollary 4.1 ([8, Corollary 1.6).] *Let R be a semiprimary ring, let e, f be primitive idempotents in R with $eRf \neq 0$. Suppose that ACC holds on right annihilator ideals. Then the following are equivalent.*

(a) ${}_RRf/r_{Rf}(eR)$ *is quasi-injective.*

(b) $eR/l_{eR}(Rf)_R$ *is quasi-injective.*

(c) (eR, Rf) *is a c-pair.*

Further, in [8], the author characterized indecomposable projective simple-quasi-injective modules and indecomposable quasi-projective R-simple-injective modules, which is a generalized result of Theorem 3 (I).

Theorem 4.3 ([8, Theorem 1.7).]

(I) *Let R be a right perfect ring and let f be a primitive idempotent in R. The following are equivalent.*

(a) ${}_RRf$ *is simple-quasi-injective.*

(b) *There exists a primitive idempotent e in R such that*

(i) $S({}_RRf) \cong T({}_RRe)$,

(ii) eRf_{fRf} *is colocal,*

(iii) *R satisfies* $\alpha_l[e, f, f]$.

(II) *Let R be a left perfect ring and let e, f be primitive idempotents in R. The following are equivalent.*

(a) $eR/l_{eR}(Rf)_R$ *is R-simple-injective.*

(b) (i) $S({}_RRf)$ *is simple and essential in* ${}_RRf$ *with* $S({}_RRf) \cong T({}_RRe)$,

(ii) eRf_{fRf} *is colocal,*

(iii) R *satisfies* $\alpha_r[e, e, f]$.

Furthermore in [8] the author generalized Theorem 3 (II) to c-pairs. We note that, in the following theorem, the equivalence between (c) and (d) was already given by Hoshino and Sumioka in [14].

Theorem 4.4 ([8, Theorem 2.4).] *Let e, f be primitive idempotents in R and let g be an idempotent of R. Suppose that (eR, Rf) is a c-pair and fRf is a left perfect ring. Then the following are equivalent.*

(a) (i) $eR/l_{eR}(Rf)_R$ *is* $gR/r_{gR}(eRg)$*-injective.*

(ii) ${}_RRf/r_{Rf}(eR)$ *is* $Rg/l_{Rg}(gRf)$*-injective.*

(b) (i) $eR/l_{eR}(Rf)_R$ *is* $gR/r_{gR}(eRg)$*-simple-injective.*

(ii) ${}_RRf/r_{Rf}(eR)$ *is* $Rg/l_{Rg}(gRf)$*-simple-injective.*

(c) $|\,(gRf/r_{gRf}(eRg))_{fRf}\,| < \infty$.

(d) $|\,{}_{eRe}(eRg/l_{eRg}(gRf))\,| < \infty$.

(e) *ACC holds on* $\{\, r_{gRf}(I) \mid {}_{eRe}I \subseteq eRg \,\}$

($\Leftrightarrow$ *DCC holds on* $\{\, l_{eRg}(I') \mid I'_{fRf} \subseteq gRf \,\}$).

As a corollary we obtain the following corollary. We note that, in the following theorem, the equivalence between (c) and (d) was already given by Hoshino and Sumioka in [14].

Corollary 4.2 ([14, Lemma 2.5; 8, Corollary 2.5).] *Let e, f be primitive idempotents in R. Suppose that (eR, Rf) is a c-pair and fRf is a left perfect ring. Then the following are equivalent.*

(a) *Both* $eR/l_{eR}(Rf)_R$ *and* ${}_RRf/r_{Rf}(eR)$ *are injective.*

(b) *Both* $eR/l_{eR}(Rf)_R$ *and* ${}_RRf/r_{Rf}(eR)$ *are R-simple-injective.*

(c) $|\, Rf/r_{Rf}(eR)_{fRf} \,| < \infty$.

(d) $|\, {}_{eRe}eR/l_{eR}(Rf) \,| < \infty$.

(e) *ACC holds on* $\{\, r_{Rf}(I) \mid {}_{eRe}I \subseteq eR \,\}$.

Last we give another corollary.

Corollary 4.3 ([8, Corollary 2.6).] *Let e, f be primitive idempotents in R. Suppose that (eR, Rf) is an i-pair and fRf is a left perfect ring.*

Then the following are equivalent.

(a) *Both eR_R and $_RRf$ are injective.*
(b) *Both eR_R and $_RRf$ are R-simple-injective.*
(c) $|\, Rf_{fRf} \,| < \infty$.
(d) $|\, _{eRe}eR \,| < \infty$.
(e) *ACC holds on* $\{\, r_{Rf}(I) \mid {}_{eRe}I \subseteq eR \,\}$.

References

1. F. W. Anderson and K. R. Fuller, "Rings and categories of modules (second edition)," Graduate Texts in Math. 13, Springer-Verlag (1991)
2. Y. Baba and K. Oshiro, *On a Theorem of Fuller,* J. Algebra **154** (1993), no.1, 86-94.
3. Y. Baba, *Injectivity of quasi-projective modules, projectivity of quasi-injective modules, and projective cover of injective modules,* J. Algebra **155** (1993), no.2, 415-434.
4. Y. Baba and K. Iwase, *On quasi-Harada rings,* J. Algebra **185** (1996), 415-434.
5. Y. Baba, *Some classes of QF-3 rings,* Comm. Alg. **28** (2000), no.6, 2639-2669.
6. Y. Baba, *On Harada rings and quasi-Harada rings with left global dimension at most* 2, Comm. Alg. **28** (2000), no.6, 2671-2684.
7. Y. Baba, *On self-duality of Auslander rings of local serial rings,* Comm. Alg. **30** (2002), no.6, 2583-2592.
8. Y. Baba, *On quasi-projective modules and quasi-injective modules,* Scientiae Mathematicae Japonicae **63** (2006), 113-120.
9. H. Bass, *Finitistic dimension and a homological generalization of semiprimary rings,* Trans. Amer. Math. Soc. **95** (1960), 466-486.
10. K. R. Fuller, *On indecomposable injectives over artinian rings,* Pacific J. Math **29** (1968), 343-354.
11. M. Harada, *Non-small modules and non-cosmall modules,* in "Ring Theory , Proceedings of 1978 Antwerp Conference" (F. Van Oystaeyen, Ed.), pp. 669-690, Dekker, New York 1979.
12. M. Harada, "Factor categories with applications to direct decomposition of modules," Lecture Note in Pure and Appl. Math., Vol. 88, Dekker, New York, (1983).
13. M. Hoshino and T. Sumioka, *Injective pairs in perfect rings,* Osaka J. Math. **35** (1998), no.3, 501-508.
14. M. Hoshino and T. Sumioka, *Colocal pairs in perfect rings,* Osaka J. Math. **36** (1999), no.3, 587-603.
15. T. Kato, *Self-injective rings,* Tohoku Math. J. **19** (1967), 485-494.
16. K. Koike, *Good self-duality of quasi-Harada rings and locally distributive rings,* J. Algebra **302** (2006), 613–645.
17. M. Morimoto and T. Sumioka, *Generalizations of theorems of Fuller,* Osaka J. Math. **34** (1997), 689-701.

18. M. Morimoto and T. Sumioka, *On dual pairs and simple-injective modules,* J. Algebra **226** (2000), no.1, 191-201.
19. M. Morimoto and T. Sumioka, *Semicolocal pairs and finitely cogenerated injective modules,* Osaka J. Math. **37** (2000), no.4, 801-820.
20. K. Oshiro, *Semiperfect modules and quasi-semiperfect mofules,* Osaka J. Math. **20** (1983), 337-372.
21. K. Oshiro, *Lifting modules, extending modules and their applications to QF-rings,* Hokkaido Math. J. **13** (1984), 310-338.
22. M. Rayer, "Small and Cosmall Modules," Ph.D. Dissertation, Indiana University, 1971.
23. L. E. T. Wu and J. P. Jans, *On quasi-projectives,* Illinois J. Math. **11** (1967), 439-448.

SOME NEW NEAR-RINGS FROM OLD RINGS*

Yong Uk Cho

Department of Mathematics Education, Silla University,
Pusan 617–736, Korea
E-mail: yucho@silla.ac.kr

In this paper, from the given ring $(R, +, \cdot)$, we will construct that a new kind of near-rings, that is, (e, t)-near-ring $(R, +, *)$ with given addition in R and new multiplication $*$ which is expressed in terms of the original multiplication and addition by defining $a * b$ to be a polynomial in a and b, where e and t are fixed (central) orthogonal idempotents in R. Next, we will give some examples and study related substructures of (e, t)-near-rings and those of base rings.

Keywords: near-rings, (e, t)-near-rings, LSD, RSD, right permutable, prime ideals

1. Introduction

Throughout this paper, our near-ring will be considered a right (abelian) near-ring, that is, a non-empty set R together with two binary operations "+" and "$*$" such that $(R, +)$ is a (an abelian) group, $(R, *)$ is a semigroup and $(a + b) * c = a * c + b * c$ for all a, b, c in R. A near-ring R is said to be zero-symmetric (resp. constant) if $a * 0 = 0$ (resp. $a * 0 = a$), for all $a \in R$.

A ring R is said to be right (left) permutable if $abc = acb$ $(abc = bac)$, right (left) self distributive if $abc = acbc$ $(abc = abac)$ for all a, b, c in R, medial if $abcd = acbd$ for all a, b, c, d in R.

In this paper, we will introduce some new kinds of near-rings from the given base rings, in particular, right permutable (or commutative, right self distributive, medial, periodic, left strongly π-regular) near-rings, and investigate the relation between the base ring properties and this new kind of near-rings properties.

To construct a new class of near-rings from the base ring $(R, +, .)$, the procedure is to begin with a ring $(R, +, .)$ and a new multiplication denoted

*2000 Mathematics Subject Classification. 16Y30.

by $*$ on $(R,+)$ so as to obtain a near-ring $(R,+,*)$. The new multiplication is expressed in terms of the original addition and multiplication by defining $a*b$ to be a polynomial in a and b with fixed (central) orthogonal idempotents e and t in R, for all a,b in R. We will also consider some relations of substructures of base rings and those of new kinds of near-rings, for example, subnear-rings, R-subgroups and ideals, in particular, prime like ideals of a new near-rings.

J. Clay [1] classified almost all abelian near-rings on given finite abelian groups of order ≤ 7. Our purpose is to introduce a special method for constructing a new class of abelian near-rings from arbitrary rings with (central) orthogonal idempotents.

An ideal P of a near-ring R is called *completely prime,* if whenever $ab \in P$, then $a \in P$ or $b \in P$ for $a,b \in R$. An ideal P of a near-ring R is called 3-*prime,* if whenever $aRb \in P$, then $a \in P$ or $b \in P$ for a, $b \in R$.

An ideal P of a near-ring R is called *equiprime,* if for $x \in R-P$, $a,b \in R$ $xra - xrb \in P$ implies $a-b \in P$ for every $r \in R$, this concept is defined by Groenewald [3,4] . We refer to Pilz [5] for other undefined basic concepts and notations.

2. Structures of New Near-Rings

We consider that an algebraic system $(R,+,\cdot)$ is a given ring with (central) orthogonal idempotents e and t. On this ring R, we can define a new operation $*: R \times R \longrightarrow R$ by $a*b = ae + abt$, for all $a,b \in R$. We will call this operation $*$ to an (e,t)-*operation* on the base ring $(R,+,\cdot)$.

Now, we will show that the algebraic system $(R,+,*)$ is an abelian near-ring.

Lemma 2.1. *Let $(R,+,\cdot)$ be a ring (resp. commutative or right permutable ring) with central orthogonal idempotents (resp. orthogonal idempotents) e and t. If $*$ is an (e,t)-operation on R, then $(R,*)$ is a semigroup.*

Proof.

Let $a,b,c \in R$. Then

$(a*b)*c = (ae+abt)*c = (ae+abt)e + (ae+abt)ct = aee+abte+aect+abtct = ae^2+0+0+abct^2 = ae+abct$. On the other hand,

$a*(b*c) = a*(be+bct) = ae+a(be+bct)t = ae+ab0+abtt = ae+abct$. Hence $(a*b)*c = a*(b*c)$. □

Proposition 2.1. *Let $(R,+,\cdot)$ be a ring with central orthogonal idempo-*

tents e and t. If $*$ is an (e,t)-operation on R, then $(R,+,*)$ is an abelian near-ring.

Proof. Together with the Lemma 2.1, it suffices to show that $*$ is right distributive under $+$. Indeed, for any $a,b,c \in R$, we have that $(a+b)*c = (a+b)e+(a+b)ct = ae+be+act+bct$. On the other hand, $a*c+b*c = ae+act+be+bct = ae+be+act+bct$. Hence $(a+b)*c = a*c+b*c$. □

From the Proposition 2.1, $(R,+,*)$ is called the *(e,t)-near-ring* obtained from the given base ring $(R,+,\cdot)$. From now on, we denote this (e,t)-near-ring $(R,+,*)$ by R^*, and denote the base ring $(R,+,\cdot)$ by R.

We will investigate some related properties of the (e,t)-near-ring and the given base ring, and then give some examples of the (e,t)-near-ring.

Remark 2.1. On the given ring $(R,+,\cdot)$ with central orthogonal idempotents e and t, if we define a new operation $*: R\times R \longrightarrow R$ by $a*b = ea+tab$, for all $a,b \in R$, then (e,t)-near-ring $(R,+,*)$ becomes a left near-ring on the base ring $(R,+,\cdot)$.

Proposition 2.2. *Let $R := (R,+,\cdot)$ be a given ring with central orthogonal idempotents e and t. Then we have the following properties.*

(1) If R is right permutable, then so is R^*.
(2) If R is medial, then so is R^*.
(3) If R is RSD, then so is R^*.
(4) If R is LSD, then so is R^*.

Proof.

(1) Assume that R is right permutable. Then for any $a,b,c \in R$, we have that $a*b*c = (ae+abt)*c = (ae+abt)e+(ae+abt)ct = aee+abte+aect+abtct = ae+0+0+abct = ae+abct$. Also, $a*c*b = (ae+act)*b = (ae+act)e+(ae+act)bt = aee+acte+aebt+actbt = ae+0+0+acbt = ae+acbt$. Since R is right permutable, this last equality is $ae+abct$. Hence $a*b*c = a*c*b$, that is, R^* is right permutable.

(2) Assume that R is medial and let $a,b,c,d \in R$. Then using the fact in the proof of Lemma 1 or above (1), we have that $a*b*c*d = (a*b*c)*d = (ae+abct)*d = (ae+abct)e+(ae+abct)dt = aee+abcte+aedt+abctdt = ae+0+0+abctdt = ae+abcdt$ and

$a*c*b*d = (a*c*b)*d = (ae+acbt)*d = (ae+acbt)e+(ae+acbt)dt = aee + acbte + aedt + acbtdt = ae + 0 + 0 + acbtdt = ae + acbdt$. Since R is medial, $abcd = acbd$. Hence $a*b*c*d = a*c*b*d$, namely, R^* is medial.

(3) Suppose that R is RSD, that is, for any $a, b, c \in R$, $abc = acbc$. Then $a*b*c = (ae+abt)*c = (ae+abt)e+(ae+abt)ct = aee+abte+aect+abtct = ae+0+0+abct = ae+abct$, and $a*c*b*c = (ae+act)*b*c = ((ae+act)e+(ae+act)bt)*c = ((ae+act)e+(ae+act)bt)e+((ae+act)e+(ae+act)bt)ct = aee+actee+aebte+actbte+aeect+actect+aebtct+actbtct = ae+0+0+0+0+0+0+acbct = ae+acbct$. Since R is RSD, $a*b*c = a*c*b*c$. Therefore, R^* is RSD.

(4) This proof is similar to that of (3). □

Proposition 2.3. *Let $(R, +, \cdot)$ be a right permutable ring with orthogonal idempotents e and t. If $*$ is an (e,t)-operation on R, then $(R, +, *)$ is an abelian right permutable near-ring.*

Proof. The proof of this proposition is clear from the proofs of Lemma 2.1, Propositions 2.1 and 2.2. □

There are a lot of examples of (e,t)-near-rings as following:

Example 2.1. Let $\mathbb{Z}_6 = \{0, 1, 2, 3, 4, 5\}$ be the ring of integers modulo 6. Taking orthogonal idempotents $e = 3$ and $t = 4$ in $\mathbb{Z}_6$, we obtained $(3,4)$-near-ring $\mathbb{Z}_6^*$ as follows:

*	0	1	2	3	4	5
0	0	0	0	0	0	0
1	3	1	5	3	1	5
2	0	2	4	0	2	4
3	3	3	3	3	3	3
4	0	4	2	0	4	2
5	3	5	1	3	5	1

For examples, $2*3 = 2\cdot3+2\cdot3\cdot4 = 6+24 = 30 = 0(mod 6), 2*5 = 2\cdot3+2\cdot5\cdot4 = 6+40 = 46 = 4(mod 6), 5*2 = 5\cdot3+5\cdot2\cdot4 = 55 = 1(mod 6)$.

Obviously, this $(3,4)$-near-ring $\mathbb{Z}_6^*$ is right permutable and medial from Proposition 3, but not commutative, because $2*5 = 4 \neq 1 = 5*2$. Furthermore, $\mathbb{Z}_6^*$ is not zero-symmetric, because $3*0 = 3 \neq 0$.

Remark 2.2. In [5, Appendix, p.409], there are 60 kinds of scheme as near-rings from the additive group $\mathbb{Z}_6 = \{0, 1, 2, 3, 4, 5\}$ modulo 6, and the following table of endomorphisms on $(\mathbb{Z}_6, +)$:

$+$	α_0	α_1	α_2	α_3	α_4	α_5
0	0	0	0	0	0	0
1	0	1	2	3	4	5
2	0	2	4	0	2	4
3	0	3	0	3	0	3
4	0	4	2	0	4	2
5	0	5	4	3	2	1

On the ground of this endomorphism table, above $(3, 4)$-near-ring $\mathbb{Z}_6^*$ on $\mathbb{Z}_6$ is written as scheme of the form:

$$(3,\ 1,\ 5,\ 3,\ 1,\ 5)$$

This scheme is a new one which is not appeared at 60 kinds of schema as near-rings in Pilz [5].

There are 6^6 kinds of scheme which are mappings on

$$\{\alpha_0, \alpha_1, \alpha_2, \alpha_3, \alpha_4, \alpha_5\}$$

Thus, in this all scheme, there exists a scheme that is not a near-ring, for example, a scheme

$$(1,\ 3,\ 5,\ 2,\ 3,\ 5)$$

is not a near-ring. Indeed this scheme (1, 3, 5, 2, 3, 5) is transformed as following Example 2:

Example 2.2. Let $\mathbb{Z}_6 = \{0, 1, 2, 3, 4, 5\}$ with addition modulo 6, from the scheme (1, 3, 5, 2, 3, 5) we define multiplication as follows:

$\cdot$	0	1	2	3	4	5
0	0	0	0	0	0	0
1	1	3	5	2	3	5
2	2	0	4	4	0	4
3	3	3	3	0	3	3
4	4	0	2	2	0	2
5	5	3	1	4	3	1

This scheme is closed and satisfies right distributive law, but does not satisfy associative law. Thus (1, 3, 5, 2, 3, 5) is not a near-ring, because $(1 \cdot 2) \cdot 3 = 5 \cdot 3 = 4 \neq 3 = 1 \cdot 4 = 1 \cdot (2 \cdot 3)$.

References

1. Clay, J. R., *The near-rings on groups of low order*, in Math. Z., **104**, (1968), pp.364–371.
2. Mason, G., *Strongly regular near-rings*, in Proc. Edinburgh Math. Soc., **23**, (1980), pp.27–35.
3. Groenewald, N.J., *Diffrent prime ideals in near-rings*, in Comm.Algebra, **10**, (1991), pp.2667–2675.
4. Groenew ald, N.J., *Prime near-rings and special radicals*, in East-West J. of Math., **3**, (2001), pp.147–162.
5. Pilz, G., *Near-Rings*, Revised Edition, North-Holland Publishing Company, Amsterdam, New York, Oxford, 1983.
6. Reddy, Y. V. and Murty, C. V. L. N., *On strongly regular near-rings*, in Proc. Edinburgh Math. Soc., **27**, (1984), pp.61–64.

ON NON-δ-M-COSINGULAR COMPLETELY $\oplus$-δ_M-SUPPLEMENTED MODULES

Derya Keskin Tütüncü

Hacettepe University, Department of Mathematics,
06800 Beytepe Ankara, Turkey
E-mail: keskin@hacettepe.edu.tr

Fatma Kaynarca

Afyonkarahisar Kocatepe University, Department of Mathematics,
03200 A.N.S. Campus Afyonkarahisar, Turkey
E-mail: fatmakaynarca@hotmail.com

Muhammet Tamer Koşan

Gebze Institute of Technology, Department of Mathematics,
41400 Gebze Kocaeli, Turkey
E-mail: mtkosan@gyte.edu.tr

In this paper we prove that any $\oplus$-δ_M-supplemented module $N \in \sigma[M]$ with SSP is completely $\oplus$-δ_M-supplemented. It is also proved that any non-δ-M-cosingular $\oplus$-δ_M-supplemented module $N \in \sigma[M]$ is (D_3) if and only if N has the SIP. Let $N \in \sigma[M]$ be any module such that $\overline{Z}_{\delta_M}(N)$ has a coclosure in N. Then we prove that N is (completely) $\oplus$-δ_M-supplemented if and only if $N = \overline{Z}^2_{\delta_M}(N) \oplus K$ for some submodule K of N such that $\overline{Z}^2_{\delta_M}(N)$ and K both are (completely) $\oplus$-δ_M-supplemented.

Keywords: δ_M-small submodule, δ_M-coclosure, (non-)δ-M-cosingular module, δ_M-supplemented module, $\oplus$-δ_M-supplemented module.

1. Introduction

Throughout this paper all rings are associative with identity and all modules are unitary right modules.

Let M be any R-module. Any R-module N is *generated by* M or *M-generated* if there exists an epimorphism $M^{(\Lambda)} \longrightarrow N$ for some indexed set Λ. An R-module N is said to be *subgenerated* by M if N is isomorphic to a submodule of an M-generated module. We denote by $\sigma[M]$ the full subcategory of the right R-modules whose objects are all right R-modules subgenerated by M. A module N in $\sigma[M]$ is called *M-singular* if $N \cong L/K$ for an $L \in \sigma[M]$ and K is essential in L.

Let M be a module and $A \leq N \in \sigma[M]$. A is called a *small submodule* of N if whenever $N = A + X$ for $X \leq N$ we have $N = X$ and it is denoted by $A \ll N$. A is called a *δ_M-small submodule* of N if whenever $N = A + X$ and N/X is M-singular for $X \leq N$ we have $N = X$ and we denote it by $A \ll_{\delta_M} N$. Note that δ_M-small submodules are the generalization of δ-small submodules in the category of *Mod-R* defined by Zhou in[9]. Clearly, if A is a small submodule of $N \in \sigma[M]$, then $A \ll_{\delta_M} N$.

Let N be any module and $A \leq B \leq N$. A is called a *coessential submodule* of B in N if $B/A \ll N/A$. A submodule A of N is said to be *coclosed* in N if it has no proper coessential submodule in N. Let $A \leq B \leq N$. If $B/A \ll N/A$ and A is a coclosed submodule of N, then we say that A is a *coclosure* of B in N. Now, let $A \leq B \leq N \in \sigma[M]$. A is called a *$\delta_M$-coessential submodule* of B in N if $B/A \ll_{\delta_M} N/A$ and we denote it by " $A \subseteq_{ce}^{\delta_M} B$ in N ". A submodule A of N is said to be *δ_M-coclosed* in N if it has no proper δ_M-coessential submodule in N and we denote it by " $A \subseteq_{cc}^{\delta_M} N$ ". Let $A \leq B \leq N \in \sigma[M]$. If $A \subseteq_{ce}^{\delta_M} B$ in N and $A \subseteq_{cc}^{\delta_M} N$, then we say that A is a *δ_M-coclosure* of B in N. By the proof of [4, Lemma 1.6], A is a δ_M-coclosure of B in N if and only if A is a minimal δ_M-coessential submodule of B in N (because the class of M-singular modules is closed under homomorphic images).

Let $N \in \sigma[M]$. N is called an *M-small module in $\sigma[M]$* if N is a small submodule of $L \in \sigma[M]$. N is called a *δ-M-small module in $\sigma[M]$* if $N \ll_{\delta_M} L \in \sigma[M]$. Note that a module N in $\sigma[M]$ is a δ-M-small module if and only if $N \ll_{\delta_M} \widehat{N}$, where $\widehat{N}$ is the M-injective hull of N (or equivalently injective hull of N in $\sigma[M]$)(see[6]). Note that the class of δ-M-small modules is closed under homomorphic images.

Let $N \in \sigma[M]$. In[7], Talebi and Vanaja define the submodule $\overline{Z}_M(N)$ as follows:

$$\overline{Z}_M(N) = \cap\{Ker g \mid g : N \longrightarrow S, S \in \mathcal{S}\}$$

where $\mathcal{S}$ is the class of M-small modules. They call any module $N \in \sigma[M]$ *M-cosingular* (*non-M-cosingular*) if $\overline{Z}_M(N) = 0$ ($\overline{Z}_M(N) = N$). In[6], inspired by this definition Özcan defines the submodule $\overline{Z}_{\delta_M}(N)$ of N as

$$\overline{Z}_{\delta_M}(N) = \cap\{Ker g \mid g : N \longrightarrow T, T \in \mathcal{DM}\}$$

where $\mathcal{DM}$ is the class of δ-M-small modules. Clearly, $\overline{Z}_{\delta_M}(N) \subseteq \overline{Z}_M(N)$. Any module $N \in \sigma[M]$ is called a *δ-M-cosingular* (*non-δ-M-cosingular*) *module* if $\overline{Z}_{\delta_M}(N) = 0$ ($\overline{Z}_{\delta_M}(N) = N$). Clearly, every δ-M-small module is δ-M-cosingular. Note that the class of non-δ-M-cosingular modules is closed under homomorphic images just like the class of non-M-cosingular modules.

Let $N \in \sigma[M]$. We set $\overline{Z}^0_{\delta_M}(N) = N$, $\overline{Z}^1_{\delta_M}(N) = \overline{Z}_{\delta_M}(N)$ and define inductively $\overline{Z}^\alpha_{\delta_M}(N)$ for any ordinal α. Thus, if α is not a limit ordinal we set $\overline{Z}^\alpha_{\delta_M}(N) = \overline{Z}_{\delta_M}(\overline{Z}^{\alpha-1}_{\delta_M}(N))$, while if α is a limit ordinal we set $\overline{Z}^\alpha_{\delta_M}(N) = \cap_{\beta<\alpha}\overline{Z}^\beta_{\delta_M}(N)$. Hence there is a descending sequence

$$N = \overline{Z}^0_{\delta_M}(N) \supseteq \overline{Z}^1_{\delta_M}(N) \supseteq \overline{Z}^2_{\delta_M}(N) \supseteq \dots$$

of submodules of N (see[6]).

Let M be any module. M has the *summand sum property (SSP)* if the sum of any two direct summands of M is a direct summand of M and M has the *summand intersection property (SIP)* if the intersection of any two direct summands of M is a direct summand of M.

Let N be any module and $X \leq N$. A *supplement* of X in N is a submodule K of N with $N = X + K$ and $X \cap K \ll K$. A submodule K of N is called a *supplement submodule* of N provided there exists a submodule X of N such that K is a supplement of X in N. If every submodule of N has a direct summand supplement in N, then N is called *$\oplus$-supplemented.* We call any module N *completely $\oplus$-supplemented* if every direct summand of N is $\oplus$-supplemented. Now, let $N \in \sigma[M]$ and $X \leq N$. A *δ_M-supplement* of X in N is a submodule K of N with $N = X + K$ and $X \cap K \ll_{\delta_M} K$. A submodule K of N is called a *δ_M-supplement submodule* of N provided there exists a submodule X of N such that K is a δ_M-supplement of X in N. If every submodule of N has a direct summand δ_M-supplement in N, then N is called *$\oplus$-δ_M-supplemented.* We call any module N in $\sigma[M]$ *completely $\oplus$-δ_M-supplemented* if every direct summand of N is $\oplus$-δ_M-supplemented. Any module M is called a *(D_3)-module* if for every direct summands K and L of M with $M = K + L$, $K \cap L$ is a direct summand of M.

In Section 2, we study completely $\oplus$-δ_M-supplemented modules. In particular, we prove that any $\oplus$-δ_M-supplemented module with the SSP is completely $\oplus$-δ_M-supplemented. Let $N \in \sigma[M]$ be a non-δ-M-cosingular $\oplus$-δ_M-supplemented module. We prove that M is (D_3) if and only if N has the SIP.

In Section 3, we decompose any $\oplus$-(δ_M-) supplemented module $N \in \sigma[M]$ in terms of the submodule $\overline{Z}^2_{\delta_M}(N)$. In particular, we prove that if $N \in \sigma[M]$ is any module such that $\overline{Z}_{\delta_M}(N)$ has a coclosure in N, then N is (completely) $\oplus$-δ_M-supplemented if and only if $N = \overline{Z}^2_{\delta_M}(N) \oplus K$ for some submodule K of N such that K and $\overline{Z}^2_{\delta_M}(N)$ are (completely) $\oplus$-δ_M-supplemented. We also characterize lifting modules in terms of $\overline{Z}^2_{\delta_M}(N)$.

2. Non-δ-M-Cosingular $\oplus$-δ_M-Supplemented Modules

The following lemma can be obtained from the proof of [9, Lemma 1.3].

Lemma 2.1. *Let $N \in \sigma[M]$.*

(1) *Let $K \leq L \leq N$. Then $L \ll_{\delta_M} N$ if and only if $K \ll_{\delta_M} N$ and $K \subseteq^{\delta_M}_{ce} L$ in N.*
(2) *Let $K, L \leq N$. Then $K + L \ll_{\delta_M} N$ if and only if $K \ll_{\delta_M} N$ and $L \ll_{\delta_M} N$.*
(3) *If $f : N \longrightarrow L$ is a homomorphism and $K \ll_{\delta_M} N$, then $f(K) \ll_{\delta_M} L$.*
(4) *If $K \leq L$, L is a direct summand of N and $K \ll_{\delta_M} N$, then $K \ll_{\delta_M} L$.*

Lemma 2.2. *Let $N \in \sigma[M]$.*

(1) *If $A \subseteq^{\delta_M}_{cc} N$, then for all $X \leq A$, $X \ll_{\delta_M} N$ implies that X is a small submodule of A and hence $X \ll_{\delta_M} A$.*
(2) *If $A \subseteq^{\delta_M}_{ce} B$ in N and $B \subseteq^{\delta_M}_{ce} C$ in N, then $A \subseteq^{\delta_M}_{ce} C$ in N.*
(3) *If $\overline{Z}_{\delta_M}(N)$ has a δ_M-coclosure in N, then $\overline{Z}_{\delta_M}(N)$ has a coclosure in N.*

Proof. (1) Let $A \subseteq^{\delta_M}_{cc} N$ and $X \leq A$ with $X \ll_{\delta_M} N$. Let $A = X + L$. We want to show that $L \subseteq^{\delta_M}_{ce} A$ in N. Let $N/L = A/L + H/L$ for any submodule H of N with $L \leq H$ and N/H M-singular. Then $N = A + H = X + H$ implies that $N = H$ since $X \ll_{\delta_M} N$. Therefore $L \subseteq^{\delta_M}_{ce} A$ in N. Since $A \subseteq^{\delta_M}_{cc} N$, $L = A$, as required.

(2) Let $N/A = C/A + L/A$ for any submodule L of N with $A \leq L$ and N/L M-singular. Then $N = C + L$ implies $N = L + B$ since $B \subseteq^{\delta_M}_{ce} C$ in N. Therefore $N = L$ since $A \subseteq^{\delta_M}_{ce} B$ in N.

(3) Assume that $A \subseteq^{\delta_M}_{ce} \overline{Z}_{\delta_M}(N)$ in N and $A \subseteq^{\delta_M}_{cc} N$. Then A is a coclosed submodule of N. Now we prove that $\overline{Z}_{\delta_M}(N)/A \ll N/A$. Since $A \subseteq^{\delta_M}_{ce} \overline{Z}_{\delta_M}(N)$ in N, $\overline{Z}_{\delta_M}(N)/A$ is a δ-M-small module. Assume that $N/A = \overline{Z}_{\delta_M}(N)/A + L/A$ for any submodule L of N with $A \leq L$. Then $N = \overline{Z}_{\delta_M}(N) + L$. Consider the natural epimorphism $\pi : N \longrightarrow N/L$. So, $\pi(N) = N/L = \pi(\overline{Z}_{\delta_M}(N)) \leq \overline{Z}_{\delta_M}(N/L)$. Therefore N/L is non-δ-M-cosingular. Now consider the epimorphism $\alpha : N/A \longrightarrow N/L$ with $Ker(\alpha) = L/A$. Then $\alpha(N/A) = \alpha(\overline{Z}_{\delta_M}(N)/A) = N/L$. Since $\overline{Z}_{\delta_M}(N)/A$

is δ-M-small, N/L is δ-M-cosingular. Thus $N = L$. This means that A is a coclosure of $\overline{Z}_{\delta_M}(N)$ in N. □

Proposition 2.1. *Let $N \in \sigma[M]$ be a $\oplus$-δ_M-supplemented (D_3)-module. Then N is completely $\oplus$-δ_M-supplemented.*

Proof. Let $N = A \oplus A'$ and $X \leq A$. Since N is $\oplus$-δ_M-supplemented, $N = K \oplus K' = K + X$ and $K \cap X \ll_{\delta_M} K$ for some submodules K and K' of N. Then $A = X + (A \cap K)$. Since N is (D_3), $A \cap K$ is a direct summand of K and A. $X \cap (A \cap K) = X \cap K \ll_{\delta_M} K$ implies that $X \cap K \ll_{\delta_M} A \cap K$ by Lemma 2.1. Therefore A is $\oplus$-δ_M-supplemented. □

Theorem 2.1. *Let $N \in \sigma[M]$.*

(1) *Assume that N is $\oplus$-δ_M-supplemented and E is a submodule of N. If for every direct summand T of N, $(E+T)/E$ is a direct summand of N/E, then N/E is $\oplus$-δ_M-supplemented. In particular, if N has the SSP, then N is completely $\oplus$-δ_M-supplemented.*

(2) *Assume that N is (D_3). Let D and E be direct summands of N such that $N/(D \cap E)$ is non-δ-M-cosingular. If N/E is $\oplus$-δ_M-supplemented, then $(D + E)/E$ is a direct summand of N/E.*

(3) *Assume that N is non-δ-M-cosingular (D_3). If N is (completely) $\oplus$-δ_M-supplemented, then it has the SSP.*

Proof. (1) Consider $A/E \leq N/E$. By hypothesis, A has a direct summand δ_M-supplement B in N. Then $(B + E)/E$ is a δ_M-supplement of A/E in N/E (by Lemma 2.1) which is a direct summand of N/E by hypothesis. Any direct summand of N is isomorphic to N/E for some direct summand E of N. Therefore if N has the SSP, then N is completely $\oplus$-δ_M-supplemented.

(2) By hypothesis, $N/E = K/E + (D + E)/E$ where K/E is a direct summand of N/E and $(D+E)/E \cap K/E \ll_{\delta_M} N/E$. That is $(E+(D\cap K))/E$ is a δ-M-small submodule of N/E and hence is a δ-M-small module. Then $(D \cap K)/(D \cap E) \cong (E + (D \cap K))/E$ is also a δ-M-small module and since $N/(D \cap E)$ is non-δ-M-cosingular, $D \cap K = D \cap E$ and hence $N/E = (D + E)/E \oplus K/E$.

(3) Since N is (D_3) $\oplus$-δ_M-supplemented, N is completely $\oplus$-δ_M-supplemented by Proposition 2.1. Suppose D and E are direct summands of N. Then N/E (being isomorphic to a direct summand of N) is $\oplus$-δ_M-supplemented and $N/(D \cap E)$ is non-δ-M-cosingular. By (2), $(D + E)/E$ is a direct summand of N/E and hence $D + E$ is a direct summand of N. □

Proposition 2.2. *Let M have the SSP with (D_3). Then M has the SIP.*

Proof. By [1, Lemma 19(2)]. □

Corollary 2.1. *Let $N \in \sigma[M]$ be a non-δ-M-cosingular $\oplus$-δ_M-supplemented module. Then N is (D_3) if and only if N has the SIP.*

Proof. Clear by Theorem 2.1 and Proposition 2.2. □

In Corollary 2.1, the non-δ-M-cosingularity condition is essential as we see in the following example.

Example 2.1. (see also, [1, Example 1]) Let F be a field and let R denote the ring

$$R = \{ \begin{bmatrix} a & x & 0 & 0 \\ 0 & b & 0 & 0 \\ 0 & 0 & b & y \\ 0 & 0 & 0 & a \end{bmatrix} \mid a, b, x, y \in F\}.$$

Then the right R-module R_R is (D_3) since every quasi-projective module is (D_3). The Jacobson radical J of R consists of all matrices in R with a zero diagonal, and $R/J \cong FxF$. Therefore J is nonzero and hence R is not a right V-ring. By [7, Corollary 2.6], R_R is not non-R-cosingular, and hence it is not non-δ-R-cosingular. We also know that R is a QF-ring. Hence R_R is $\oplus$-supplemented and hence $\oplus$-δ_M-supplemented. Consider the right ideals $K = \{ \begin{bmatrix} 0 & 0 & 0 & 0 \\ 0 & b & 0 & 0 \\ 0 & 0 & b & x \\ 0 & 0 & 0 & 0 \end{bmatrix} \mid b, x \in F\}$ and $N = \{ \begin{bmatrix} 0 & b & 0 & 0 \\ 0 & b & 0 & 0 \\ 0 & 0 & b & x \\ 0 & 0 & 0 & 0 \end{bmatrix} \mid b, x \in F\}$ of R. Clearly, K and N are direct summands of R_R. But since $N \cap K = \{ \begin{bmatrix} 0 & 0 & 0 & 0 \\ 0 & 0 & 0 & 0 \\ 0 & 0 & 0 & x \\ 0 & 0 & 0 & 0 \end{bmatrix} \mid x \in F\}$ is a nilpotent right ideal of R, it is not a direct summand of R_R. Thus R_R does not have the SIP.

3. $\oplus$-δ_M-Supplemented Modules and $\overline{Z}^2_{\delta_M}$

Proposition 3.1. *Let $N \in \sigma[M]$ and $U \leq N$ such that $f(U) \leq U$ for each homomorphism $f : N \longrightarrow N$. If N is $\oplus$-δ_M-supplemented, then N/U is*

$\oplus$-δ_M-supplemented. If moreover, U is a direct summand of N, then U is also $\oplus$-δ_M-supplemented.

Proof. Let T be any direct summand of N. Then $N = T \oplus T'$ for some submodule T' of N. Then it is easy to see that $U = (U \cap T) \oplus (U \cap T')$. Thus we have

$$(T+U) \cap (T'+U) \leq [(T+U+T') \cap U] + [(T+U) \cap T'].$$

Hence

$$(T+U) \cap (T'+U) \leq U + [(T+U \cap T + U \cap T') \cap T'].$$

Therefore $(T+U) \cap (T'+U) \leq U$ and $(T+U)/U \oplus (T'+U)/U = N/U$. By Theorem 2.1, N/U is $\oplus$-δ_M-supplemented.

Now suppose that U is a direct summand of N. Let V be a submodule of U. Then there exist submodules K and K' of N such that $N = K \oplus K' = K+V$ and $K \cap V \ll_{\delta_M} K$. Thus $U = V + (K \cap U)$. Since $U = (U \cap K) \oplus (U \cap K')$, $K \cap U$ is a direct summand of U. Now $V \cap (K \cap U) = V \cap K \ll_{\delta_M} K$ implies that $V \cap K \ll_{\delta_M} U \cap K$ by Lemma 2.1. Therefore U is $\oplus$-δ_M-supplemented. □

Theorem 3.1. *Any finite direct sum of $\oplus$-δ_M-supplemented modules in $\sigma[M]$ is $\oplus$-δ_M-supplemented.*

Proof. Let n be any positive integer and N_i a $\oplus$-δ_M-supplemented module in $\sigma[M]$ for each $1 \leq i \leq n$. Let $N = N_1 \oplus \cdots \oplus N_n$. To prove that N is $\oplus$-δ_M-supplemented it is sufficient by induction on n to prove this in the case when $n = 2$. Thus suppose $n = 2$.

Let L be any submodule of N. Let H be a δ_M-supplement of $N_2 \cap (N_1 + L)$ in N_2 such that H is a direct summand of N_2. Then $N = H + (N_1 + L)$ and $H \cap (N_1 + L) \ll_{\delta_M} H$. Let K be a δ_M-supplement of $N_1 \cap (L + H)$ in N_1 such that K is a direct summand of N_1. Then $N = (H \oplus K) + L$ and by Lemma 2.1, $(H \oplus K) \cap L \leq H \cap (K + L) + K \cap (H + L) \ll_{\delta_M} N$. So, $(H \oplus K) \cap L \ll_{\delta_M} H \oplus K$ again by Lemma 2.1. Therefore $H \oplus K$ is a direct summand δ_M-supplement of L in N. □

Theorem 3.2. *Let $N \in \sigma[M]$ and $\alpha \geq 1$ any ordinal. Then*

$$\overline{Z}^{\alpha+1}_{\delta_M}(N) = \cap\{X \mid X \subseteq^{\delta_M}_{ce} \overline{Z}^{\alpha}_{\delta_M}(N) \text{ in } N\} = \cap\{X \mid \overline{Z}^{\alpha}_{\delta_M}(N)/X \ll N/X\}$$

and the family of all (δ_M)-coessential submodules of $\overline{Z}^{\alpha}_{\delta_M}(N)$ in N is closed under finite intersections.

Proof. By the same proof as the proof of [7, Theorem 3.3]. $\square$

Corollary 3.1. *Let $N \in \sigma[M]$. If $\overline{Z}_{\delta_M}(N)$ has a (δ_M)-coclosure in N, then $\overline{Z}^2_{\delta_M}(N)$ is the unique (δ_M)-coclosure of $\overline{Z}_{\delta_M}(N)$ and hence is the largest non-δ-M-cosingular submodule of N.*

Proof. We give the proof when $\overline{Z}_{\delta_M}(N)$ has a δ_M-coclosure. When $\overline{Z}_{\delta_M}(N)$ has a coclosure the same proof is repeated by taking coclosure of $\overline{Z}_{\delta_M}(N)$ in N instead of δ_M-coclosure of $\overline{Z}_{\delta_M}(N)$ in N. Let S and T both be δ_M-coclosures of $\overline{Z}_{\delta_M}(N)$ in N. By Theorem 3.2, $S \cap T \subseteq_{ce}^{\delta_M} \overline{Z}_{\delta_M}(N)$ in N and so $S = T$ since S and T are both δ_M-coclosed. Therefore δ_M-coclosure of $\overline{Z}_{\delta_M}(N)$ is unique. If $X \subseteq_{ce}^{\delta_M} \overline{Z}_{\delta_M}(N)$ in N, then since S is the minimal δ_M-coessential submodule of $\overline{Z}_{\delta_M}(N)$ in N, $S \leq X$ and so $S \leq \overline{Z}^2_{\delta_M}(N)$ by Theorem 3.2. Thus $\overline{Z}^2_{\delta_M}(N)$ is the unique δ_M-coclosure of $\mathbb{Z}$ in N, because $S = \overline{Z}^2_{\delta_M}(N)$. Now if $X \subseteq_{ce}^{\delta_M} \overline{Z}^2_{\delta_M}(N)$ in N, then since $\overline{Z}^2_{\delta_M}(N) \subseteq_{ce}^{\delta_M} \overline{Z}_{\delta_M}(N)$ in N, we have $X \subseteq_{ce}^{\delta_M} \overline{Z}_{\delta_M}(N)$ in N by Lemma 2.2. By Theorem 3.2, $\overline{Z}^3_{\delta_M}(N) = \overline{Z}^2_{\delta_M}(N)$ and so $\overline{Z}^2_{\delta_M}(N)$ is the largest non-δ-M-cosingular submodule of N. $\square$

Let $N \in \sigma[M]$. Note that for every direct summand A of N, $\overline{Z}^2_{\delta_M}(A) = \overline{Z}^2_{\delta_M}(N) \cap A$ since it can be easily seen that $\overline{Z}_{\delta_M}(\oplus_{i \in I} A_i) = \oplus_{i \in I} \overline{Z}_{\delta_M}(A_i)$ (see, for example, [7, Proposition 2.1(4)]). Then for each decomposition $N = N_1 \oplus N_2$ of N we have that $\overline{Z}^2_{\delta_M}(N) = [\overline{Z}^2_{\delta_M}(N) \cap N_1] \oplus [\overline{Z}^2_{\delta_M}(N) \cap N_2]$. Therefore if N is $\oplus$-δ_M-supplemented, then $N/\overline{Z}^2_{\delta_M}(N)$ is $\oplus$-δ_M-supplemented and if moreover $\overline{Z}^2_{\delta_M}(N)$ is a direct summand of N, then $\overline{Z}^2_{\delta_M}(N)$ is also $\oplus$-δ_M-supplemented by Proposition 3.1. Now we can give the following Theorem:

Theorem 3.3. *Let $N \in \sigma[M]$ be any module such that $\overline{Z}_{\delta_M}(N)$ has a coclosure in N. Then N is $\oplus$-δ_M-supplemented if and only if $N = \overline{Z}^2_{\delta_M}(N) \oplus K$ for some submodule K of N such that $\overline{Z}^2_{\delta_M}(N)$ and K both are $\oplus$-δ_M-supplemented.*

Proof. *Sufficiency:* By Theorem 3.1.

Necessity: Since N is $\oplus$-δ_M-supplemented, there exist submodules K and K' of N such that $N = K \oplus K' = K + \overline{Z}^2_{\delta_M}(N)$ and $K \cap \overline{Z}^2_{\delta_M}(N) = \overline{Z}^2_{\delta_M}(K) \ll_{\delta_M} K$. Then $\overline{Z}^2_{\delta_M}(K)$ is δ-M-small and so δ-M-cosingular. On the other hand, by Corollary 3.1, $\overline{Z}^2_{\delta_M}(N)$ is a non-δ-M-cosingular submodule of N. So, $\overline{Z}^2_{\delta_M}(K)$ is non-δ-M-cosingular. Hence $\overline{Z}^2_{\delta_M}(K) = 0$. Therefore $N = K + \overline{Z}^2_{\delta_M}(N) = K \oplus \overline{Z}^2_{\delta_M}(N)$. Now by the above remark, K and $\overline{Z}^2_{\delta_M}(N)$ are $\oplus$-δ_M-supplemented submodules of N. □

Note that Theorem 3.3 is true if $\overline{Z}_{\delta_M}(N)$ has a δ_M-coclosure in N by Lemma 2.2.

Corollary 3.2. *Let $N \in \sigma[M]$ be any module such that $\overline{Z}_{\delta_M}(N)$ has a coclosure (or δ_M-coclosure) in N. Then N is $\oplus$-supplemented if and only if $N = \overline{Z}^2_{\delta_M}(N) \oplus K$ for some submodule K of N such that $\overline{Z}^2_{\delta_M}(N)$ and K both are $\oplus$-supplemented.*

Proof. *Sufficiency:* By [2, Theorem 1.4].

Necessity: Since N is $\oplus$-supplemented, then it is $\oplus$-δ_M-supplemented. So, by Theorem 3.3, $N = \overline{Z}^2_{\delta_M}(N) \oplus K$ for some submodule K of N (Note that $\overline{Z}^2_{\delta_M}(N)$ is non-δ-M-cosingular and $\overline{Z}^2_{\delta_M}(K) = 0$). By [3, Proposition 2.5], K and $\overline{Z}^2_{\delta_M}(N)$ are both $\oplus$-supplemented. □

Theorem 3.4. *Let $N \in \sigma[M]$ be any module such that $\overline{Z}_{\delta_M}(N)$ has a coclosure in N. Then N is completely $\oplus$-δ_M-supplemented if and only if $N = \overline{Z}^2_{\delta_M}(N) \oplus K$ for some submodule K of N such that $\overline{Z}^2_{\delta_M}(N)$ and K both are completely $\oplus$-δ_M-supplemented.*

Proof. Assume N is completely $\oplus$-δ_M-supplemented. Then N is $\oplus$-δ_M-supplemented and by Theorem 3.3, $N = \overline{Z}^2_{\delta_M}(N) \oplus K$ for some submodule K of N. Since all direct summands of $\overline{Z}^2_{\delta_M}(N)$ and K are also direct summands of N, $\overline{Z}^2_{\delta_M}(N)$ and K are completely $\oplus$-δ_M-supplemented. Conversely, let $N = \overline{Z}^2_{\delta_M}(N) \oplus K$ for some submodule K of N with K and $\overline{Z}^2_{\delta_M}(N)$ completely $\oplus$-δ_M-supplemented. By Theorem 3.1, N is $\oplus$-δ_M-supplemented. Suppose $N = D \oplus D'$. Then $N = \overline{Z}^2_{\delta_M}(N) \oplus K$ and $\overline{Z}^2_{\delta_M}(N) = \overline{Z}^2_{\delta_M}(D) \oplus \overline{Z}^2_{\delta_M}(D')$ implies $D = \overline{Z}^2_{\delta_M}(D) \oplus T$ and $D' = \overline{Z}^2_{\delta_M}(D') \oplus T'$. $T \oplus T' \cong K$ and K is completely $\oplus$-δ_M-supplemented implies both T and T' are $\oplus$-δ_M-supplemented. $\overline{Z}^2_{\delta_M}(N)$ is completely $\oplus$-δ_M-supplemented implies both $\overline{Z}^2_{\delta_M}(D)$ and $\overline{Z}^2_{\delta_M}(D')$ are $\oplus$-δ_M-supplemented. Hence both D and D' are $\oplus$-δ_M-supplemented by Theorem 3.1. □

Note that Theorem 3.4 is true if $\overline{Z}_{\delta_M}(N)$ has a δ_M-coclosure in N by Lemma 2.2.

Corollary 3.3. *Let $N \in \sigma[M]$ be any module such that $\overline{Z}_{\delta_M}(N)$ has a coclosure (or δ_M-coclosure) in N. Then N is completely $\oplus$-supplemented if and only if $N = \overline{Z}^2_{\delta_M}(N) \oplus K$ for some submodule K of N such that $\overline{Z}^2_{\delta_M}(N)$ and K both are completely $\oplus$-supplemented.*

Proof. By the proof of Theorem 3.4 and [2, Theorem 1.4]. □

Let M be any module. M is called a *lifting module* if for every submodule A of M, there is a direct summand X of M such that $X \leq A$ and $A/X \ll M/X$. The module M is called *amply supplemented* if for all submodules K and L of M with $M = K + L$, K contains a supplement of L in M. By [8, 41.12], every lifting module is amply supplemented and by [4, Proposition 1.5], every submodule of an amply supplemented module has a coclosure. Also it is easy to see that every lifting module is (completely) $\oplus$-supplemented. Now we can characterize lifting modules in terms of $\overline{Z}^2_{\delta_M}(N)$ as follows:

Theorem 3.5. *Let $N \in \sigma[M]$. Then N is lifting if and only if $N = \overline{Z}^2_{\delta_M}(N) \oplus K$ for some submodule K of N such that K and $\overline{Z}^2_{\delta_M}(N)$ are lifting and K and $\overline{Z}^2_{\delta_M}(N)$ are relatively projective.*

Proof. Assume N is lifting. Then N is $\oplus$-supplemented and hence by Corollary 3.2, $N = \overline{Z}^2_{\delta_M}(N) \oplus K$ for some submodule K of N where $\overline{Z}^2_{\delta_M}(K) = 0$ and $\overline{Z}^2_{\delta_M}(N)$ is non-δ-M-cosingular. Let $f : \overline{Z}^2_{\delta_M}(N) \longrightarrow \overline{Z}^2_{\delta_M}(N)/B$ be the natural epimorphism and $g : K \longrightarrow \overline{Z}^2_{\delta_M}(N)/B$ any homomorphism. Define $L = \{x + y \mid x \in K,\ y \in \overline{Z}^2_{\delta_M}(N) \text{ and } g(x) = -f(y)\}$. Since $N = L + \overline{Z}^2_{\delta_M}(N)$, we may assume that L is not small in N. Since N is lifting, there is a direct summand X of N such that $N = X \oplus Y$, $X \leq L$ and $L/X \ll N/X$. Then $N = X + \overline{Z}^2_{\delta_M}(N)$ and so Y is an epimorphic image of $\overline{Z}^2_{\delta_M}(N)$ and hence is non-δ-M-cosingular. By [6, Theorem 2.18] (or by the proof of [7, Theorem 3.5]), $Y \leq \overline{Z}^2_{\delta_M}(N)$. Hence $\overline{Z}^2_{\delta_M}(N) = Y \oplus (X \cap \overline{Z}^2_{\delta_M}(N))$. Then there is a direct summand Z of N such that $X = Z \oplus (X \cap \overline{Z}^2_{\delta_M}(N))$ and $N = Z \oplus \overline{Z}^2_{\delta_M}(N)$. Let $\pi : N \longrightarrow \overline{Z}^2_{\delta_M}(N)$ be the projection map corresponding to $N = Z \oplus \overline{Z}^2_{\delta_M}(N)$ and let $h = \pi_{|K}$. Then $fh = g$ and so K is $\overline{Z}^2_{\delta_M}(N)$-projective. Now we prove that $\overline{Z}^2_{\delta_M}(N)$

is K-projective. Since $\overline{Z}^2_{\delta_M}(K) = 0$, $\overline{Z}^2_{\delta_M}(K/L) = 0$ for any $L \leq K$, by [6, Theorem 2.18]. Since $\overline{Z}^2_{\delta_M}(N)$ is non-δ-M-cosingular, each of its epimorphic images is also and so $\mathrm{Hom}(\overline{Z}^2_{\delta_M}(N), K/L) = 0$. Thus $\overline{Z}^2_{\delta_M}(N)$ is K-projective. Conversely, let $N = \overline{Z}^2_{\delta_M}(N) \oplus K$ for some submodule K of N with K and $\overline{Z}^2_{\delta_M}(N)$ relatively projective lifting modules. Then N is lifting by [5, Theorem 3.6]. □

Acknowledgments

The first and second authors would like to thank the organization committee of CJK5 for their financial supports.

References

1. M. Alkan and A. Harmanci, *On Summand Sum and Summand Intersection Property of Modules*, Turkish Journal of Mathematics (2002) **26**, 131-147.
2. A. Harmanci, D. Keskin and P.F. Smith, *On ⊕–supplemented Modules*, Acta Math. Hungar. (1999) **83(1-2)**, 161-169.
3. A. Idelhadj and R. Tribak, *On Some Properties of ⊕-supplemented Modules*, Int. J. Math. Math. Sci. (2003) **69**, 4373-4387.
4. D. Keskin, *On Lifting Modules*, Comm. Alg. (2000) **28(7)**, 3427-3440.
5. Y. Kuratomi, *On Direct Sums of Lifting Modules and Internal Exchange Property*, Comm. Alg. (2005) **33(6)**, 1795-1804.
6. A. Ç. Özcan, *The Torsion Theory Cogenerated by δ-M-small Modules and GCO-Modules*, Comm. Alg. (2007) **35**, 1-11.
7. Y. Talebi and N. Vanaja, *The Torsion Theory Cogenerated by M–small Modules*, Comm. Alg. (2002) **30(3)**, 1449-1460.
8. R. Wisbauer, *Foundations of Module and Ring Theory*, Gordon and Breach, Philadelphia 1991.
9. Y. Zhou, *Generalizations of Perfect, Semiperfect and Semiregular Rings*, Alg. Coll. (2000) **7(3)**, 305-318.

Commutative rings and zero-divisor semigroups of regular polyhedrons

Gaohua Tang

Department of Mathematics, Guangxi teacher's Education University, Nanning 530023, China
E-mail: tanggh@gxtc.edu.cn

Huadong Su

Department of Mathematics, Guangxi teacher's Education University, Nanning 530023, China
E-mail: huadongsu@sohu.com

Yangjiang Wei

Department of Mathematics, Guangxi teacher's Education University, Nanning 530023, China
E-mail: weiyangjiang2004@yahoo.com.cn

In this paper, we study commutative rings and commutative zero-divisor semigroups determined by graphs. We have completely determined the commutative rings and commutative zero-divisor semigroups of all regular polyhedrons.
2000 Mathematics Subject Classification: 20M14, 13M05, 05C90
Key Words: Zero-divisor graph, zero-divisor semigroup, commutative ring, regular polyhedron.

1. Introduction

Given a commutative ring R with multiplicative identity 1, the zero-divisor graph of R is the graph where the vertices are the nonzero zero-divisors of R, and where there is an undirected edge between two distinct vertices x and y if and only if $xy = 0$. The zero-divisor graph of R is denoted by $\Gamma(R)$. This definition of $\Gamma(R)$ first appeared in [7], where many of the most basic features of $\Gamma(R)$ are investigated. For a given connected simple graph G if there exists a commutative ring R such that $\Gamma(R) \cong G$, then we say that G has *corresponding rings*, and we call R a ring determined by the graph G.

Similarly, for any commutative semigroup S with zero element 0, there is an undirected zero-divisor graph $\Gamma(S)$ associated to S. The vertex set of $\Gamma(S)$ is the set of all nonzero zero-divisors of S and for distinct vertices x and y of $\Gamma(S)$, there is an edge connecting x and y if and only if $xy = 0$. In [1] and [2], some fundamental properties and possible structures of $\Gamma(S)$ were studied. For any commutative semigroup S, let T be the set of all zero-divisors of S. Then T is an ideal of S and in particular, it is also a semigroup with the property that all elements of T are zero-divisors. We call such semigroups *zero-divisor semigroups*. Obviously we have $\Gamma(S) \cong \Gamma(T)$. For a given connected simple graph G if there exists a zero-divisor semigroup S such that $\Gamma(S) \cong G$, then we say that G has *corresponding semigroups*, and we call S a semigroup determined by the graph G.

Zero-divisor graphs of commutative rings or commutative semigroups have been studied in several articles, such as [1], [2], [3], [4], [5], [6], [8] and [9]. As we know, there are exactly five regular polyhedrons, namely, tetrahedron, cube, octahedron, dodecahedron and icosahedron(see the following G1 to G5). In this paper, we have completely determined the commutative rings and commutative zero-divisor semigroups of all regular polyhedrons(Theorem 2.1, Corollary 2.1 and Theorem 2.5).

In this paper, all rings are commutative rings with multiplicative identity 1; all semigroups are multiplicative commutative zero-divisor semigroups with zero element 0, where $0x = 0$ for all $x \in S$; and all graphs in this paper are undirected simple and connected. If x is a vertex in a graph G, let $N(x)$ be the vertices in G adjacent to x and $\overline{N(x)} = N(x) \cup \{x\}$. For other graph theoretical notions and notations adopted in this paper, please refer to [11].

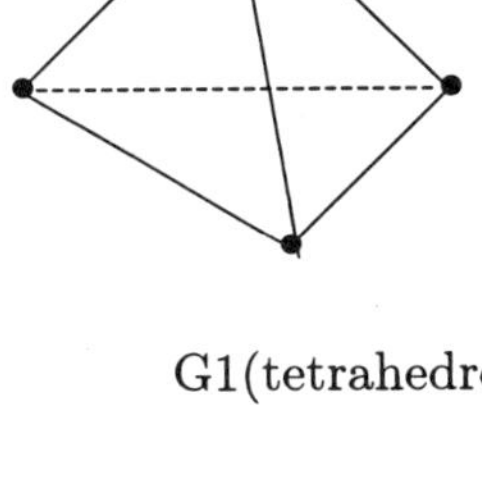

$\Longrightarrow$

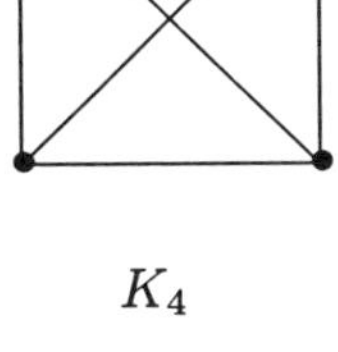

G1(tetrahedron) K_4

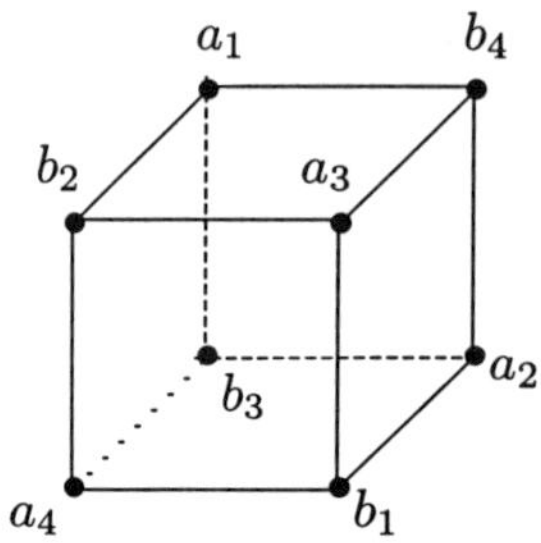

G2(cube)

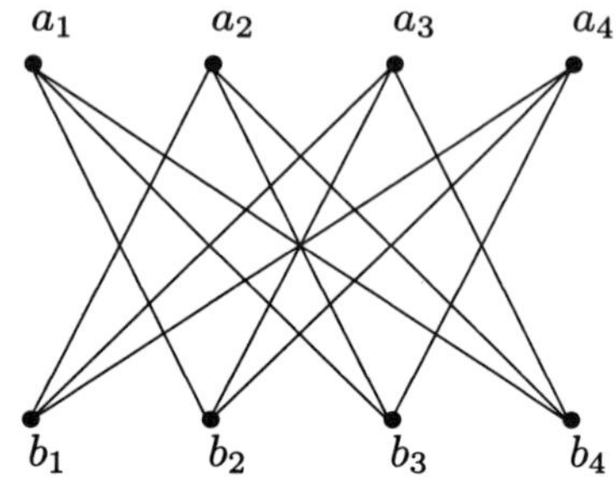

$H_{4,4}$

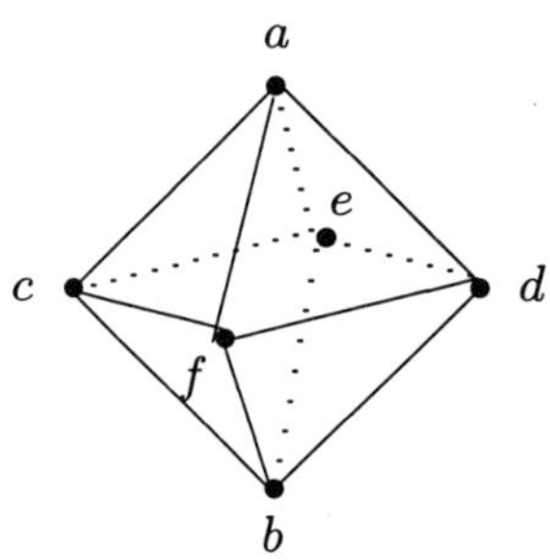

G3(octahedron)

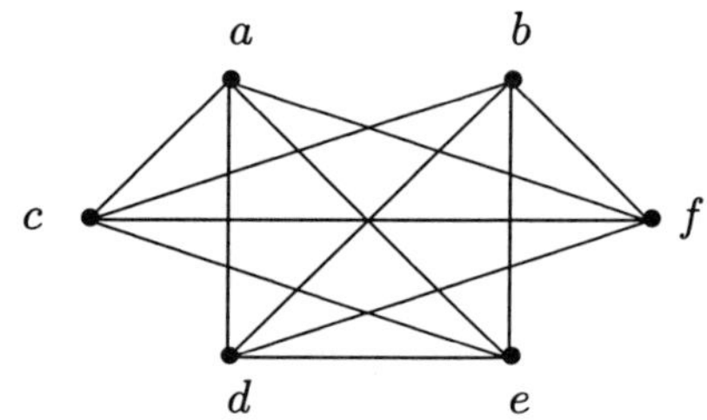

$K_{2,2,2}$

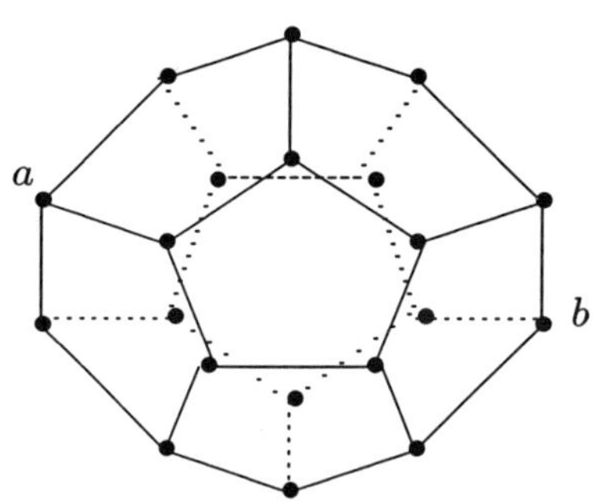

G4(dodecahedron)

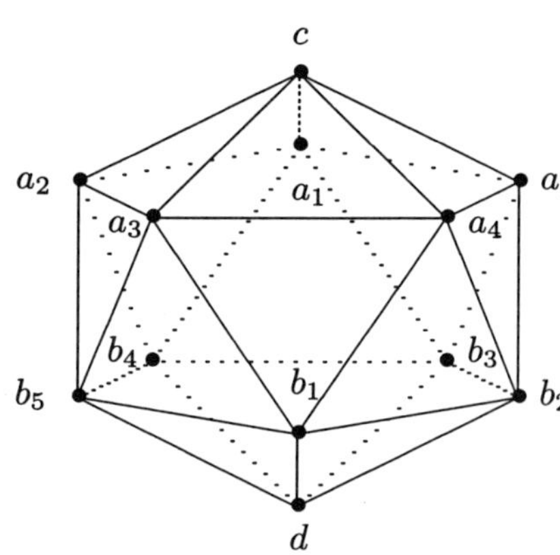

G5(icosahedron)

2. Main Results

Theorem 2.1. *(1) There are twelve corresponding semigroups and two corresponding rings to tetrahedron;*

(2) There are no corresponding semigroups and no corresponding rings to dodecahedron;

(3) There are no corresponding semigroups and no corresponding rings to icosahedron.

Proof: (1) For tetrahedron, it is a complete graph K_4. In [5], Wu have show a formula $K(n)$to calculate the number of non-isomorphic zero-divisor semigroups corresponding to the complete graph K_n and particularly $K(4) = 12$. Thus, there are twelve corresponding semigroups to tetrahedron. On the other hand, by [7, Example 2.1(b)], there are two corresponding rings Z_{25} and $Z_5[X]/(X^2)$ corresponding to K_4.

(2) From the graph of dodecahedron(see G4), we find that the distance from vertex a to vertex b is 5, i.e., $d(a,b) = 5$. So there are no corresponding rings to dodecahedron by [7, Theorem 2.1] and there are no corresponding semigroups to dodecahedron by [1, Theorem 1].

(3) Assume that there is a commutative semigroup S such that $\Gamma(S)$ is isomorphic to icosahedron, we consider the value of cd(see G5). If $cd = c$ then $b_1c = b_1cd = 0$, a contradiction. If $cd = d$ then $a_1d = a_1cd = 0$, a contradiction. If $cd = a_i$ for some i, from G5, we see that there exists some b_j such that $b_ja_i \neq 0$, but $b_jd = 0$, which implies a contradiction. If $cd = b_i$ for some i, from the graph of icosahedron, we see that there exists some a_j such that $b_ia_j \neq 0$, but $a_jc = 0$, a contradiction too. So there are no corresponding semigroups to icosahedron and therefore there are no corresponding rings to icosahedron. □

Theorem 2.2. *For any $n \geq 3$, let the connected simple graph*

$$H_{n,n} = \{a_1, \cdots, a_n; b_1, \cdots, b_n\}$$

be the subgraph of the complete bipartite graph $K_{n,n}$, which is obtained by deleting the edges $a_i - b_i (i = 1, 2, \cdots, n)$ from $K_{n,n}$. Then $H_{n,n}$ has no corresponding rings and no corresponding semigroups.

Proof: Since $N(a_1) \cup N(b_1) = \{b_2, \cdots, b_n\} \cup \{a_2, \cdots, a_n\}$, $|N(a_1) \cup N(b_1)| = 2n - 2$. But for any $v \in H_{n,n}$, $|\overline{N(v)}| = n < 2n - 2$ for $n \geq 3$. So there are no corresponding semigroups associated to $H_{n,n}$ by [1, Theorem 1(4)] and therefore there are no corresponding rings to $H_{n,n}$. □

Corollary 2.1. *The cube has no corresponding rings and no corresponding semigroups*

Proof: Since the graph of cube is $H_{4,4}$, there are no corresponding rings and no corresponding semigroups to cube by Theorem 2.2. □

Lemma 2.1. [8, Theorem 3.1], *Let R be a commutative ring. If $\Gamma(R)$ is a complete r-partite graph with $r \geq 3$, then at most one part has more than one vertex.* □

For zero-divisor semigroups of complete r-partite graphs, we have:

Theorem 2.3. *There are corresponding zero-divisor semigroups to the complete 3-partite graph $K_{m,n,p}$, for any positive integers m, n and p. In general, there are corresponding zero-divisor semigroups to the complete r-partite graph if $r \geq 3$.*

Proof: Let G be a complete 3-partite graph on $A \cup B \cup C$, and $A = \{a_1, \cdots, a_m\}$, $B = \{b_1, \cdots, b_n\}$, $C = \{c_1, \cdots, c_p\}$. We let $S = A \cup B \cup C \cup \{0\}$, and define an operation on S:

(1) $a_i a_j = a_1$, for all $i \neq j$, $a_i^2 = a_i$, for any $1 \leq i \leq m$.
(2) $b_i b_j = b_1$, for all $i \neq j$, $b_i^2 = b_i$, for any $1 \leq i \leq n$.
(3) $c_i c_j = c_1$, for all $i \neq j$, $c_i^2 = c_i$, for any $1 \leq i \leq p$.
(4) $a_i b_j = 0, a_i c_k = 0, b_j c_k = 0$, for all possible i, j, k.

Since (4), we only need to check the associative law holds in A, B and C respectively. By symmetry, we need only to check the associative law in A, namely:

$$(a_i a_j) a_k = a_i (a_j a_k), \qquad for\ \ all\ \ 1 \leq i, j, k \leq n. \qquad (\triangle)$$

Case1. If $i = j = k$, the equality obviously holds.
Case2. If there are at least two elements different among a_i, a_j and a_k, the right side of $(\triangle)$ is a_1, and the left side of $(\triangle)$ is a_1 too. Thus the equality holds.

In general, for complete r-partite graph, we can analogously define an operation on r parts. This completes our proof. □

Theorem 2.4. *Let $A = \{a_1, \cdots, a_m\}, B = \{b_1, \cdots, b_n\}, C = \{c_1, \cdots, c_p\}$ be the three parts of the complete 3-partite graph $K_{m,n,p}$. If $S = A \cup B \cup C \cup \{0\}$ is a commutative zero-divisor semigroup of the graph $K_{m,n,p}$, then*

(1) $A \cup \{0\}$, $B \cup \{0\}$ and $C \cup \{0\}$ are ideals of S;
(2) There are at most one element a_i in A such that $a_i^2 = 0$;
(3) There are at least one idempotent elements in A if $|A| \geq 2$.

Note: In the next, if there is one element a_i in A such that $a_i^2 = 0$, we always assume $a_1^2 = 0$. In this case, we conclude that $a_1 a_i = a_1$, for all $i \neq 1$. Similar discussions for B and C.

Proof: (1) Obviously.

(2) If $a_i^2 = a_j^2 = 0$, for some $i \neq j$, let us consider the value of $a_i a_j$. If $a_i a_j = a_k$, for some $k \neq j$, then $a_j a_k = 0$, a contradiction. If $a_i a_j = a_j$, then $a_i a_j = a_i^2 a_j = 0$, also a contradiction.

(3) Assume that there are no idempotent elements in A. Since $2 \leq |A| < \infty$, we can find a finite sequence $a_{i_1}, a_{i_2}, \cdots, a_{i_k}$ such that $a_{i_1}^2 = a_{i_2}, a_{i_2}^2 = a_{i_3}, \cdots, a_{i_{k-1}}^2 = a_{i_k}, a_{i_k}^2 = a_{i_1}$, therefore $(a_{i_1} a_{i_2} \cdots a_{i_k})^2 = a_{i_1}^2 a_{i_2}^2 \cdots a_{i_k}^2 = a_{i_2} a_{i_3} \cdots a_{i_k} a_{i_1} = a_{i_1} a_{i_2} \cdots a_{i_k}$ which is an idempotent, a contradiction. Therefore, our proof is finished. □

Theorem 2.5. *The octahedron has no corresponding rings but has twenty corresponding semigroups.*

Proof: Since octahedron is $K_{2,2,2}$, it has no corresponding rings by Lemma 2.1. Now, we discuss the corresponding semigroups to octahedron($K_{2,2,2}$). Let $A = \{a, b\}, B = \{c, d\}, C = \{e, f\}$ be the three parts of the graph $K_{2,2,2}$, and let $S = \{0, a, b, c, d, e, f\}$ be a zero-divisor semigroup corresponding to the graph $K_{2,2,2}$. Then, by Theorem 2.4(2), we have four cases for our discussion.

Case1. Assume $a^2 = c^2 = e^2 = 0$, then we obtain $ab = a, cd = c$ and $ef = e$. By Theorem 2.4(3), we have $b^2 = b$, $d^2 = d$ and $f^2 = f$. Hence, we have one table on $S(table1)$.

Case2. Assume $a^2 = c^2 = 0$, $e^2 \neq 0$ and $f^2 \neq 0$, then we have $ab = a, cd = c$. $b^2 = b$, $d^2 = d$.

Subcase2.1. Let $e^2 = e, f^2 = f$. If $ef = e$, then we have a table on $S(table2)$. If $ef = f$, the table is isomorphic to table2.

·	a	b	c	d	e	f
a	0	a	0	0	0	0
b	a	b	0	0	0	0
c	0	0	0	c	0	0
d	0	0	c	d	0	0
e	0	0	0	0	0	e
f	0	0	0	0	e	f

*table*1

·	a	b	c	d	e	f
a	0	a	0	0	0	0
b	a	b	0	0	0	0
c	0	0	0	c	0	0
d	0	0	c	d	0	0
e	0	0	0	0	e	e
f	0	0	0	0	e	f

*table*2

Subcase2.2. Let $e^2 = e, f^2 = e$. Then $ef = e$ or f. So we have two new tables on $S(table3-4)$.

$\cdot$	a	b	c	d	e	f
a	0	a	0	0	0	0
b	a	b	0	0	0	0
c	0	0	0	c	0	0
d	0	0	c	d	0	0
e	0	0	0	0	e	e
f	0	0	0	0	e	e

$table3$

$\cdot$	a	b	c	d	e	f
a	0	a	0	0	0	0
b	a	b	0	0	0	0
c	0	0	0	c	0	0
d	0	0	c	d	0	0
e	0	0	0	0	e	f
f	0	0	0	0	f	e

$table4$

Case3. Assume $a^2 = 0$, $c^2 \neq 0$, $d^2 \neq 0$, $e^2 \neq 0$ and $f^2 \neq 0$. Then we have $ab = a, b^2 = b$. By Theorem 2.4(3), we have the following three cases.
Subcase3.1. Assume $c^2 = c, d^2 = d, e^2 = e$ and $f^2 = f$. Then, by symmetry, we have only one table on $S(table5)$.

$\cdot$	a	b	c	d	e	f
a	0	a	0	0	0	0
b	a	b	0	0	0	0
c	0	0	c	c	0	0
d	0	0	c	d	0	0
e	0	0	0	0	e	e
f	0	0	0	0	e	f

$table5$

$\cdot$	a	b	c	d	e	f
a	0	a	0	0	0	0
b	a	b	0	0	0	0
c	0	0	c	c	0	0
d	0	0	c	d	0	0
e	0	0	0	0	e	e
f	0	0	0	0	e	e

$table6$

Subcase3.2. Assume $c^2 = c, d^2 = d, e^2 = e$ and $f^2 \neq f$. Then, $f^2 = e$. In this case, $ef = e$ or f, then we have two tables on $S(table6-7)$.

$\cdot$	a	b	c	d	e	f
a	0	a	0	0	0	0
b	a	b	0	0	0	0
c	0	0	c	c	0	0
d	0	0	c	d	0	0
e	0	0	0	0	e	f
f	0	0	0	0	f	e

$table7$

$\cdot$	a	b	c	d	e	f
a	0	a	0	0	0	0
b	a	b	0	0	0	0
c	0	0	c	c	0	0
d	0	0	c	c	0	0
e	0	0	0	0	e	e
f	0	0	0	0	e	e

$table8$

Subcase3.3. Assume $c^2 = c, d^2 \neq d, e^2 = e$ and $f^2 \neq f$. Then, $d^2 = c$

$f^2 = e$. In this case, we have three new tables on $S(table8-10)$.

.	a	b	c	d	e	f
a	0	a	0	0	0	0
b	a	b	0	0	0	0
c	0	0	c	c	0	0
d	0	0	c	c	0	0
e	0	0	0	0	e	f
f	0	0	0	0	f	e

*table*9

.	a	b	c	d	e	f
a	0	a	0	0	0	0
b	a	b	0	0	0	0
c	0	0	c	d	0	0
d	0	0	d	c	0	0
e	0	0	0	0	e	f
f	0	0	0	0	f	e

*table*10

Case4. Assume $a^2 \neq 0$, $b^2 \neq 0$, $c^2 \neq 0$, $d^2 \neq 0$, $e^2 \neq 0$ and $f^2 \neq 0$. By Theorem 2.4(3), we have four cases.

Subcase4.1. Assume $a^2 = a, b^2 = b, c^2 = c, d^2 = d, e^2 = e$ and $f^2 = f$. By symmetry, we have only one table on $S(table11)$.

.	a	b	c	d	e	f
a	a	a	0	0	0	0
b	a	b	0	0	0	0
c	0	0	c	c	0	0
d	0	0	c	d	0	0
e	0	0	0	0	e	e
f	0	0	0	0	e	f

*table*11

.	a	b	c	d	e	f
a	a	a	0	0	0	0
b	a	b	0	0	0	0
c	0	0	c	c	0	0
d	0	0	c	d	0	0
e	0	0	0	0	e	e
f	0	0	0	0	e	e

*table*12

Subcase4.2. Assume $a^2 = a, b^2 = b, c^2 = c, d^2 = d, e^2 = e$ and $f^2 \neq f$. Then, $f^2 = e$. In this case, $ef = e$ or f, we have two tables on $S(table12-13)$.

.	a	b	c	d	e	f
a	a	a	0	0	0	0
b	a	b	0	0	0	0
c	0	0	c	c	0	0
d	0	0	c	d	0	0
e	0	0	0	0	e	f
f	0	0	0	0	f	e

*table*13

.	a	b	c	d	e	f
a	a	a	0	0	0	0
b	a	b	0	0	0	0
c	0	0	c	c	0	0
d	0	0	c	c	0	0
e	0	0	0	0	e	e
f	0	0	0	0	e	e

*table*14

Subcase4.3. Assume $a^2 = a, b^2 = b, c^2 = c, d^2 \neq d, e^2 = e$ and $f^2 \neq f$. Then, $d^2 = c$ and $f^2 = e$. In this case, similar to subcase3.3, we have three tables on $S(table14-16)$.

$\cdot$	a	b	c	d	e	f
a	a	a	0	0	0	0
b	a	b	0	0	0	0
c	0	0	c	c	0	0
d	0	0	c	c	0	0
e	0	0	0	0	e	f
f	0	0	0	0	f	e

*table*15

$\cdot$	a	b	c	d	e	f
a	a	a	0	0	0	0
b	a	b	0	0	0	0
c	0	0	c	d	0	0
d	0	0	d	c	0	0
e	0	0	0	0	e	f
f	0	0	0	0	f	e

*table*16

Subcase4.4. Assume $a^2 = a, b^2 \neq b, c^2 = c, d^2 \neq d, e^2 = e$ and $f^2 \neq f$. Then, $b^2 = a$, $d^2 = c$ and $f^2 = e$. In this case, we have four tables on $S(table17-20)$.

$\cdot$	a	b	c	d	e	f
a	a	a	0	0	0	0
b	a	a	0	0	0	0
c	0	0	c	c	0	0
d	0	0	c	c	0	0
e	0	0	0	0	e	e
f	0	0	0	0	e	e

*table*17

$\cdot$	a	b	c	d	e	f
a	a	a	0	0	0	0
b	a	a	0	0	0	0
c	0	0	c	c	0	0
d	0	0	c	c	0	0
e	0	0	0	0	e	f
f	0	0	0	0	f	e

*table*18

$\cdot$	a	b	c	d	e	f
a	a	a	0	0	0	0
b	a	a	0	0	0	0
c	0	0	c	d	0	0
d	0	0	d	c	0	0
e	0	0	0	0	e	f
f	0	0	0	0	f	e

*table*19

$\cdot$	a	b	c	d	e	f
a	a	b	0	0	0	0
b	b	a	0	0	0	0
c	0	0	c	d	0	0
d	0	0	d	c	0	0
e	0	0	0	0	e	f
f	0	0	0	0	f	e

*table*20

The associativity of the above twenty tables have been verified by a computer procedure. □

The numbers of non-isomorphic commutative rings and zero-divisor semigroups of all regular polyhedrons are listed in the following table.

regular polyhedrons	corresponding rings	corresponding semigroups
tetrahedron	2	12
cube	0	0
octahedron	0	20
dodecahedron	0	0
icosahedron	0	0

Acknowledgments

This research was supported by National Natural Science Foundation of China (10771095), Guangxi Natural Science Foundation(0575052, 0832107), Innovation Project of Guangxi Graduate Education(2006106030701M05, 2007106030701M15) and Scientific Research Foundation of Guangxi Educational Committee(200508164).

References

1. F. DeMeyer, L. DeMeyer, *Zero divisor graphs of semigroups*, J. Algebra 283(2005) 190-198.
2. F. DeMeyer, T. McKenzie, K. Schneider, *The zero-divisor graph of a commutative semigroup*, Semigroup Forum 65(2002) 206-214.
3. T. S. Wu, *On directed zero-diuvisor graphs of finite rings*, Discrete Math. 296(1)(2005) 73-86.
4. T. S. Wu, L. Chen, *Simple graphs and commutative zero-divisor semigroups*, Algebra Colloq.(to appear)
5. T. S. Wu, F. Cheng, *The structures of zero-divisor semigroups with graph* $K_n \circ K_2$, Semigroup Forum (to appear)
6. T. S. Wu, D. C. Lu, *Zero-divisor semigroups and some simple graphs*, Communications in Algebra 34(8)(2006) 3043-3052.
7. D. F. Anderson, P. S. Livingston, *The zero-divisor graph of a commutative ring*, J. Algebra 217(1999) 434-447
8. S. Akbari, H. R. Maimani, S. Yassemi, *When a zero-divisor graph is planar or a completer r-partite graph*, J. Algeba 270(1)(2003) 169-180.
9. R. J. Wilson, Introduction to Graph Theory, Fourth Edition, Pearson Education Limited, 1996.

REMARKS ON QF-2 RINGS, QF-3 RINGS AND HARADA RINGS

Ken-ichi Iwase

Tennoji senior high-school attached to Osaka kyoiku university, Tennoji, Osaka, JAPAN

In the procedings of the 1978 antwerp conference, M.Harada studied those rings whose non-small left modules contains non-zero injective submodules. K.Oshiro called perfect rings with this condition "*left Harada* rings". These rings are two sided artinian, right QF-2, and right and left QF-3 rings containing QF rings and *Nakayama* rings, and moreover, these rings have left and also right ideal theoretic characterizations.

The purpose of this paper is to study the following well known theorems (*see Anderson-Fuller* [1]) :

Theorem I. Right or left artinian QF-2 rings are QF-3.

Theorem II. For a right or left artinian ring R, R is QF-3 if and only if its injective
hull $E(R_R)$ is projective.

Theorem III. Every Nakayama ring R with a simple projective right ideal is expressed
as a factor ring of an upper triangular matrix ring over a division ring.

In Theorems I, II, we are little anxious whether the assumption "right or left artinian" is natural or not. This assumption also appears in the following well known theorem due to Fuller [6] :

Let R be a right or left artinian ring and let e be a primitive idempotent in R. Then
eR_R is injective if and only if there exists a primitive idempotent f in R such that
$S(eR) \cong fR/fJ$ and $S(Rf) \cong Re/Je$, where $S(X)$ and J mean the socle of X and
the Jacobson radical of R, respectively.

In Baba-Oshiro [2], this theorem is improved for a semiprimary ring with "ACC or DCC" for right annihilator ideals, where ACC and DCC mean the ascending chain condition and the descending chain condition, respectively. As the condition ACC or DCC for right annihilator ideals is equivalent to the condition ACC or DCC for left annihilator ideals, the replacement of "right or left artinian" with "semiprimary ring with ACC or DCC for annihilator right ideals" is quite natural.

In this paper, from this view point, we improve Theorem I as follows: Semiprimary QF-2 rings with ACC or DCC for right annihilator ideals are QF-3. For Theorem II, we show that, for a left perfect ring R with ACC or DCC for right annihilator ideals, R is QF-3 if its injective hull $E(R_R)$ is projective. For Theorem III, using the structure theorem of left Harada rings, we improve the theorem as follows: Left Harada rings with a simple projective right ideal is expressed as a factor ring of an upper triangular matrix ring over a division ring.

Keywords: Harada ring; Nakayama rings; QF-3 rimgs; ACC or DCC on right annihilator ideals.

1. Introduction

Throughout this paper, we let R be an associative ring with identity and modules are unitary R-modules. For an R-module M, We denote the injective hull,the Jacobson radical, the top $M/J(M)$, the socle, and loewy length by $E(M)$, $J(M)$, $T(M)$, $S(M)$, and $L(M)$. We define the i-th socle $S_i(M)$ of M by $S_i(M)/S_{i-1}(M) = S(M/S_{i-1}(M))$ inductively, where we let $S_1(M) = S(M)$. Further we put $J := J(R_R)$ and we let $Pi(R)$ be a complete set of orthogonal primitive idempotents in R. And, for a subset S of R, we denote the left (right) annihilator of S in R by $l_R(S)$ $(r_M(S))$, respectively.

@A ring is called a *Quasi-Frobenius* (abbreviated *QF*) ring if it is one sided artinian one sided self-injective. Let R be a semiperfect ring and $e, f \in Pi(R)$. $(eR; Rf)$ is called an *injective pair* (abbreviated i-pair) if $S(eR_R) \cong T(fR_R)$ and $S(_RRf) \cong T(_RRe)$ are satisfied. A ring is called a *basic* ring if for distinct e and $f \in Pi(R)$, $eR \ncong fR$. Let R be a basic *QF*-ring. For a complete basic set of indecomposable primitive idempotents $e_1, e_2, .., e_m$ in R, i.e., $Pi(R) = \{e_1, .., e_n\}$ and $e_iR \ncong e_jR$ for every distinct $i, j \in \{1, 2, .., n\}$, if there is a permutation σ of $\{1, .., m\}$ such that $(e_jR, Re_{\sigma(j)})$ is an i-pair for every $j \in \{1, 2, .., m\}$, this permutation σ is called a *Nakayama permutation.* A ring R is called a *right* (*left*) *QF-2* ring if every its indecomposable projective right (left) module has a simple socle. And R is called a *QF-2* ring if it is both left and right *QF-2*. A faithful right R-module M is called *minimal faithful* if, for any faithful right R-module N, there exists a direct summand M' of N such that $M' \cong M$. A ring R is called a *right* (*left*) *QF-3* ring if R has a minimal faithful right (left) R-module, and is called a *QF-3 ring* if it is both left and right QF-3. An artinian ring is called *right* (*left*) *Nakayama* ring if its right (left) indecomposable projective module is uniserial, and is called a *Nakayama* ring if it is both right and left *Nakayama.* We call a module M *non-small* if M is not small in $E(M)$, and otherwise we call M small. For a condition (*) : Every non-small left R-module contains a non-zero injective submodule, a perfect ring with condition (*) is called a *left Harada* ring, and is called a *Harada* ring if it is both right and left *Harada. Left Harada* rings are characterized as follows:

Theorem A. ([12]) The following (a),(b) and (c) are equivalent.

(a) R is a left Harada ring.
(b) R satisfies the following (1) and (2).

(1) R is a two-sided artinian ring.
(2) There exists a complete set $\{e_{ij}\}_{i=1,j=1}^{mn(i)}$ of primitive idempotents in R such that

(i) $e_{i1}R_R$ is injective for $i = 1, \ldots, m$,
(ii) $e_{ij}R_R \cong e_{i,j-1}R_R$ or $\cong e_{i,j-1}J_R$ for $i = 1, \ldots, m,\ j = 2, \ldots, n(i)$.

When R is a basic left *Harada* ring, we note that condition (2)-(ii) above can be replaced by
$(ii)'$ $e_{ij}R \cong e_{i1}J^{j-1}$ for any $i = 1, ., m,\ j = 2, ., n(i)$.

We note the following:

Remark

(1) Left *Harada* rings are Morita invariant.

(2) Left *Harada* rings are two sided artinian, *right QF-2*, and *right and left QF-3*, @@@but not left *QF-2* in general.

For example, we consider the ring $Q = K[x, y]/(x^2, y^2)$, where K is a field. Then Q is a local QF algebra.

Putting $J = J(Q)$, $S = S(Q)$ and $\overline{Q} = Q/S$, we make the following ring: $R = \begin{pmatrix} Q & \overline{Q} \\ J & \overline{Q} \end{pmatrix}$. Then R is a left *Harada* ring, but not a left *QF-2* ring.

(3) Let R be a two sided artinian. Then R is a left *Harada* ring if and only @@@@if R is a right *QF-2* and right and left *QF-3* ring with the property that, for @@any indecomposable projective module P_R and any projective submodule A of @@P, P/A is a uniserial module.

Actually, (1) and (2) are well known (see [12]). To see (3), we may show that any submodule B with $P \subsetneq B \subsetneq A$ is projective. Since B is a cyclic hollow module, there is a projective cover $Q \xrightarrow{f} B$ with Q indecomposable projective. Since A is not a singular module, we see that $\ker f = 0$. Hence B is projective.

2. Improve versions of Theorem I and Theorem II

Recall that a right R-module M is called *uniform*f every non-zero submodule of M is essential. We note that, if R is left perfect, M_R is uniform if and only if M_R is colocal.

The *uniform dimension*of a module M is the infimum of those cardinal numbers c such that $\#I \leq c$ for every independent set $\{N_i\}_{i \in I}$ of non-zero submodules of M. We denote the uniform dimension of M by unif.dim$M$, where $\#I$ means the number of elements of I.

Proposition 2.1. (*c.f.* [3, Proposition.3.1.2]) *Let R be a ring. We consider the following four conditions.*

(*a*) *R is right QF-3.*
(*b*) *R contains a faithful injective right ideal.*
(*c*) *For any projective right R module P_R, $E(P_R)$ is projective.*
(*d*) *$E(R_R)$ is projective.*

Then the following hold.

(1) (*a*) $\Rightarrow$ (*b*) *holds. Further, if R is a left perfect ring, then* (*b*) $\Rightarrow$ (*a*) *also holds.*
(2) *If R is left perfect, then* (*b*) $\Leftrightarrow$ (*c*) *holds.*
(3) *If ACC or DCC holds on right annihilator ideals, then* (*d*) $\Rightarrow$ (*b*) *does.*
(4) (*c*) $\Rightarrow$ (*d*) *holds in general.*

Proof. (2) This follows from [19, Theorem1.3. (4) $\Leftrightarrow$ (6)]

(1) (*a*) $\Rightarrow$ (*b*).Let M_R be a minimal faithful right R module. Since R_R is faithful, then there exists right ideal $I_R \lessdot\!\!\oplus R$ such that $I_R \cong M_R$. On the other hand, since $E(R_R)$ is faithful, $I_R \lessdot\!\!\oplus E(R_R)$. Then I_R is injective, that is, R contains a faithful injective right ideal I_R.

(*b*) $\Rightarrow$ (*a*).We assume that there exists a faithful injective right ideal I of R. Then $I = \oplus_{i=1}^{n} e_i R$ for some primitive idempotents $e_1, \ldots, e_n$ in R since R is left perfect. We may assume that $\{e_i\}_{i=1}^{k}$ is a basic set of

$\{e_i\}_{i=1}^n$. Then we note that $\{S(e_iR_R)\}_{i=1}^k$ is a set of pairwise non-isomorphic essential simple socle since R is left perfect. Let N be a faithful right R-module. For each $i = 1, \ldots, k$, we have a monomorphism $\varphi_i : e_iR \to N$ since $S(e_iR_R)$ is an essential simple socles. Then $\sum_{i=1}^k \mathrm{Im}\varphi_i = \oplus_{i=1}^k \mathrm{Im}\varphi_i$ since $S(e_1R_R), \ldots, S(e_kR_R)$ are pairwise non-isomorphic. So we have a split monomorphism : $\oplus_{i=1}^k e_iR \to N$. Hence $\oplus_{i=1}^k e_iR$ is a minimal faithful right R-module.

(3) $(d) \Rightarrow (b)$.We assume that ACC or DCC holds on right annihilator ideal in R. First we show the following Claim.

Claim.

(1) *For any right ideal A of R,* $\mathrm{Hom}_R(r_Rl_R(A)/A,\ E(R_R)) = 0$.
(2) *For any non-zero right ideals A, B of R with $A + B = A \oplus B$, $r_Rl_R(A) < r_Rl_R(A \oplus B)$.*
(3) *unif.dim $R_R < \infty$.*

Proof of Claim *(1) Assume that* $\mathrm{Hom}_R(r_Rl_R(A)/A, E(R_R)) \neq 0$ *for some right ideal A of R. Let $0 \neq \varphi \in$* $\mathrm{Hom}_R(r_Rl_R(A)/A, E(R_R))$. *$\varphi$ is extended to $\tilde{\varphi} \in$* $\mathrm{Hom}_R(R/A, E(R_R))$. *And, since $E(R_R)$ is projective, there exists $\psi \in$* $\mathrm{Hom}_R(E(R_R), R)$ *with $\psi\varphi \neq 0$. We put $a := \psi\tilde{\varphi}(\overline{1})$. Then $aA = 0$. So $a{\cdot}r_Rl_R(A) = 0$ since $a \in l_R(A) = l_Rr_Rl_R(A)$. Hence $\psi\varphi = (a)_L|_{r_Rl_R(A)/A} = 0$, a contradiction.*

(2) Assume that $r_Rl_R(A) = r_Rl_R(A \oplus B)$. Then $A \oplus B \leq r_Rl_R(A)$. On the other hand, $\mathrm{Hom}_R((A\oplus B)/A,\ E(R_R)) \neq 0$ *since $(A\oplus B)/A \cong B$ is isomorphic to a submodule of $E(R_R)$. Hence there exists an epimorphism:* $\mathrm{Hom}_R(r_Rl_R(A)/A, E(R_R)) \to \mathrm{Hom}_R((A\oplus B)/A, E(R_R))$ *because $E(R_R)$ is injective. This contradicts with (1).*

(3) This follows from (2).

Claim is shown.

From Claim (3), we see that $E(R_R)$ has a finite decomposition $E(R_R) = \oplus_{i=1}^n E_i$ of indecomposable submodules. Then, for each $i = 1, \ldots, n$, E_i has the exchange property because $\mathrm{End}(E_i)$ *is a local ring. On the other hand, since $E(R_R)$ is projective, we have a (split) monomorphism $\iota : E(R_R) \to R^{(S)}$ for some set S. So $E_i \cong e_iR$ for some primitive idempotent e_i in R. Now we may consider that $\{E_1, E_2, \ldots, E_m\}$ is an irredundant subset of $\{E_i\}_{i=1}^n$. Then $\sum_{i=1}^m e_iR = \oplus_{i=1}^m e_iR$ is a faithful injective right ideal of R.* □

Remark : By Proposition 2.1, when R is a left perfect ring, we have $(a) \Leftrightarrow (b) \Leftrightarrow (c)$ and $(a) \Rightarrow (d)$, but in general $(d) \Rightarrow (a)$ is not true.

For example, for the set $\mathbb{Q}$ of rational numbers and the set $\mathbb{Z}$ of integers, we consider $R = \begin{pmatrix} \mathbb{Q} & \mathbb{Q} \\ 0 & \mathbb{Z} \end{pmatrix}$. Then ${}_RR$ is noetherian and has a faithful injective right ideal, so that $E(R_R)$ is projective, but R does not have minimal faithful right R module. (c.f. [18, Theorem 6.2 (Vinsonhaler)])

The following Theorem is due to K.R.Fuller.

Theorem B. ([1, Theorem 31.3]) Let R be a right or left artinian ring and let $f \in Pi(R)$. Then ${}_RRf$ is injective if and only if there is a primitive idempotent e in R such that $S({}_RRf) \cong T({}_RRe)$ and $S(eR_R) \cong T(fR_R)$.

Using this Theorem B, he showed that every right or left artinian QF-2 ring is QF-3. Y. Baba and K. Oshiro improved Theorem B in [2] as follows:

Theorem C. ([2]) Let R be a semiprimary ring which satisfies ACC or DCC for right annihilator ideais and let $e, f \in Pi(R)$. Then the following conditions are equivalent:

(1) ${}_RRf$ is injective with $S({}_RRf) \cong T({}_RRe)$.
(2) eR_R is injective with $S(eR_R) \cong T(fR_R)$.

Now we show the following.

Theorem 2.1. *If R is a semiprimary QF-2 ring with ACC or DCC for right annihilator ideals, then R is QF-3.*

Proof. It suffices to show that R is right *QF-3*. We may assume that $Pi(R) = \{e_{ij}\}_{i=1,j=1}^{m\ \ n(i)}$, where for any $i,\ j \in \{1, \ldots, m\}$,

(i) $S(e_{i1}R_R) \not\cong S(e_{j1}R_R)$ if $i \neq j$,
(ii) $S(e_{i1}R_R) \cong S(e_{i2}R_R) \cong \cdots \cong S(e_{in(i)}R_R)$, and
(iii) $L(e_{i1}R_R) \geq L(e_{i2}R_R) \geq \cdots \geq L(e_{in(i)}R_R)$.

We may show that each $e_{i1}R_R$ is injective by Proposition 2.1. Let $e \in \{e_{i1}\}_{i=1}^m$. We have $f \in Pi(R)$ with fR_R a projective cover of $S(eR_R)$ since R is *QF*-2. Then $0 \neq S(eR_R)f \subseteq Rf$. Further we claim that $J\,S(eR_R)f = 0$. Assume that $J\,S(eR_R)f \neq 0$. There exist $g \in Pi(R)$ and $x \in gJ$ with $x\,S(eR_R)f \neq 0$. We consider the left multiplication map $(x)_L : eR \to gR$. And we see that it is a monomorphism from the assumption $J\,S(eR_R)f \neq 0$

because $S(eR_R)$ is the essential simple socle. On the other hand, $\mathrm{Im}(x)_L \subseteq gJ$ because $x \in J$. Therefore $L(eR_R) < L(gR_R)$. This contradicts with the maximality of $L(eR_R)$. So $0 \neq S(eR_R)f \subseteq S({}_RRf)$, i.e., $e\,S({}_RRf) \neq 0$. Hence we have the projective cover ${}_RRe \to S({}_RRf)$ since R is QF-2. Thus $(eR; Rf)$ is an i-pair. Since R satisfies ACC or DCC for right annihilators, we see that eR_R is injective from Theorem C. □

3. An improve version of Theorem III

For our purpose, we need the following structure theorem due to Oshiro ([15]-[17]):

Theorem D. Let R be a basic left Harada ring. Then R can be constructed as an upper staircase factor ring of a block extension of its frame QF-subring $F(R)$.

In order to understand this structure theorem, we must review the sketch of the proof of Theorem D (for details, see Baba-Oshiro's Lecture Note).

Let F be a basic QF-ring with $Pi(F) = \{e_1, ., e_y\}$. We put $A_{ij} := e_iFe_j$ for any i, j, and, in particular, put $Q_i := A_{ii}$ for any i. Then we may represent F as

$$F = \begin{pmatrix} A_{11} & A_{12} & \cdots & A_{1y} \\ A_{21} & A_{22} & \cdots & A_{2y} \\ \cdots & \cdots & \cdots & \cdots \\ A_{y1} & A_{y2} & \cdots & A_{yy} \end{pmatrix} = \begin{pmatrix} Q_1 & A_{12} & \cdots & A_{1y} \\ A_{21} & Q_2 & \cdots & A_{2y} \\ \cdots & \cdots & \cdots & \cdots \\ A_{y1} & \cdots & A_{y,y-1} & Q_y \end{pmatrix}.$$

For $k(1), ., k(y) \in \mathbb{N}$, the *block extension* $F(k(1), ., k(y))$ of Fs defined as follows: For each $i,\ s \in \{1, ., y\},\ j \in \{1, ., k(i)\},\ t \in \{1, ., k(s)\}$, let

$$P_{ij,st} = \begin{cases} Q_i & \text{if } i = s,\ j \leq t, \\ J(Q_i) & \text{if } i = s,\ j > t, \\ A_{is} & \text{if } i \neq s, \end{cases}$$

and

$$P(i,s) = \begin{pmatrix} P_{i1,s1} & P_{i1,s2} & \cdots & P_{i1,sk(s)} \\ P_{i2,s1} & P_{i2,s2} & \cdots & P_{i2,sk(s)} \\ \vdots & \vdots & \ddots & \vdots \\ P_{ik(i),s1} & P_{ik(i),s2} & \cdots & P_{ik(i),sk(s)} \end{pmatrix}.$$

Consequently, when $i = s$, we have the $k(i) \times k(i)$ matrix

$$P(i,i) = \begin{pmatrix} Q_i & \cdots & \cdots & Q_i \\ J(Q_i) & \ddots & & \vdots \\ \vdots & \ddots & \ddots & \vdots \\ J(Q_i) & \cdots & J(Q_i) & Q_i \end{pmatrix}$$

which we denote by $Q(i)$, and, when $i \neq s$, we have the $k(i) \times k(s)$ matrix

$$P(i,s) = \begin{pmatrix} A_{is} & \cdots & A_{is} \\ & \cdots & \\ A_{is} & \cdots & A_{is} \end{pmatrix}$$

Furthermore, we set

$$P = F(k(1),\,.\,,k(y)) = \begin{pmatrix} P(1,1) & P(1,2) & \cdots & P(1,y) \\ P(2,1) & P(2,2) & \cdots & P(2,y) \\ \cdots & \cdots & \cdots & \cdots \\ P(y,1) & P(y,2) & \cdots & P(y,y) \end{pmatrix}$$

$$= \begin{pmatrix} Q(1) & P(1,2) & \cdots & P(1,y) \\ P(2,1) & Q(2) & \cdots & P(2,y) \\ \cdots & \cdots & \cdots & \cdots \\ P(y,1) & P(y,2) & \cdots & Q(y) \end{pmatrix}.$$

Since F is a basic QF-ring, we see that P is a basic left *Harada* ring with matrix size $k(1)+\cdots+k(y)$. We say that $F(k(1),\,.\,,k(y))$ is a *block extension* of F for $\{k(1),\,.\,,k(y)\}$. indexblock extension In more detail, this matrix representation is given by

$$P = F(k(1),\,.\,,k(y)) = \begin{pmatrix} P_{11,11} & \cdots & P_{11,1k(1)} & \cdots & P_{11,y1} & \cdots & P_{11,yk(y)} \\ \vdots & & \vdots & & \vdots & & \vdots \\ P_{1k(1),11} & \cdots & P_{1k(1),1k(1)} & \cdots & P_{1k(1),y1} & \cdots & P_{1k(1),yk(y)} \\ \vdots & & \vdots & & \vdots & & \vdots \\ P_{y1,11} & \cdots & P_{y1,1k(1)} & \cdots & P_{y1,y1} & \cdots & P_{y1,yk(y)} \\ \vdots & & \vdots & & \vdots & & \vdots \\ P_{yk(y),11} & \cdots & P_{yk(y),1k(1)} & \cdots & P_{yk(y),y1} & \cdots & P_{yk(y),yk(y)} \end{pmatrix}.$$

If we set

$$p_{ij} = \langle 1 \rangle_{ij,ij} \;,$$

where this means an element of P which the $(ij,\ ij)$-position $= 1$, and another positions are 0.
For each $i = 1, .., y,\ j = 1, .., k(i)$, then $\{p_{ij}\}_{i=1,j=1}^{y\ \ k(i)}$ is a well-indexed set of a complete set of orthogonal primitive idempotents of $P = F(k(1), .., k(y))$.

For $Pi(P)$, we note that

$$p_{ij}P_P \cong p_{i1}J(P)_P^{j-1}$$

for any $i = 1, .., y$ and $j = 1, .., k(i)$.

Given the situation above, the following are equivalent:

(1) F is a QF ring with a Nakayama permutation:

$$\begin{pmatrix} e_1 & \cdots & e_y \\ e_{\sigma(1)} & \cdots & e_{\sigma(y)} \end{pmatrix}.$$

(2) $P = F(k(1), .., k(y))$ is a basic left *Harada* ring of type $(*)$ with a well-indexed set $Pi(P) = \{p_{ij}\}_{i=1,j=1}^{y\ \ k(i)}$.

Let R be a basic *Harada* ring. We call R a basic *Harada* ring of *type* $(*)$ if there is a permutation σ of $\{1, 2, .., m\}$ such that $(e_{j1}R, Re_{\sigma(j)n(\sigma(j))})$ is an i-pair for every $j \in \{1, 2, .., m\}$.

From now on, we assume that the Nakayama permutation of F is

$$\begin{pmatrix} e_1 & \cdots & e_y \\ e_{\sigma(1)} & \cdots & e_{\sigma(y)} \end{pmatrix},$$

and we take the block extension $P = F(k(1), .., k(y))$ of F. Let $i \in \{1, .., y\}$ and consider the i-pair $(e_iF; Fe_{\sigma(i)})$. Put $S(A_{ij}) := S({}_{Q_i}A_{ij}) = S({A_{ij}}_{Q_j})$. Then we define an upper staircase left $Q(i)$- right $Q(\sigma(i))$-subbimodule $S(i, \sigma(i))$ of $P(i, \sigma(i))$ with tiles $S(A_{ij})$ as follows:

(I) Suppose that $i = \sigma(i)$: Then we see from above argument that $S(A_{ij})$ is simple as both a left and a right ideal of $Q_i = A_{ii}$. Put $Q := Q_i$, $J := J(Q_i)$ and $S := S(Q_i)$. Then, in the $k(i) \times k(i)$ matrix ring,

$$Q(i) = P(i,i) = \begin{pmatrix} Q & \cdots & \cdots & Q \\ J & \ddots & & \vdots \\ \vdots & \ddots & \ddots & \vdots \\ J & \cdots & J & Q \end{pmatrix},$$

we define an upper staircase left $Q(i)$- right $Q(i)$-subbimodule $S(i,i) = S(i,\sigma(i))$ of $Q(i)$ as follows:

$$S(i,i) = \begin{pmatrix} 0\cdots 0 & & S \\ & & \\ & & \\ 0 & & \end{pmatrix}. \qquad (\text{ the } (1,1) \text{ position} = 0),$$

where, for the form of $S(i,i)$, we assume that
(1) the $(1,1)$-position $= 0$,
(2) when Q is a division ring, that is, $Q = S$,

$$S(i,i) = \begin{pmatrix} 0\cdots 0 & & & S \\ & \ddots & & \\ & & \ddots & \\ & & & \ddots \\ 0 & & & 0 \end{pmatrix}$$

Then, since S is an ideal of Q, we see that $S(i,i) = S(i,\sigma(i))$ is an ideal of Q_i.

We let $\overline{Q(i)} = \overline{P(i,\sigma)} = P(i,\sigma)/S(i,\sigma(i))$ for the subbimodule $S(i,\sigma(i))$. In $Q(i)$, we replace Q or J of the (p,q)-position by $\overline{Q} = Q/S$ or $\overline{J} = J/S$, respectively, when the (p,q)-position of $S(i,i)$ is S. Then we may represent $\overline{Q(i)}$ with the matrix ring which is made by these replacements.

For example,

$$\overline{Q(i)} = \begin{pmatrix} Q & \overline{Q} & \overline{Q} & \overline{Q} & \overline{Q} & \overline{Q} \\ J & \overline{Q} & \overline{Q} & \overline{Q} & \overline{Q} & \overline{Q} \\ J & \overline{J} & \overline{Q} & \overline{Q} & \overline{Q} & \overline{Q} \\ J & J & J & Q & \overline{Q} & \overline{Q} \\ J & J & J & J & \overline{Q} & \overline{Q} \\ J & J & J & J & \overline{J} & \overline{Q} \end{pmatrix} = \begin{pmatrix} Q & Q & Q & Q & Q & Q \\ J & Q & Q & Q & Q & Q \\ J & J & Q & Q & Q & Q \\ J & J & J & Q & Q & Q \\ J & J & J & J & Q & Q \\ J & J & J & J & J & Q \end{pmatrix} \Big/ \begin{pmatrix} 0 & S & S & S & S & S \\ 0 & S & S & S & S & S \\ 0 & S & S & S & S & S \\ 0 & 0 & 0 & 0 & S & S \\ 0 & 0 & 0 & 0 & S & S \\ 0 & 0 & 0 & 0 & S & S \end{pmatrix}$$

(II) Now suppose that $i \neq \sigma(i)$: Put $S := S_{i\sigma(i)} = S({}_{Q_i}A_{i\sigma(i)}) = S(A_{i\sigma(i)}{}_{Q_{\sigma(i)}})$. Then S is a left Q_i- right $Q_{\sigma(i)}$-subbimodule of $A = A_{i\sigma(i)}$.

In the left $Q(i)$- right $Q(\sigma(i))$-bimodule

$$P(i,\sigma(i)) = \begin{pmatrix} A \cdots A \\ \cdots \\ A \cdots A \end{pmatrix} \qquad (\, k(i) \times k(\sigma(i))\text{-matrix}\,),$$

we define an upper staircase subbimodule $S(i,\sigma(i))$ of $P(i,\sigma(i))$ with tiles S of $P(i,\sigma(i))$ as follows:

$$S(i,\sigma(i)) = \begin{pmatrix} 0 \cdots 0 \,\rfloor & & S \\ & \rfloor & \\ & & \rule{0pt}{0pt}\underline{\qquad\quad} \\ 0 & & \end{pmatrix} \qquad (\text{ the } (1,1)\text{-position} = 0\,)$$

and put $\overline{P(i,\sigma)} := P(i,\sigma(i))/S(i,\sigma(i))$. We may represent $\overline{P(i,\sigma)}$ as

$$\overline{P(i,\sigma)} = \begin{pmatrix} A \cdots A \,\rfloor & & \overline{A} \\ & \rfloor & \\ & & \underline{\qquad\quad} \\ A & & \end{pmatrix}.$$

Next we define a subset X of $P = F(k(1), .\,, k(y))$ by

$$X = \begin{pmatrix} X(1,1) & X(1,2) & \cdots & X(1,y) \\ X(2,1) & X(2,2) & \cdots & X(2,y) \\ \cdots & \cdots & \cdots & \cdots \\ X(y,1) & X(y,2) & \cdots & X(y,y) \end{pmatrix},$$

where $X(i,j)$ $(\subseteq Q_i)$ and $X(i,j)$ $(\subseteq P(i,j))$ are defined by

$$X(i,i) = \begin{cases} 0 & \text{if } i \neq \sigma(i), \\ S(i,i) & \text{if } i = \sigma(i), \end{cases}$$

$$X(i,j) = \begin{cases} 0 & \text{if } j \neq \sigma(i), \\ S(i,j) & \text{if } j = \sigma(i). \end{cases}$$

Then we see that X is an ideal of $P = F(k(1), .\,, k(y))$. The factor ring $F(k(1), .\,, k(y))/X$ is then called an *upper staircase factor ring*indexupper staircase factor ring of $P = F(k(1), .\,, k(y))$. If, in the representation

$$P = F(k(1), .\,, k(y)) = \begin{pmatrix} P(1,1) & P(1,2) & \cdots & P(1,y) \\ P(2,1) & P(2,2) & \cdots & P(2,y) \\ \cdots & \cdots & \cdots & \cdots \\ P(y,1) & P(y,2) & \cdots & P(y,y) \end{pmatrix},$$

we replace $P(i,\sigma(i))$ with $\overline{P(i,\sigma(i))}$ and put $\overline{P} := F(k(1),\,.\,,k(y))/X$, then it is convenient to represent $\overline{P}$ as follows:

$$\overline{P} = \begin{pmatrix} P(1,1) & \cdots & \overline{P(1,\sigma(1))} & \cdots & \cdots & \cdots & P(1,y) \\ \cdots & \cdots & \cdots & \cdots & \cdots & \cdots & \cdots \\ \cdots & \cdots & \cdots & \cdots & \cdots & \cdots & \cdots \\ P(i,1) & \cdots & \cdots & \overline{P(i,\sigma(i))} & \cdots & \cdots & \cdots \\ \cdots & \cdots & \cdots & \cdots & \cdots & \cdots & \cdots \\ \cdots & \cdots & \cdots & \cdots & \cdots & \cdots & \cdots \\ \cdots & \cdots & \cdots & \cdots & \cdots & \cdots & \cdots \\ P(y,1) & \cdots & \cdots & \cdots & \overline{P(y,\sigma(y))} & \cdots & P(y,y) \end{pmatrix}.$$

From the form of $\overline{P}$ together with $k \geq 1$, where the k appears in the matrices above (**I**), (**II**), we can see that $\overline{P} = F(k(1),\,.\,,k(y))/X$ is a basic left *Harada* ring. Moreover, by the upper staircase form of $S(i,\sigma(i))$, we have left *Harada* rings $P = P_1 = F(k(1),\,.\,,k(y))$, P_2, P_3, ., P_{l-1}, $P_l = \overline{P}$ and canonical surjective ring homomorphisms $\varphi_i : P_i \to P_{i+1}$ with kerφ_i a simple ideal of P_i as follows:

$$P_1 \xrightarrow{\varphi_1} P_2 \xrightarrow{\varphi_2} P_3 \xrightarrow{\varphi_3} \cdots \xrightarrow{\varphi_{l-2}} P_{l-1} \xrightarrow{\varphi_{l-1}} P_l = \overline{P} = F(k(1),\,.\,,k(y))/X.$$

The following is the fundamental structure theorem (see Oshiro [17]).

Theorem E.

For a given basic QF-ring F, every upper staircase factor ring P/X of a block extension $P = F(k(1),\,.\,,k(y))$ is a basic left *Harada* ring, and, for any basic left Harada ring R, there is a basic QF-subring $F(R)$ which is called the frame QF-subring, R is represented in this form by $F(R)$.

Using theorem E, we show the following:

Theorem 3.1. *Let R be a basic indecomposable left Harada ring. If R has a simple projective right R-module, then R can be represented as an upper triangular matrix ring over a division ring as follows:*

$$R \cong \begin{pmatrix} \mathrm{D} & & & & \\ & \ddots & & 0 & \\ & & \mathrm{D} & & \\ & 0 & & \ddots & \\ & & & & \mathrm{D} \end{pmatrix}.$$

Proof. We arrange $Pi(R)$ as a well-indexed set $\{e_{ij}\}_{i=1,j=1}^{m\ \ n(i)}$. We may assume that $e_{mn(m)}R_R$ is simple. We recall the maps $\sigma, \rho : \{1, ., m\} \to \mathbb{N}$ which satisfy the condition that $(e_{i1}R; Re_{\sigma(i)\rho(i)})$ is an i-pair for any $i = 1, \ldots, m$. Now $e_{mn(m)}R_R$ is simple. We see that $\sigma(m) = m$ and $\rho(m) = n(m)$. By Theorem E there exists $\{e_{i_j1}\}_{j=1}^{y}$ ($\subseteq \{e_{i1}\}_{i=1}^{m}$) such that $\{e_{i_j1}\}_{j=1}^{y} = \{e_{\sigma(i_j)1}\}_{j=1}^{y}$ and the frame QF-subring of R is

$$F(R) = \begin{pmatrix} e_{i_11}Re_{i_11} & \cdots & e_{i_11}Re_{i_y1} \\ \cdots & \cdots & \cdots \\ e_{i_y1}Re_{i_11} & \cdots & e_{i_y1}Re_{i_y1} \end{pmatrix}.$$

And for $F(R)$, there exist $k(1), k(2), ., k(y) \in \mathbb{N}$ such that R can be represented as an upper staircase factor ring of the block extension $P = F(R)(k(1), ., k(y))$.

On the other hand, by using Theorem 3,

$$R \cong \overline{P} = \begin{pmatrix} P(1,1) & \cdots & \cdots & \overline{P(1,\sigma(1))} & \cdots & P(1,m) \\ & & \cdots & \cdots & & \\ P(i,1) & \cdots & \overline{P(i,\sigma(i))} & \cdots & & \\ & & \cdots & \cdots & & \\ P(m,1) & & \cdots & \cdots & & \overline{P(m,\sigma(m))} \end{pmatrix},$$

where for $\{e_{i1}\}_{i=1}^{m}$ we put T such that $R \cong T(n(1), ., n(m))$ and $P(k,l)$ as above sketch of Theorem 3 for every $k, l \in \mathbb{N}$. In this case, $k(1) + . + k(y) = n(1) + . + n(m)$. Here we note that $\sigma(m) = m$, $\rho(m) = n(m)$, and $e_{mn(m)}R$ appears as the last row of the m-th row block of P. Since $e_{mn(m)}R_R$ is simple, we see that $e_{i_y1}Re_{i_y1} \cong e_{mn(m)}Re_{mn(m)}$ is a division ring,

$$\overline{P(y,y)} = \begin{pmatrix} D & & & \\ & \ddots & & \\ & & D & 0 \\ & 0 & & \ddots \\ & & & & D \end{pmatrix}$$

and since R is a basic ring, then

$$(P(y,1) \cdots P(y,y-1)) = 0,$$

where we put $D := \mathrm{End}(e_{i_y1}R_R) \cong e_{i_y1}Re_{i_y1}$. Then we claim that $P(j,y) = 0$ for any $j \in \{1,\ldots,y-1\}$. Since $j \neq y$, $\sigma(j) \neq y$. It follows that, if $P(j,y) \neq 0$, then $P(j,y)P(y,\sigma(j)) \neq= 0$, which is impossible. Because for any $j \in \{1,\ldots,y-1\}$, $e_{i_j1}Re_{i_y1} \cdot e_{i_y1}Re_{\sigma(i_j)1} \supseteq S(e_{i_j1}R)e_{\sigma(i_j)1} \neq 0$. Since R is an indecomposable ring, it follows that $y = 1$, $F(R) = D$. And hence

$$R = \overline{P(1,1)} \cong$$

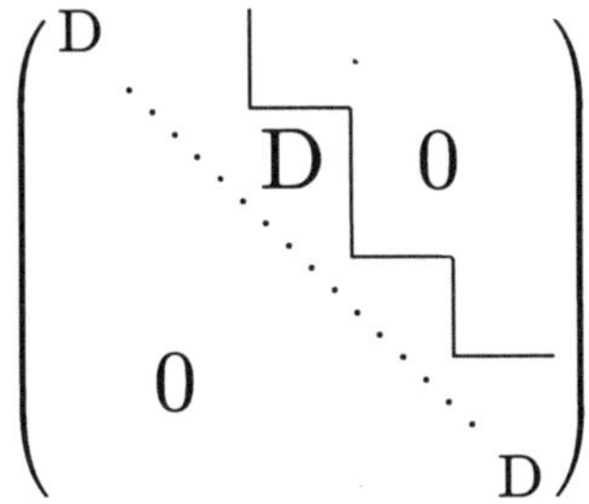

□

By Theorem 3.1 we have the following corollary.

Corollary 3.1. (*c.f.* [1, Theorem 32.8]) *Let R be a basic indecomposable Nakayama ring. If R has a simple projective right R-module, then R can be represented as a factor ring of an upper triangular matrix ring over a division ring.*

Acknowledgements

The auther thanks Professor K.Oshiro, Professor T.Sumioka and Professor Y.Baba for giving me many useful suggestions.

References

1. F. W. Anderson and K. R. Fuller, "Rings and Categories of Modules (second edition)," Graduate Texts in Math. 13, Springer-Verlag, Heidelberg/New York/Berlin (1991).
2. Y. Baba and K. Oshiro, On a Theorem of Fuller, *J. Algebra* 154, 1, 86–94(1993).
3. Y. Baba and K.Oshiro, Classical Artinian Rings and Related Topics, *Lecture note*, preprint.
4. Y. Baba and K. Iwase, On quasi-Harada rings, *J. Algebra* 185, 415–434 (1996).
5. C. Faith, "Algebra II. Ring Theory", Grundlehren Math. Wiss. 192, Springer-Verlag, Heidelberg/New York/Berlin (1976).
6. K. R. Fuller, On indecomposable injectives over artinian rings, *Pacific J. Math* 29, 115–135 (1968).

7. M. Harada, Non-small modules and non-cosmall modules, in "Ring Theory. Proceedings of 1978 Antwerp Conference" (F. Van Oystaeyen, Ed.) Dekker, New York, 669–690(1979).
8. M. Harada, On one-sided *QF-2* rings I, *Osaka J. Math.* 17, 421–431 (1980).
9. M. Harada, On one-sided *QF-2* rings II, *Osaka J. Math.* 17, 433–438 (1980).
10. M. Harada, "Factor Categories with Applications to Direct Decomposition of Modules," Lect. Notes Pure Appl. Math. 88, Dekker, New York (1983).
11. M. Morimoto and T. Sumioka, Generalizations of theorems of Fuller, *Osaka J. Math.* 34, 689–701 (1997).
12. K. Oshiro, Lifting modules, extending modules and their applications to *QF*-rings, *Hokkaido Math. J.* 13, 310–338 (1984).
13. K. Oshiro, lifting modules, extending modules and their applications to generalized uniserial rings, *Hokkaido Math. J.* 13, 339–346 (1984).
14. K. Oshiro, Structure of Nakayama rings, *Proceedings 20th Symp. Ring Theory*, Okayama, 109–133(1987).
15. K. Oshiro, On Harada rings I, *Math. J. Okayama Univ.* 31, 161–178 (1989).
16. K. Oshiro, On Harada rings II, *Math. J. Okayama Univ.* 31, 179–188 (1989).
17. K. Oshiro, On Harada rings III, *Math. J. Okayama Univ.* 32, 111–118 (1990).
18. H. Tachikawa, "Quasi-Frobenius Rings and Generalizations", Lect. Notes Math. 351, Springer-Verlag, New York/Heidelberg/Berlin (1973).
19. R. Colby and E. Rutter.Jr, Generalizations of QF-3 algebras, *Transactions of the American Math. Society* vol.153, 371-386 (1971)

ON THE RELATIVE (QUASI-)DISCRETENESS OF MODULES

Derya Keskin Tütüncü

Department of Mathematics, Hacettepe University,
06800 Beytepe Ankara, Turkey
E-mail: keskin@hacettepe.edu.tr

Nil Orhan Ertaş

Department of Mathematics, Süleyman Demirel University,
Isparta, Turkey
E-mail: orhannil@yahoo.com

In this paper we investigate the relative (quasi-)discreteness of any module M with respect to any module N. Let $0 \longrightarrow N' \longrightarrow N \longrightarrow N'' \longrightarrow 0$ be an exact sequence and let M be an N-amply supplemented module. Assume that $\mathcal{B}(M/T, N)$ is closed under supplement submodules for every factor module M/T of M. Then M is N-lifting(quasi-discrete) if and only if M is N'- and N''-lifting(quasi-discrete).

Keywords: $\mathcal{A}$-lifting module, $\mathcal{A}$-$\oplus$-supplemented module.

1. Introduction

In recent years, several authors (c.f. [1]-[5]) have studied discrete modules and other related concepts as interesting generalizations of the concept of projectivity and lifting property of modules. A comprehensive account of the theory of discrete modules can be found in[5] and.[4]

Although relative projectivity has been studied in detail, it appears that not much work has yet been done to study the discreteness of an arbitrary module M with respect to another module N, namely the N-discreteness of M. In this paper, we propose a definition for M to be N-(quasi-)discrete and derive some consequences and properties. We show that if M is $\oplus_{i=1}^{n} N_i$-lifting ((quasi-)discrete), then M is N_i-lifting ((quasi-)discrete) for every $i = 1, \cdots, n$.

Throughout this paper, R denotes a ring with identity and all modules are unital right modules unless otherwise specified. For R-modules A and B, we write $A \ll B$ ($A \leq_d B$) to denote that A is a small submodule (a direct summand) of B. Let M be a module and $A \leq B \leq M$. If $B/A \ll M/A$, then A is called a *coessential* submodule of B in M. If B has no proper coessential submodule in M, then B is called a *coclosed* submodule of M. For the other definitions we refer to[1],[4] and[5].

2. Preliminaries

Consider the following conditions for a module M.

(D_1) For every submodule K of M, there exists a direct summand N of M with $N \subseteq K$ and $K/N \ll M/N$.

(D_2) Every submodule K of M with $M/K \cong T \leq_d M$ is itself a direct summand of M.

(D_3) If A and B are direct summands of M with $M = A + B$, then $A \cap B \leq_d M$.

Then the module M is called *lifting, quasi-discrete or discrete*, respectively, if M satisfies (D_1), (D_1) and (D_3) or (D_1) and (D_2).

In,[3] Oshiro generalized the concepts of lifting, quasi-discrete and discrete modules only for submodules taken from a given family $\mathcal{A}$. This was done by considering the properties

$\mathcal{A}-(C_1)$ For all $A \in \mathcal{A}$ there exists a direct summand N of M such that $N \subseteq A$ and $A/N \ll M/N$.

$\mathcal{A}-(C_2)$ For any $A \in \mathcal{A}$ with $A \leq_d M$, any sequence $M \longrightarrow M/A \longrightarrow 0$ splits.

$\mathcal{A}-(C_3')$ For any $A \in \mathcal{A}$ and $B \leq M$ with $A, B \leq_d M$ and $M = A + B$, if $A \cap B \ll M$, then $A \cap B = 0$.

Let M be a module. Throughout this paper $\mathcal{A}$ will be a family of submodules of M which is closed under coessential submodules (namely, if $A \in \mathcal{A}$ and $A/B \ll M/B$, then $B \in \mathcal{A}$) and (α):

(α) For $A \in \mathcal{A}$ and $B \leq M$, $M/A \cong M/B$ implies that $B \in \mathcal{A}$.

Oshiro called the module M $\mathcal{A}$-*lifting*, $\mathcal{A}$-*quasi-semiperfect* or $\mathcal{A}$-*semiperfect*, respectively, if M satisfies $\mathcal{A}-(C_1)$, $\mathcal{A}-(C_1)$ and $\mathcal{A}-(C_3')$, or $\mathcal{A}-(C_1)$ and $\mathcal{A}-(C_2)$. Clearly, $\mathcal{A}$-lifting property is equivalent to the following property:

For all $A \in \mathcal{A}$ there exists a decomposition $M = M_1 \oplus M_2$ such that $M_1 \leq A$ and $M_2 \cap A \ll M_2$.

Now we define the following properties

$\mathcal{A}-(D_1)$ (or $\mathcal{A}$-lifting) M satisfies $\mathcal{A}-(C_1')$.

$\mathcal{A}-(D_2)$ For all $A \in \mathcal{A}$ with $M/A \cong N \leq_d M$, then $A \leq_d M$.

$\mathcal{A}-(D_3)$ For any $A \in \mathcal{A}$ and $B \leq M$ with $A, B \leq_d M$ and $M = A + B$, $A \cap B \leq_d M$.

$\mathcal{A}-(D_5)$ For any $A \in \mathcal{A}$ and $B \leq_d M$ with $M = A + B$, every homomorphism from $A/(A \cap B)$ to $B/(A \cap B)$ can be lifted to a homomorphism from $A/(A \cap B)$ to B.

It is easy to check that $\mathcal{A}-(C_2)$ and $\mathcal{A}-(D_2)$ are the same and $\mathcal{A}-(D_3)$ implies $\mathcal{A}-(C_3')$. If we take $\mathcal{A}$ as the class of all submodules of any module M, where M is lifting, then $\mathcal{A}-(C_3')$ implies $\mathcal{A}-(D_3)$ (see, for example, [5, Lemma 4.27]). We also note that the condition $\mathcal{A}-(D_5)$ implies the condition $\mathcal{A}-(C_5')$ in[3] for any family $\mathcal{A}$ of submodules of any module M.

Let M be a module and $\mathcal{A}$ a family of submodules of M which is closed under coessential submodules and (α). Then the module M is said to be $\mathcal{A}$-*quasi-discrete* or $\mathcal{A}$-*discrete*, respectively, if M satisfies $\mathcal{A}-(D_1)$ and $\mathcal{A}-(D_3)$ or $\mathcal{A}-(D_1)$ and $\mathcal{A}-(D_2)$. When $\mathcal{A}$ consists of all submodules of M then M is $\mathcal{A}$-lifting, $\mathcal{A}$-quasi-discrete, $\mathcal{A}$-discrete or $\mathcal{A}-(D_5)$, respectively, iff M is lifting, quasi-discrete, discrete or (D_5).

Let M be a module. M is called $\mathcal{A}$-$\oplus$-*supplemented* if every submodule N of M with $N \in \mathcal{A}$ has a direct summand supplement in M.

3. Relative (Quasi-)Discreteness

Theorem 3.1. *Let M be $\mathcal{A}-(D_2)$. Then M is $\mathcal{A}-(D_3)$.*

Proof. Let $A \in \mathcal{A}$, $B \leq M$ such that $A, B \leq_d M$ and $M = A + B$. Then $M = A \oplus A_1 = B \oplus B_1$ for some submodules A_1 and B_1 of M. Put $X = A \cap B$. Clearly, $B \cap (B_1 + X) = X$. Now $M/(B_1 \oplus X) \cong M/A$ implies that $B_1 \oplus X \in \mathcal{A}$ since $A \in \mathcal{A}$. Also, $M/(B_1 \oplus X) \cong A_1 \leq_d M$ implies that $B_1 \oplus X \leq_d M$ since M is $\mathcal{A}-(D_2)$. Thus $X = A \cap B \leq_d M$. Hence M is $\mathcal{A}-(D_3)$. □

Corollary 3.1. *Every $\mathcal{A}$-discrete module is $\mathcal{A}$-quasi-discrete.*

Lemma 3.1. *Let M be a module. M is $\mathcal{A}$-lifting if and only if every submodule A of M with $A \in \mathcal{A}$ can be written as $A = B \oplus S$ such that $B \leq_d M$ and $S \ll M$.*

Proof. Let M be $\mathcal{A}$-lifting and $A \in \mathcal{A}$. Then there exists a decomposition $M = M_1 \oplus M_2$ such that $M_1 \leq A$ and $M_2 \cap A \ll M_2$. So $A = M_1 \oplus (A \cap M_2)$ and the result follows with $B = M_1$ and $S = A \cap M_2$.

Conversely, let $A \in \mathcal{A}$. By hypothesis, $A = A_1 \oplus S$ with $A_1 \leq_d M$ and $S \ll M$. Then $M = A_1 \oplus A_2$ for some submodule A_2 of M. Therefore $A = A_1 \oplus (A \cap A_2)$. Let $\pi : M = A_1 \oplus A_2 \longrightarrow A_2$ be the projection. Then $\pi(A) = \pi(A_1 + S) = \pi(S) = \pi(A_1 + (A \cap A_2)) = \pi(A \cap A_2) = A \cap A_2$. Since $S \ll M$, $\pi(S) = A \cap A_2 \ll A_2$ by [5, Lemma 4.2(3)]. Thus M is $\mathcal{A}$-lifting□

Lemma 3.2. *Let M be $\mathcal{A}$-lifting. Then every submodule of M in $\mathcal{A}$ has a direct summand supplement in M and every coclosed submodule of M in $\mathcal{A}$ is a direct summand of M.*

Proof. Let $A \in \mathcal{A}$. Then there exists a decomposition $M = T \oplus K$ such that $T \leq A$ and $K \cap A \ll K$. Clearly, $M = A + K$. So, K is a direct summand supplement of A in M.

Now let $P \in \mathcal{A}$ be a coclosed submodule of M. By Lemma 3.1, $P = P_1 \oplus S$ with $P_1 \leq_d M$ and $S \ll M$. By [1, Lemma 1.1], $S \ll P$ and hence $S = 0$. Therefore P is a direct summand of M. □

Corollary 3.2. *Every $\mathcal{A}$-lifting module is $\mathcal{A}$-$\oplus$-supplemented.*

Lemma 3.3. *Let M be a module. Then M is $\mathcal{A} - (D_3)$ if and only if for every direct summands A and B of M with $A \in \mathcal{A}$ and $M = A + B$, there exists a submodule B' of M such that $B' \leq B$ and $M = A \oplus B'$.*

Proof. Assume M is $\mathcal{A}-(D_3)$. Let $A, B \leq_d M$ with $A \in \mathcal{A}$ and $M = A+B$. Then $A \cap B \leq_d M$ and hence $M = (A \cap B) \oplus L$ for some submodule L of M. Now $B = (A \cap B) \oplus (B \cap L)$. Put $B' = B \cap L$. Therefore $M = A \oplus B'$ and $B' \leq B$.

Conversely, let $A, B \leq_d M$ with $A \in \mathcal{A}$ and $M = A + B$. Then by hypothesis, there exists a submodule B' of M such that $B' \leq B$ and $M = A \oplus B'$. Then $B = B' \oplus (A \cap B)$. Hence $A \cap B \leq_d M$ since $B \leq_d M$□

Proposition 3.1. *Let M be a module. Assume that for every direct summands A and B of M with $A \in \mathcal{A}$ and $M = A \oplus B$, B is A-projective. Then M is $\mathcal{A} - (D_3)$.*

Proof. Let $A, B \leq_d M$ with $A \in \mathcal{A}$ and $M = A + B$. $M = A \oplus A_1$ for some submodule A_1 of M. By hypothesis, A_1 is A-projective. By [4, 41.14], there exists a submodule B' of M such that $B' \leq B$ and $M = A \oplus B'$. By Lemma 3.3, M is $\mathcal{A} - (D_3)$. □

Let M be any module. Consider the following property
(*) If $A \in \mathcal{A}$ and $B \leq M$ with $M = A + B$, then $B \in \mathcal{A}$.

Proposition 3.2. *Let M be a module and $\mathcal{A}$ satisfies (*). M is $\mathcal{A}$-quasi-discrete if and only if for every decomposition $M = A \oplus B$ with $A, B \in \mathcal{A}$, A and B are relatively projective and M is $\mathcal{A}$-lifting.*

Proof. Assume M is $\mathcal{A}$-quasi-discrete. Let $M = A \oplus B$ with $A \in \mathcal{A}$. Let $M = A + T$ for any submodule T of M. By (*), $T \in \mathcal{A}$. Then there exists a direct summand T' of M such that $T' \leq T$ and $T/T' \ll M/T'$ since M is $\mathcal{A}$-lifting. Then $M = A + T'$. By Lemma 3.3, there exists a submodule T^* of M such that $T^* \leq T'$ and $M = A \oplus T^*$. Therefore B is A-projective by [4, 41.14]. By the same way, A is B-projective.

Conversely, assume that M is $\mathcal{A}$-lifting and for every decomposition $M = A \oplus B$ with $A \in \mathcal{A}$, A and B are relatively projective. Let $M = A \oplus B$ with $A \in \mathcal{A}$. By Proposition 3.1, M is $\mathcal{A} - (D_3)$. Hence M is $\mathcal{A}$-quasi-discrete. □

Theorem 3.2. *Let M be $\mathcal{A}$-lifting. Then M is $\mathcal{A} - (D_3)$ if and only if M is $\mathcal{A} - (D_5)$.*

Proof. Firstly, we prove that if M is $\mathcal{A} - (D_5)$, then M is $\mathcal{A} - (D_3)$ without assuming M is $\mathcal{A}$-lifting. Let $A, B \leq_d M$, $A \in \mathcal{A}$ and $M = A + B$. $M = A \oplus A_1 = B \oplus B_1$ for some submodules A_1, B_1 of M. Put $X = A \cap B$ and $\overline{M} = M/X$. Then $\overline{M} = \overline{A} \oplus \overline{B} = \overline{A} \oplus \overline{A_1} = \overline{B} \oplus \overline{B_1}$. Let π and π_1 be the projections $\overline{M} = \overline{B} \oplus \overline{B_1} \longrightarrow \overline{B}$ and $\overline{M} = \overline{B} \oplus \overline{B_1} \longrightarrow \overline{B_1}$, respectively. Then it is clear that $\overline{A} = \{\pi(\overline{a}) + \pi_1(\overline{a}) \mid a \in A\}$, $\pi_1(\overline{A}) = \overline{B_1}$ and the map $f : \overline{B_1} \longrightarrow \overline{B}$ defined by $f(\pi_1(\overline{a})) = \pi(\overline{a})$ is well-defined. As we do in the proof of Theorem 3.1, $B_1 + X \in \mathcal{A}$. Since M is $\mathcal{A} - (D_5)$, the homomorphism $f : (B_1 + X)/X \longrightarrow B/X$ can be lifted to a homomorphism $g : (B_1 + X)/X \longrightarrow B$, where $\pi_X : B \longrightarrow B/X$ is the natural epimorphism. Let $\alpha : B_1 + X \longrightarrow (B_1 + X)/X$ be the natural epimorphism. Then we have the homomorphism $g\alpha : B_1 + X \longrightarrow B$. Put $P = \{b_1 + (g\alpha)(b_1) \mid b_1 \in B_1\}$. We prove that $A = P \oplus X$. Let $x \in X \cap P$. Then $x = b_1 + (g\alpha)(b_1)$ for some $b_1 \in B_1$ and hence $b_1 \in B \cap B_1 = 0$. Therefore $X \cap P = 0$. Let $b_1 + (g\alpha)(b_1) \in P$, where $b_1 \in B_1$. Since $\pi_X g\alpha(b_1) = f\alpha(b_1)$, $g\alpha(b_1) + X = f(b_1 + X)$. Since $\pi_1(\overline{A}) = \overline{B_1}$, there exists $a \in A$ such that $b_1 + X = \pi_1(a + X)$.

Hence $g\alpha(b_1) + X = f(\pi_1(a + X)) = \pi(a + X)$. Let $a = b + b_1'$, where $b \in B$ and $b_1' \in B_1$. Then $\overline{a} = \overline{b} + \overline{b_1'}$, $\pi_1(\overline{a}) = \overline{b_1'} = \overline{b_1}$ and $\pi(\overline{a}) = \overline{b}$. Now $g\alpha(b_1)+X = b+X$ implies that $b_1+g\alpha(b_1)+X = b_1+b+X = b_1'+b+X$. So $b_1 + g\alpha(b_1) - a = x$ for some $x \in X$. Thus $b_1 + g\alpha(b_1) \in A$. This means that $P+X \subseteq A$. Conversely, let $a \in A$ and assume that $a = b+b_1'$, where $b \in B$, $b_1' \in B_1$. Then $\overline{a} = \pi(\overline{a})+\pi_1(\overline{a}) = f(\pi_1(\overline{a}))+\pi_1(\overline{a})$. There exists an element $b_1 \in B_1$ such that $\pi_1(\overline{a}) = \overline{b_1}$. Therefore $\overline{a} = \pi_X g(\overline{b_1}) + \pi_1(\overline{a}) = g(\overline{b_1}) + \pi_1(\overline{a})+X = g\alpha(b_1)+\pi_1(\overline{a})+X = g\alpha(b_1)+b_1+X$. Now $a-g\alpha(b_1)-b_1 \in X$, namely $a \in P + X$. Thus $A = P \oplus X$. Therefore $X = A \cap B \leq_d M$. Hence M is $\mathcal{A} - (D_3)$.

Now assume that M is $\mathcal{A}$-lifting and $\mathcal{A} - (D_3)$, namely M is $\mathcal{A}$-quasi-discrete. We show that M is $\mathcal{A}-(D_5)$. Let $A \in \mathcal{A}$, $B \leq_d M$ and $M = A+B$. Put $X = A \cap B$ and $\overline{M} = M/X$. Let $f : A/X \longrightarrow B/X$ be a homomorphism and $\pi : B \longrightarrow B/X$ the natural epimorphism. Let $C = \{x \in M \mid \overline{x} = \overline{a} + f(\overline{a}) \text{ for some } a \in A\}$. Then C is a submodule of M containing X. It is not hard to see that $\overline{M} = \overline{C} \oplus \overline{B}$. Since $M/C \cong \overline{B} \cong M/A$, $C \in \mathcal{A}$. Since M is $\mathcal{A}$-lifting, there exists a direct summand C_1 of M such that $C_1 \leq C$ and $C/C_1 \ll M/C_1$. Then $C_1 \in \mathcal{A}$ since $\mathcal{A}$ is closed under coessential submodules. Now $M = C_1 + B$. Since M is $\mathcal{A}-(D_3)$, $M = (C_1 \cap B) \oplus T$ for some submodule T of M. Note that $C_1 \cap B \subseteq X$ and $B = (C_1 \cap B) \oplus (T \cap B)$. Define the homomorphism $\varphi : A/X \longrightarrow B$ by $\varphi(a + X) = y$, where $a \in A$, $f(a+X) = x+y+X = y+X$ and $x \in C_1 \cap B$, $y \in T \cap B$. Clearly, φ lifts f. □

Proposition 3.3. *Let M be $\mathcal{A} - (D_5)$. Suppose that if $A \in \mathcal{A}$ and $A \leq B$, then $B \in \mathcal{A}$. Then for all $A \in \mathcal{A}$ and all $B \leq_d M$ with $M = A + B$, M/B is M/A-projective.*

Proof. Let $A \in \mathcal{A}$, $B \leq_d M$ and $M = A + B$. Let $f : M/B \longrightarrow (M/A)/(T/A)$ be any homomorphism and $\pi_{T/A} : M/A \longrightarrow (M/A)/(T/A)$ the natural epimorphism where $T/A \leq M/A$. Let $\eta : (M/A)/(T/A) \longrightarrow M/T$ be the obvious isomorphism. Then we have the homomorphism $\eta f : M/B \longrightarrow M/T$. Clearly, $M = T + B$ and by hypothesis on $\mathcal{A}$, $T \in \mathcal{A}$. Let $\alpha : T/(T \cap B) \longrightarrow M/B$ and $\beta : M/T \longrightarrow B/(T \cap B)$ be the obvious isomorphisms. So we have the homomorphism $\varphi = \beta\eta f\alpha : T/(T \cap B) \longrightarrow B/(T \cap B)$. Let $\pi_{T\cap B} : B \longrightarrow B/(T \cap B)$ be the natural epimorphism. Then there exists a homomorphism $g : T/(T \cap B) \longrightarrow B$ such that $\pi_{T\cap B}\, g = \varphi$. Let $\psi = \pi_A i g \alpha^{-1} : M/B \longrightarrow M/A$, where $i : B \longrightarrow M$ is the inclusion map and $\pi_A : M \longrightarrow M/A$ is the natural epimorphism. It is easy to see that $\pi_{T/A}\psi = f$. Therefore M/B is M/A-projective. □

Corollary 3.3. *Let M be a (D_3)-module. Then whenever $M = A \oplus B$, A and B are relatively projective.*

Let M and N be modules. In,[2] it is defined the class
$\mathcal{B}(M,N) = \{A \leq M \mid \exists X \leq N, \exists f \in Hom(M, N/X), \mathrm{Ker}f/A \ll M/A\}$.

By [2, Lemma 2.2], this class $\mathcal{B}(M,N)$ is closed under coessential submodules. The next lemma shows that $\mathcal{B}(M,N)$ is closed under the property (α).

Lemma 3.4. *The class $\mathcal{B}(M,N)$ is closed under the property (α).*

Proof. Let $A \in \mathcal{B}(M,N)$, $B \leq M$ and $M/A \cong M/B$. Let η be the isomorphism from M/B to M/A. Since $A \in \mathcal{B}(M,N)$, there exist a submodule X of N and a homomorphism $f : M \longrightarrow N/X$ with $\mathrm{Ker}f/A \ll M/A$. Define the homomorphism $\beta : M \longrightarrow N/X$ by $m \mapsto n + X$, where $m \in M$, $\eta(m+B) = m_1 + A$ for some $m_1 \in M$ and $f(m_1) = n + X$ for some $n \in N$. β is well-defined since $A \leq \mathrm{Ker}f$. Assume $b \in B$. Then $\eta(b+B) = 0 + A$ and $f(0) = 0 + X$ and so $\beta(b) = 0 + X$. Therefore $B \leq \mathrm{Ker}\beta$. Since $\eta^{-1}(\mathrm{Ker}f/A) = \mathrm{Ker}\beta/B$, $\mathrm{Ker}\beta/B \ll M/B$. Thus $B \in \mathcal{B}(M,N)$. □

Let M and N be modules. In,[2] it is defined for M to be *N-lifting*, *N-quasi-discrete* or *N-discrete*, respectively, when M is $\mathcal{A}$-lifting, $\mathcal{A}$-quasi-discrete or $\mathcal{A}$-discrete for $\mathcal{A} = \mathcal{B}(M,N)$. It is easy to see that M is lifting (quasi-)discrete if and only if M is M-lifting (quasi-)discrete if and only if M is N-lifting (quasi-)discrete for every module N.

Example 3.1. (i) Clearly $\mathcal{B}(M,0) = \{M\}$ for any module M. Therefore every module M is 0-lifting ((quasi-)discrete). This means that the $\mathbb{Z}$-modules $\mathbb{Z}_{\mathbb{Z}}$ and $\mathbb{Q}_{\mathbb{Z}}$ are 0-lifting ((quasi-)discrete). But we know that they are not lifting.
(ii) Let S be any simple projective module and M any module. Then clearly the class $\mathcal{B}(M,S)$ is contained in the class of all direct summands of M. Therefore M is S-lifting.

Lemma 3.5. *Let $f : N' \longrightarrow N$ be a monomorphism and $g : N \longrightarrow N''$ be an epimorphism. Then $\mathcal{B}(M,N')$ and $\mathcal{B}(M,N'') \subseteq \mathcal{B}(M,N)$.*

Proof. Let $A \in \mathcal{B}(M,N')$. There exist a submodule $X \leq N'$ and a homomorphism $\alpha : M \longrightarrow N'/X$ such that $\mathrm{Ker}\alpha/A \ll M/A$. Define the homomorphism $\beta : N'/X \longrightarrow N/f(X)$ by $n' + X \mapsto f(n') + f(X)$, where

$n' \in N$. Clearly, $\mathrm{Ker}\beta = 0$. Then $\mathrm{Ker}\beta\alpha = \alpha^{-1}(\mathrm{Ker}\beta) = \mathrm{Ker}\alpha$. Therefore $A \in \mathcal{B}(M, N)$.

Let $B \in \mathcal{B}(M, N'')$. There exist a submodule $T \leq N''$ and a homomorphism $h : M \longrightarrow N''/T$ such that $\mathrm{Ker}h/B \ll M/B$. Define the homomorphism $\eta : N''/T \longrightarrow N/g^{-1}(T)$ by $n'' + T \mapsto n + g^{-1}(T)$, where $n'' \in N''$ and $g(n) = n''$ for some $n \in N$. Clearly, $\mathrm{Ker}\eta = 0$. Then $\mathrm{Ker}\eta h = h^{-1}(\mathrm{Ker}\eta) = \mathrm{Ker}h$. Therefore $B \in \mathcal{B}(M, N)$. □

Theorem 3.3. *Let M be a module. Let $f : N' \longrightarrow N$ be a monomorphism and $g : N \longrightarrow N''$ an epimorphism. If M is N-lifting (respectively (quasi-)discrete), then M is N'- and N''-lifting (respectively (quasi-)discrete).*

Proof. It is clear by Lemma 3.5. □

Corollary 3.4. *If M is $N = \oplus_{i=1}^{n} N_i$-lifting ((quasi-)discrete), then M is N_i-lifting ((quasi-)discrete) for every $i = 1, \cdots, n$.*

Let M be a module. Following,[2] M is called *N-amply supplemented* if for any submodules A, B of M with $A \in \mathcal{B}(M, N)$ and $M = A + B$ there exists a supplement P of A in M such that $P \leq B$ and M is called *N-weakly supplemented* if for any submodule A of M with $A \in \mathcal{B}(M, N)$, there exists a submodule B of M such that $M = A + B$ and $A \cap B \ll M$. Then every N-amply supplemented module is N-weakly supplemented. It is not hard to see that every epimorphic image of an N-weakly supplemented module is again N-weakly supplemented.

Lemma 3.6. *[2, Lemma 3.12] Let M be a module and $K \in \mathcal{B}(M, N)$. Consider the following conditions:*

(1) K is a supplement submodule of M.
(2) K is coclosed in M.
(3) For all $A \leq K$, $A \ll M$ implies that $A \ll K$.

Then (1)$\Longrightarrow$(2)$\Longrightarrow$(3) hold. If M is N-weakly supplemented, then (3)$\Longrightarrow$(1).

Theorem 3.4. *Let $0 \longrightarrow N' \xrightarrow{f} N \xrightarrow{g} N'' \longrightarrow 0$ be an exact sequence and let M be an N-amply supplemented module. Assume that $\mathcal{B}(M/T, N)$ is closed under supplement submodules for every factor module M/T of M (namely, every supplement submodule of M/T belongs to $\mathcal{B}(M/T, N)$ for every factor module M/T of M). Then M is N-lifting if and only if it is both N'-lifting and N''-lifting.*

Proof. If M is N-lifting, then it is both N'-lifting and N''- lifting by Theorem 3.3. Conversely, suppose that M is both N'-lifting and N''-lifting. We want to show that M is N-lifting. By [2, Proposition 3.15], it is enough to show that every coclosed (namely, supplement) submodule K of M with $K \in \mathcal{B}(M, N)$ is a direct summand of M. Let $K \in \mathcal{B}(M, N)$ and K a coclosed submodule of M. There exist a submodule $L \leq N$ and a homomorphism $\varphi : M \longrightarrow N/L$ such that $\mathrm{Ker}\varphi/K \ll M/K$. Let $L' = \varphi^{-1}((L + \mathrm{Ker}g)/L)$. Then $\mathrm{Ker}\varphi \leq L'$ and so $K \leq L'$. By [2, Lemma 3.13], M/K is N-amply supplemented and by hypothesis, $\mathcal{B}(M/K, N)$ is closed under supplement submodules. Define the homomorphism $\eta : M \longrightarrow N''/g(L)$ by $m \mapsto g(n) + g(L)$, where $\varphi(m) = n + L$ for some $n \in N$. Let $m \in M$ and $\eta(m) = 0 + g(L)$. Then $g(n) + g(L) = 0 + g(L)$, where $\varphi(m) = n + L$. Now $n \in L + \mathrm{Ker}g$ implies that $\varphi(m) \in (L + \mathrm{Ker}g)/L$ and hence $m \in L'$. Therefore $\mathrm{Ker}\eta \subseteq L'$. Let $l' \in L'$, namely $\varphi(l') \in (L + \mathrm{Ker}g)/L$. Assume that $\varphi(l') = x + L$, where $x \in \mathrm{Ker}g$. Then $\eta(l') = g(x) + g(L) = 0 + g(L)$. Thus $L' = \mathrm{Ker}\eta$. So, $L' \in \mathcal{B}(M, N'')$. By Proposition 3.5, $L' \in \mathcal{B}(M, N)$. By [2, Lemma 2.2], $L'/K \in \mathcal{B}(M/K, N)$. By [2, Proposition 3.14], L'/K has an s-closure in M/K, namely there exists a coclosed submodule K'/K of M/K with $K'/K \leq L'/K$ and $L'/K' \ll M/K'$. Now by [2, Lemma 2.2], $K'/K \in \mathcal{B}(M/K, N)$. Then by [1, Lemma 1.4(1)] and Lemma 3.6, K' is coclosed in M. Now $K' \in \mathcal{B}(M, N'')$ since $L'/K' \ll M/K'$. Since K' is a coclosed submodule of M in $\mathcal{B}(M, N'')$, K' is a direct summand of M by Lemma 3.2. Let $M = K' \oplus K''$ for some submodule K'' of M. Clearly, $K = (K \oplus K'') \cap K'$ and hence $M/K = ((K \oplus K'')/K) \oplus (K'/K)$. Therefore $(K \oplus K'')/K$ is a supplement submodule of M/K. By [1, Lemma 1.4(1)], $K \oplus K''$ is a supplement submodule of M. We claim that $K \oplus K'' \in \mathcal{B}(M, N')$. Clearly, $\varphi(L') \subseteq (L + \mathrm{Ker}g)/L$. Define the homomorphism $\varphi_1 : M \longrightarrow N'/f^{-1}(L)$ by $k' + k'' \mapsto n' + f^{-1}(L)$, where $k' \in K'$, $k'' \in K''$, $\varphi(k') = x + L$ with $x \in \mathrm{Ker}g$ and since $\mathrm{Ker}g = \mathrm{Im}f$, $x = f(n')$ for some $n' \in N'$. It is easy to check that $K \oplus K'' \subseteq \mathrm{Ker}\varphi_1 \subseteq \mathrm{Ker}\varphi + K''$ and $\mathrm{Ker}\varphi_1/(K \oplus K'') \ll M/(K \oplus K'')$. Therefore $K \oplus K''$ is a supplement submodule of M in $\mathcal{B}(M, N')$. Since M is N'-lifting, $K \oplus K''$ is a direct summand of M by Lemma 3.2 and hence K is a direct summand of M. □

Theorem 3.5. *Let* $0 \longrightarrow N' \xrightarrow{f} N \xrightarrow{g} N'' \longrightarrow 0$ *be an exact sequence and let* M *be an* N*-amply supplemented module. Assume that* $\mathcal{B}(M/T, N)$ *is closed under supplement submodules for every factor module* M/T *of* M*. Then* M *is* N*-quasi-discrete if and only if it is both* N'*-quasi-discrete and* N''*-quasi-discrete.*

Proof. If M is N-quasi-discrete, then it is both N'-quasi-discrete and N''-quasi-discrete by Theorem 3.3.

Conversely, suppose that M is both N'-quasi-discrete and N''-quasi-discrete. By Theorem 3.4, M is N-lifting. Let K and H be the direct summands of M with $K \in \mathcal{B}(M,N)$ and $M = K + H$. By the proof of Theorem 3.4, $K = K' \cap (K \oplus K'')$ for some direct summands K' and $K \oplus K''$ of M with $K' \in \mathcal{B}(M,N'')$, $K \oplus K'' \in \mathcal{B}(M,N')$ and $M = K' \oplus K''$. Clearly, $M = (K \oplus K'') + H$. Since M is $N' - (D_3)$, $(K \oplus K'') \cap H \leq_d M$. Now $[H \cap (K \oplus K'')] + K = (K \oplus K'') \cap (K + H) = K \oplus K''$ implies that $M = K + K'' + K' = [H \cap (K \oplus K'')] + K + K' = [H \cap (K \oplus K'')] + K'$. Since M is $N'' - (D_3)$, $H \cap (K \oplus K'') \cap K' = H \cap K \leq_d M$. Thus M is $N - (D_3)$ and hence N-quasi-discrete. □

Acknowledgment

The first author would like to thank Hacettepe University for the 05 G 602 001 Project.

References

1. D.Keskin, *On Lifting Modules*, Comm. Algebra, **28(7)**, 3427-3440 (2000).
2. D.Keskin and A.Harmanci, *A Relative Version of the Lifting Property of Modules*, Algebra Colloquium, **11(3)**, 361-370 (2004).
3. K.Oshiro, *Semiperfect Modules and Quasi-semiperfect Modules*, Osaka J. Math., **20**, 337-372 (1983).
4. R.Wisbauer, *Foundations of Module and Ring Theory*, Gordon and Breach, Philadelphia (1991).
5. S.H.Mohamed and B.J.Müller, *Continuous and Discrete Modules*, London Math. Soc. LNS 147 Cambridge Univ. Press, Cambridge (1990).

COMPATIBLE ALGEBRA STRUCTURES OF LIE ALGEBRAS

F.KUBO*

Department of Applied Mathematics, Graduate School of Engeering, Hiroshima University, Higashi Hiroshima, JAPAN
E-mail: remakubo@amath.hiroshima-u.ac.jp

A compatible algebra products of a finite-dimensional semisimple Lie algebra $\mathfrak{g}$ with a Lie bracket $[-,-]$ are studied. If $\mathfrak{g}$ is simple of type A_1 or not of type A_n of $n \geqq 2$, then the compatible algebra products must be the scalar multiples of the Lie bracket $[-,-]$. In case that $\mathfrak{g}$ is simple of type A_n of $n \geqq 2$, such a product is a sum of a scalar multiple of $[-,-]$ and a deformed one of the ordinal associative products on the full $(n+1) \times (n+1)$ matix algebra. Then we give a alternative proof to the triviality of the compatible associative algebra structures of a semisimple Lie algebra $\mathfrak{g}$.

Keywords: Lie Algebra; Poisson Algebra.

1. Introduction

Let k be an algebraically closed field of characteristic zero. Let $\mathfrak{g}$ be a Lie algebra over k. A *compatible* algebra product, being denoted by $x * y$ for $x, y \in \mathfrak{g}$, satisfies the Leibniz law

$$[x, y * z] = [x, y] * z + y * [x, z]$$

for $x, y, z \in \mathfrak{g}$, in other words, $\mathfrak{g}$ is a $*$-algebra on which every adjoint map of $\mathfrak{g}$ acts as an $*$-derivation $(\mathrm{ad}x(y * z) = \mathrm{ad}x(y) * z + y * \mathrm{ad}x(z))$.

A compatible product is an element of $\mathrm{Hom}_{\mathfrak{g}}(\mathfrak{g} \otimes \mathfrak{g}, \mathfrak{g})$ the set of all $\mathfrak{g}-$module homomorphisms of $\mathfrak{g} \otimes \mathfrak{g}$ into $\mathfrak{g}$. In fact, A mapping ϕ of $\mathfrak{g} \otimes \mathfrak{g}$ into $\mathfrak{g}$ defined by $\phi(x \otimes y) = x * y$ satisfies $\phi([x, y \otimes z] = [x, \phi(y \otimes z)]$. Our goals of this paper are followings:

1) We will give the explicit forms of $\mathrm{Hom}_{\mathfrak{g}}(\mathfrak{g} \otimes \mathfrak{g}, \mathfrak{g})$.

*The author gratefully wishes to acknowledge the support of Japan Society for the Promotion of Science under Grant-in-Aid for Scientific Research (18540035)

2) We will give a alternative proof to the triviality of a compatible associative product, that is, $\mathfrak{g} * \mathfrak{g} = 0$.

Note: In our paper [3], we have shown the second assertion in more general setting. This can be stated as follows in our contex:

If $\phi \in \mathrm{Hom}_{\mathfrak{g}}(\mathfrak{g} \otimes \mathfrak{g}, \mathfrak{g})$ and $\phi(\phi(x \otimes y) \otimes z) = \phi(x \otimes \phi(y \otimes z))$ then $\phi = 0$.

2. Typical elements of $\mathrm{Hom}_{\mathfrak{g}}(\mathfrak{g} \otimes \mathfrak{g}, \mathfrak{g})$

One typical element of $\mathrm{Hom}_{\mathfrak{g}}(\mathfrak{g} \otimes \mathfrak{g}, \mathfrak{g})$ is given by

$$L(x \otimes y) := [x, y] \tag{$*$}$$

for $x, y \in \mathfrak{g}$. We construct another type of elements of $\mathrm{Hom}_{\mathfrak{g}}(\mathfrak{g} \otimes \mathfrak{g}, \mathfrak{g})$ for the case that $\mathfrak{g}$ is of type A_n of $n \geqq 2$. Let A be an associative algebra over k, writing xy the a ssociative product of $x, y \in A$, and denote by $[A]$ the Lie algebra A whose Lie bracket $[-,-]$ is given by the ordinal associative commutator $[x, y] := xy - yx$ for $x, y \in A$. We choose an invariant form $\alpha : [A] \times [A] \to k$, so that $\alpha([x, y], z) = \alpha(x, [y, z])$ for $x, y, z \in A$ ([2, III.4]), and an element z_0 of the center of the associative algebra A. Then one can have an $[A]$−module homomorphism ϕ of $[A] \otimes [A]$ to $[A]$ defined by

$$\phi(x \otimes y) = xy - \alpha(x, y) z_0$$

for $x, y \in A$. Now we consider the case that A is the full $(n+1) \times (n+1)$ matrix algebra $M_{n+1}(k)$ and define a $[M_{n+1}(k)]$-module homomorphism of $M_{n+1}(k) \otimes M_{n+1}(k)$ to $M_{n+1}(k)$ by

$$D(x \otimes y) := xy - \frac{1}{n+1} \mathrm{Tr}(xy) E \tag{$**$}$$

for $x, y \in M_{n+1}(k)$, where E is the unit matrix in $M_{n+1}(k)$ and $\mathrm{Tr}(xy)$ is the trace of the matrix xy. Let $\mathfrak{sl}_{n+1}(k)$ be the set of all matrices having trace zero in $M_{n+1}(k)$. It is immediate that D can be regarded as a $\mathfrak{sl}_{n+1}(k)$-module homomorphism of $\mathfrak{sl}_{n+1}(k) \otimes \mathfrak{sl}_{n+1}(k)$ to $\mathfrak{sl}_{n+1}(k)$.

We can now state the main theorem (one of our goals) as follows:

Theorem *Let $\mathfrak{g}$ be a simple Lie algebra.*

(1) *If $\mathfrak{g}$ is not of type A_n of $n \geq 2$ then*

$$\mathrm{Hom}_{\mathfrak{g}}(\mathfrak{g} \otimes \mathfrak{g}, \mathfrak{g}) = kL.$$

(2) *If $\mathfrak{g}$ is of type A_n of $n \geq 2$ then*

$$\mathrm{Hom}_{\mathfrak{g}}(\mathfrak{g} \otimes \mathfrak{g}, \mathfrak{g}) = kL \oplus kD.$$

Here L is of (∗) and D is of (∗∗).

3. Notations and the usefull elements in H

Here we state some basic notion: Let $\mathfrak{g}$ be a finite-dimensional simple Lie algebra over k, H a Cartan subalgebra of $\mathfrak{g}$, $\{\alpha_1, \ldots, \alpha_n\}$ the set of simple roots and $C = (C_{ij})$ the Cartan matrix for $\mathfrak{g}$ relative to H. We choose Chevalley generators $e_1, \ldots, e_n, h_1, \ldots, h_n, f_1, \ldots, f_n$ of $\mathfrak{g}$, so that e_i is a non-zero root vector of the root α_i, f_i of $-\alpha_i$,

$$[e_i, f_i] = h_i \quad \text{and} \quad C = (\alpha_i(h_j)).$$

Let $\kappa(x, y) = \mathrm{Tr}((\mathrm{ad}_{\mathfrak{g}} x)(\mathrm{ad}_{\mathfrak{g}} y))$ be the Killing form of $\mathfrak{g}$ (or H) and we denote by $(-,-)$ the transfered Killing form to H^* the dual space of H and $< \alpha, \beta >= 2(\alpha, \beta)/(\beta, \beta)$ for α, $\beta \in H^*$ with $\beta \neq 0$. By the formula ${}^t(\alpha_1, \ldots, \alpha_n) = C^t(\lambda_1, \ldots, \lambda_n)$, we have the fundamental dominant weights $\lambda_1, \ldots, \lambda_n$, so that $< \lambda_i, \alpha_j > (= 2(\lambda_i, \alpha_j)/(\alpha_j, \alpha_j)) = \delta_{ij}$, equivalently, $\lambda_i(h_j) = \delta_{ij}$.

Now let us introduce the elements H_i's. Define the elements $H_1, \ldots, H_n$ of H by

$$(H_1, \ldots, H_n)C = (h_1, \ldots, h_n).$$

Then we have $\alpha_i(H_j) = \delta_{ij}$. This complete the following table of the correspondance between the elements of H and those of H^*:

H^*		H	: the rule of correspondance
α	$\longleftrightarrow$	t_α	: $\alpha(h) = \kappa(t_\alpha, h)$
(α, β)	$=$	$\kappa(t_\alpha, t_\beta)$	: transfer
α_i	$\longleftrightarrow$	h_i	: $h_i = 2t_{\alpha_i}/(\alpha_i, \alpha_i)$
λ_i	$\longleftrightarrow$	h_i	: $\lambda_i(h_j) = \delta_{ij}$
α_i	$\longleftrightarrow$	H_i	: $\alpha_i(H_j) = \delta_{ij}$

Here $< \alpha, \beta >= 2(\alpha, \beta)/(\beta, \beta)$, and

$$< \alpha_i, \alpha_j >= C_{ij} = \alpha_i(h_j).$$

4. Proof of the theorem

Let γ be the highest root of $\mathfrak{g}$. If $\mathfrak{g}$ is not of type A_n of $n \geqq 1$, then $\gamma = \lambda_p$ for some fundamental dominant weight λ_p, and if $\mathfrak{g}$ is of type A_n of $n \geqq 2$ then $\gamma = \lambda_1 + \lambda_n$. For the convenience of the readers, we recall the tables of [1, Table 2(p.66), Table 1(p69)] of the correspondance between the highest roots and the fundamental weights, and put it at the end of this proof.

Observing the α_i−string through γ one can easily see that $\gamma - \alpha_i$ is not a root for α_i with $<\gamma, \alpha_i>= 0$. Now take nonzero root vectors e and f of γ and $-\gamma$ respectively. Then $\mathfrak{g} \otimes \mathfrak{g} = U(\mathfrak{g})(e \otimes f)$ and we have $[e_i, e \otimes f] = 0$ for such e_i's that $\gamma(h_i) = 0$.

Let ϕ be any element of $\mathrm{Hom}_{\mathfrak{g}}(\mathfrak{g} \otimes \mathfrak{g}, \mathfrak{g})$ and put $h_0 = \phi(e \otimes f)$. Then h_0 must satisfies $\alpha_i(h_0) = 0$ for α_i with $\gamma(h_i) = 0$. By taking another basis $\{H_1, \ldots, H_n\}$ of H introduced before, we have $h_0 = \sum_{i=1}^{n} \alpha_i(h_0)H_i$. Hence if $\mathfrak{g}$ is not of type A_n of $n \geqq 2$ then $h_0 \in kH_p$, and if $\mathfrak{g}$ is of type A_1 then $h_0 \in kH_1 = kh_1$. For these two cases one therefore gets $\dim \mathrm{Hom}_{\mathfrak{g}}(\mathfrak{g} \otimes \mathfrak{g}, \mathfrak{g}) \leqq 1$. In the case that $\mathfrak{g}$ is of type A_n of $n \geqq 2$ we have $h_0 \in kH_1 + kH_n$, hence $\dim \mathrm{Hom}_{\mathfrak{g}}(\mathfrak{g} \otimes \mathfrak{g}, \mathfrak{g}) \leqq 2$. Theorem follows from the fact that the $\mathfrak{g}$-module homomorphisms L and D cover the space $\mathrm{Hom}_{\mathfrak{g}}(g \otimes \mathfrak{g}, \mathfrak{g})$ for all the types of $\mathfrak{g}$.

Type	highest root		f'd. weight
A_n :	$\alpha_1 + \cdots + \alpha_n$	$\leftrightarrow$	$\lambda_1 + \lambda_n$
A_1 :	α_1	$\leftrightarrow$	$2\lambda_1$
B_n :	$\alpha_1 + 2\alpha_2 + \cdots + 2\alpha_n$	$\leftrightarrow$	λ_2
C_n :	$2\alpha_1 + \cdots + 2\alpha_{n-1} + \alpha_n$	$\leftrightarrow$	$2\lambda_1$
D_n :	$\alpha_1 + 2\alpha_2 + \cdots + 2\alpha_{n-2} + \alpha_{n-1} + \alpha_n$	$\leftrightarrow$	λ_2
E_6 :	$\alpha_1 + 2\alpha_2 + 2\alpha_3 + 3\alpha_4 + 2\alpha_5 + \alpha_6$	$\leftrightarrow$	λ_2
E_7 :	$2\alpha_1 + 2\alpha_2 + 3\alpha_3 + 4\alpha_4 + 3\alpha_5 + 2\alpha_6 + \alpha_7$	$\leftrightarrow$	λ_1
E_8 :	$2\alpha_1 + 3\alpha_2 + 4\alpha_3 + 6\alpha_4 + 5\alpha_5 + 24\alpha_6 + 3\alpha_7 + 2\alpha_8$	$\leftrightarrow$	λ_8
F_4 :	$2\alpha_1 + 3\alpha_2 + 4\alpha_3 + 2\alpha_4$	$\leftrightarrow$	λ_1
G_2 :	$3\alpha_1 + 2\alpha_2$	$\leftrightarrow$	λ_2

□

Our problem of determining the structures of $\mathrm{Hom}_{\mathfrak{g}}(g \otimes \mathfrak{g}, \mathfrak{g})$ will be solved for the case that $\mathfrak{g}$ is semisimple as follows: Write $\mathfrak{g} = \mathfrak{g}_1 \oplus \cdots \oplus \mathfrak{g}_r$ where $\mathfrak{g}_i$ is a simple Lie ideal of $\mathfrak{g}$. It is easy to see that $\phi \in \mathrm{Hom}_{\mathfrak{g}}(\mathfrak{g} \otimes \mathfrak{g}, \mathfrak{g})$ satisfies $\phi(\mathfrak{g}_i \otimes \mathfrak{g}_j) = 0$ for $i \neq j$ and $\phi(\mathfrak{g}_i \otimes \mathfrak{g}_i) \subseteq \mathfrak{g}_i$, and then $\mathrm{Hom}_{\mathfrak{g}}(\mathfrak{g} \otimes \mathfrak{g}, \mathfrak{g}) = \oplus_{i=1}^{r} \mathrm{Hom}_{\mathfrak{g}_i}(\mathfrak{g}_i \otimes \mathfrak{g}_i, \mathfrak{g}_i)$. Hence we can express every element of $\mathrm{Hom}_{\mathfrak{g}}(\mathfrak{g} \otimes \mathfrak{g}, \mathfrak{g})$ in a linear combinations of the $\mathfrak{g}_i$−module homomorphisms of the forms L and D of (*) and (**) respectively.

5. Corollary

We now deal with the triviality of the compatible *associative* algebra structures of semisimple Lie algebras in our context.

Corollary (Kubo[3]) *Let $\mathfrak{g}$ be a semisimple Lie algebra with a Lie bracket*

$[-,-]$. *If* $\phi \in Hom_{\mathfrak{g}}(\mathfrak{g}\otimes\mathfrak{g},\mathfrak{g})$ *satisfies the associative law:* $\phi(\phi(x\otimes y)\otimes z) = \phi(x\otimes\phi(y\otimes z))$ *for* $x, y, z \in \mathfrak{g}$, *then* $\phi = 0$.

Proof. We may assume that $\mathfrak{g}$ is simple. Let $\mathfrak{g}$ be of type A_1 or not of type A_n of $n \geqq 2$. Take the root vector e_1 of the simple root α_1, f_1 of $-\alpha_1$, so that $[e_1, f_1] = h_1$. By Theorem one can write $\phi(-\otimes-) = s[-,-]$ for some $s \in k$. Since $\phi(e_1\otimes e_1) = 0$, we have $0 = \phi(f_1\otimes\phi(e_1\otimes e_1)) = \phi(\phi(f_1\otimes e_1)\otimes e_1) = s^2[[f_1, e_1], e_1] = -2s^2e_1$ by the associativity, and then $s = 0$, which says that $\phi = 0$.

Let $\mathfrak{g}$ be of type A_n of $n \geqq 2$. We identify $\mathfrak{g}$ with $\mathfrak{sl}_{n+1}(k)$ in $M_{n+1}(k)$. By Theorem we can write $\phi(x\otimes y) = s[x,y] + t(xy - (\mathrm{Tr}(xy)/(n+1))E)$ for $x, y \in \mathfrak{sl}_{n+1}(k)$. Let us denote by e_{ij} the $(n+1)\times(n+1)$ matrix with 1 in the (i,j)th place and 0 elsewhere, and $h_i = e_{ii} - e_{i+1\,i+1}$ for $1 \leqq i \leqq n$. Then $\phi(h_1\otimes h_1) = t(e_{11} + e_{22} - (2/(n+1))E)$ and $\phi(e_{12}\otimes h_1) = -(2s+t)e_{12}$. The associativities $\phi(e_{12}\otimes\phi(h_1\otimes h_1)) = \phi(\phi(e_{12}\otimes h_1)\otimes h_1)$ and $\phi(h_2\otimes\phi(h_1\otimes h_1)) = \phi(\phi(h_2\otimes h_1)\otimes h_1)$ lead us the equalities $t^2((n-1)/(n+1)) = (2s+t)^2$ and $t = 0$, respectively. We therefore have $\phi = 0$ □

Acknowledgements

The author thanks the referee of our paper[3] for his or her comments. The proof of this paper is based on them.

References

1. J.E.Humphreys, *Introduction to Lie algebras and Representation Theory* , (Springer–Verlag, New York, 1972).
2. N.Jacobson, *Lie algebras,* (Interscience, New York, 1962).
3. F.Kubo, *Finite-dimensional non-commutative Poisson algebras, J. Pure Appl.Algebra* **113**, 307 (1996).

On τ-coherent rings

Kanzo Masaike

Department of Mathematics, Tokyo Gakugei University, Japan

E-mail: masaike@u-gakugei.ac.jp

We study left-right symmetry of rings for which every fnitely generated torsion free left (or right) module is embedded into free modules. We give a necessary and sufficient condition of a ring R with flat left injective hulls under which R has a flat right injective hull. Let Q be a left coherent maximal left quotient ring of R such that every finitely presented left Q-module is torsion free. Then, Q_R is flat if and only if R is left τ-coherent, relative to Lambek torsion theory.

Let R be a ring with identity and τ the Lambek torsion theory (cf. [12]) with respect to a ring R, which is cogenerated by the injective hull $E(_RR)$ (or $E(R_R)$) of the left (right) R-module R. Levy [7] is the first to investigate rings whose finitely generated τ-torsion free left modules are contained in free modules. In this note such rings are said to be *left TF*. Indeed, properties of these rings are closely conected to the left-right symmetry of their quotient rings (cf. [3], [8], [9]). Our present purpose is to study the condition of a left TF-ring R for which R becomes right TF and apply it to study those rings which have two-sided maximal quotient rings.

We shall say that a τ-torsion free left R-module M is *τ-finitely generated*, if M has a finitely generated τ-dense submodule N, i.e., $\mathrm{Hom}_R(M/N, E(_RR)) = 0$. On the other hand, left R-module M is said to be *τ-finitely presented*, if M is isomorphic to $\oplus_{i=1}^n R/I$, where I is a τ-finitely generated submodule of $\oplus_{i=1}^n R$, a direct sum of n copies of R. Hoshino and Takashima proved in Lemma 1.2 of [6] that every τ-finitely presenteded τ-torsion free left R-module is torsionless, if and only if every τ-finitely presented τ-torsion free right R-module is torsionless. If these equivalent conditions are satisfied, R is said to be *τ-absolutely pure.*

1. Embedding in free modules.

Let M be a left R-module. Assume $(a_1, a_2, \cdots, a_n) \in \oplus_{i=1}^n R$ and $(m_1, m_2, \cdots, m_n) \in \oplus_{i=1}^n M$. Then, we define a multiplication by

$$(a_1, a_2, \cdots, a_n) * (m_1, m_2, \cdots, m_n) = a_1 m_1 + a_2 m_2 + \cdots + a_n m_n.$$

Let X, Y be subsets of $\oplus_{i=1}^n M$ and$\oplus_{i=1}^n R$ respectively. Then, we put

$$\begin{aligned} l_{R^n}(X) &= \{\boldsymbol{a} \in \oplus_{i=1}^n R \mid \boldsymbol{a} * \boldsymbol{x} = 0 \text{ for every } \boldsymbol{x} \in X\}, \\ r_{R^n}(X) &= \{\boldsymbol{a} \in \oplus_{i=1}^n R \mid \boldsymbol{x} * \boldsymbol{a} = 0 \text{ for every } \boldsymbol{x} \in X\}, \\ r_{M^n}(Y) &= \{\boldsymbol{v} \in \oplus_{i=1}^n M \mid \boldsymbol{y} * \boldsymbol{v} = 0 \text{ for every } \boldsymbol{y} \in Y\}. \end{aligned}$$

Lemma 1.1. *The following conditions are euivalent for a ring R.*

(1) R is τ-absolutely pure.

(2) If $F = \{\boldsymbol{a}_1, \boldsymbol{a}_2, \cdots, \boldsymbol{a}_k\}$ is a finite subset of $\oplus_{i=1}^n R$, then $l_{R^n}(r_{E(_RR)^n}(F)) = l_{R^n}(r_{R^n}(F))$.

Proof. Every R-homomorphism $l_{R^n}(r_{E(_RR)^n}(F)) \to E(_RR)$ is extended to $\oplus_{i=1}^n R \to E(_RR)$, which is realized by a right multiplication of an element of $\oplus_{i=1}^n E(_RR)$. From this fact we are able to deduce that $\oplus_{i=1}^n R/l_{R^n}(r_{E(_RR)^n}(F))$ is τ-torsion free and $\sum_{i=1}^t R\boldsymbol{a}_i$ is a τ-dense submodule of $l_{R^n}(r_{E(_RR)^n}(F))$. Therefore, every τ-finitely presented τ-torsion free left R-module is torsionlles, if and only if $l_{R^n}(r_{E(_RR)^n}(F)) = l_{R^n}(r_{R^n}((F))$ for every finite subset F of $\oplus_{i=1}^n R$.

A ring R is said to be *left coherent* (resp. *left τ-coherent*), if every finitely (resp. *τ-finitely*) generated left ideal is finitely (rep. τ-finitely) presented, i.e., if $\boldsymbol{a}$ is an element of $\oplus_{i=1}^n R$, then $l_{R^n}(\boldsymbol{a})$ is finitely (resp. τ-finitely) generated. In this case for every finite subset F of $\oplus_{i=1}^n R$, $l_{R^n}(F)$ is finitely (resp. τ-finitely) generated.

Theorem 1.1. *Let R be a left TF ring. Then, the following conditions are equinalent.*

(1) R is right TF.

(2) R is left τ-coherent, and for every inverse system of R-homomorphisms $\{f_\lambda : R \to M_\lambda\}_{\lambda \in \Lambda}$ with each M_λ a torsionless left R-module and coker f_λ a τ-torsion left R-modue, coker $\varprojlim f_\lambda$ is a τ-torsion left R-module, where the index set Λ is directed.

Proof. Assume (1). Let $\boldsymbol{a}$ be an element of $\oplus_{i=1}^n R$ and $L = l_{R^n}(\boldsymbol{a})$. It is evident that $\oplus_{i=1}^n R/r_{R^n}(L)$ is a torsionless right R-module and hence contained in a finitely generated free module. Namely, there exists an R-monomorphism $\theta_i : \oplus_{i=1}^n R \to R, i = 1, 2, \cdots, k$ such that $\cap_{i=1}^k$ Ker $\theta_i =$

$r_{R^n}(L)$. Hence there exsts a finite subset $F = \{\boldsymbol{b}_1, \boldsymbol{b}_2, \cdots, \boldsymbol{b}_k\}$ of $\oplus_{i=1}^n R$ such that $r_{R^n}(L) = r_{R^n}(F)$. Then, L has a finitely generated τ-dense submodule generated by F in view of the argument in Lemmma 1, since

$$L = l_{R^n}(r_{R^n}(L)) = l_{R^n}(r_{R^n}(F)) = l_{R^n}(r_{E({}_RR)^n}(F)).$$

Therefore, R is left τ-coherent. On the other hand, since R is right TF, every finitely genarated submodule of $E(R_R)$ is torsionless. Then (2) holds from Thorem 1.2 of [5].

Conversely, assume (2). Since R is τ-absolutely pure, the condition (2) implies that every finitely generated τ-torsion free right R-module is torsionless by Thorem 1.2 of [5], again. Assume K is a right R-submodule of $\oplus_{i=1}^n R$ such that $\oplus_{i=1}^n R/K$ is τ-torsion free and hence torsionless. It is evident that $r_{R^n}(l_{R^n}(K)) = K$. On the other hand, $\oplus_{i=1}^n R/l_{R^n}(K)$ is embedded into a finitely generated free left R-module, since R is a left TF ring. Therefore, there exists a finite subset F of K such that $l_{R^n}(K) = l_{R^n}(F)$. Since R is left τ-coherent, there exists a finite subset $G = \{\boldsymbol{p}_1, \boldsymbol{p}_2, \cdots, \boldsymbol{p}_t\}$ in $l_{R^n}(K)$ such that $\sum_{i=1}^t R\boldsymbol{p}_i$ is a τ-dense submodule of $l_{R^n}(K)$ and hence

$$K = r_{R^n}(l_{R^n}(K)) = r_{R^n}(\textstyle\sum_{i=1}^t R\boldsymbol{p}_i) = r(G).$$

Thus, K is embedded into $\oplus_{i=1}^t R$ and this implies R is a right TF-ring.

Proposition 1.1. *Let Q be a maximal left quotient ring of R. Then R is left TF, if and only if Q is left TF and for every $q_1, q_2, \cdots, q_t \in Q$, $\cap_{i=1}^t (q_i^{-1}R)$ is a τ-finitely generated τ-dense right ideal of R, where $q_i^{-1}R = \{r \in R \mid q_i r \in Q\}$.*

Proof. Assume R is left TF. Then, it is easily seen that Q is also left TF. Every R-homomorphism $f : \sum_{i=1}^t Rq_i + R \to R$ is extended to $Q \to Q$, so that f is realized by a right multiplication of an element of $\cap_{i=1}^n (q_i^{-1}R)$. Since $\sum_{i=1}^t Rq_i + R$ is embeded into $\oplus_{i=1}^n R$, there exist a finite subset $\{p_1, p_2, \cdots, p_n\}$ in $\cap_{i=1}^t (q_i^{-1}R)$ such that

$$l_R(\textstyle\sum_{i=1}^n p_i R) = \cap_{i=1}^t l_R(p_i) = 0.$$

Now, R is τ-absolutely pure and hence $\sum_{i=1}^n p_i R$ is a τ-dense submodule of $r_R(l_R(\sum_{i=1}^n p_i R)) = R$ from the argument in Lemma 1.1. Hence $\cap_{i=1}^t (q_i^{-1}R)$ is τ-finitely generated dense right ideal of R.

Coversely, assume Q is left TF and for every $q_1, q_2, \cdots, q_t \in Q$, $\cap_{i=1}^t (q_i^{-1}R)$ is a τ-finitely generated τ-dense right idealof R. Let M be a finitely generated τ-torsion free left R-modle. We may regard M to be a submodule of a direct product of copies of $E({}_RR) = E({}_QQ)$. The finitely generated left Q-module QM is contained in $\oplus_{i=1}^n Q$ and we may write

$$M = \sum_{i=1}^t R\boldsymbol{p}_i, \text{ where, } \boldsymbol{p}_i = (p_{i1}, p_{i2}, \cdots, p_{in}) \in \oplus_{i=1}^n Q.$$

Then, $\cap_{i=1}^t(\cap_{j=1}^n(p_{ij}^{-1}R))$ contains a finitely generated τ-dense right ideal generated by $a_1, a_2, \cdots, a_k \in R$. Let $\hat{a_i}$ be the R-homomorphims from M to $\oplus_{i=1}^n R$ defined by the right multiplication of an element a_i in R $(i = 1, 2, \cdots, k)$. Since $\cap_{i=1}^k$ Ker $\hat{a_i} = 0$, M is embedded into a direct sum of $n \times k$ copies of R.

In [2] Chase proved that the following conditios are equivalent for a right R-module M.

(a) M is flat.

(b) If $\boldsymbol{a} \in \oplus_{i=1}^n R$ and $(v_1, v_2, \cdots, v_n) \in l_{M^n}(\boldsymbol{a})$, then there exist $(b_{i1}, b_{i2}, \cdots, b_{in}) \in l_{R^n}(\boldsymbol{a})$, $i = 1, 2, \cdots, k$ and $(u_1, u_2, \cdots, u_k) \in \oplus_{i=1}^k M$ such that $v_i = \sum_{i=1}^k u_j b_{ji}$, $i = 1, 2, \cdots, n$.

Assume every finitely generated submodule of $E(R_R)$ is embedded in free module. This means that if $\boldsymbol{v} \in \oplus_{i=1}^n E(R_R)$, then there exist finite elements $\boldsymbol{b}_1, \boldsymbol{b}_2, \cdots, \boldsymbol{b}_k$ in $\oplus_{i=1}^n R$ such that $\cap_{i=1}^k r_{R^n}(\boldsymbol{b}_i) = r_{R^n}(\boldsymbol{v})$. In this case, it is known that $E(R_R)$ is flat (cf. [4]). Concerning this result, when $M = E(R_R)$, we are able to weaken the above condition (b) as follows.

Proposition 1.2. *The following conditions are equivalent for a ring R.*

(1) $E(R_R)$ is flat.

(2) If $\boldsymbol{a} \in \oplus_{i=1}^n R$ and $\boldsymbol{v} \in l_{E(_RR)^n}(\boldsymbol{a})$, then there exist finite elements $\boldsymbol{b}_1, \boldsymbol{b}_2, \cdots, \boldsymbol{b}_k \in l_{R^n}(\boldsymbol{a})$ such that $\cap_{i=1}^k r_{R^n}(\boldsymbol{b}_i) \subseteq r_{R^n}(\boldsymbol{v})$.

Proof. The implication (1) $\Rightarrow$ (2) is trivial from the above result of Chase, if we put $\boldsymbol{v} = (v_1, v_2, \cdots, v_n)$ and $\boldsymbol{b}_i = (b_{i1}, b_{i2}, \cdots, b_{in})$.

Assume (2). Let $\boldsymbol{a} \in \oplus_{i=1}^n R$ and $\boldsymbol{v} \in l_{E(_RR)^n}(\boldsymbol{a})$. Further, assume $\boldsymbol{b}_1, \boldsymbol{b}_2, \cdots, \boldsymbol{b}_k$ satisfy the condition in (2). Let N be the right R-submodule of $\oplus_{i=1}^k R$ such that

$$N = \{(\boldsymbol{b_1} * \boldsymbol{x}, \boldsymbol{b_2} * \boldsymbol{x}, \cdots, \boldsymbol{b_k} * \boldsymbol{x}) \mid \boldsymbol{x} \in \oplus_{i=1}^k R\}$$

. We are able to defin an R-homomorphism θ from N to R by

$$\theta((\boldsymbol{b_1} * \boldsymbol{x}, \boldsymbol{b_2} * \boldsymbol{x}, \cdots, \boldsymbol{b_k} * \boldsymbol{x})) = \boldsymbol{v} * \boldsymbol{x}.$$

Then, θ is extended to an R-homomorphism $\hat{\theta}$ from $\oplus_{i=1}^k R$ to $E(R_R)$. Therefore, $\hat{\theta}$ is realized by the left multiplication of an element

$(u_1, u_2, \cdots, u_k) \in \oplus_{i=1}^n E(R_R)$. Let $\boldsymbol{e_i}$ be the element of $\oplus_{i=1}^k R$ such that the i-th component is 1 and 0 elsewhere. It follows that

$$\begin{aligned} v_i = \boldsymbol{v} * \boldsymbol{e_i} &= \hat{\theta}((\boldsymbol{b}_1 * \boldsymbol{e}_i, \boldsymbol{b}_2 * \boldsymbol{e}_i, \cdots, \boldsymbol{b}_k * \boldsymbol{e}_i)) \\ &= (u_1, u_2, \cdots, u_k) * (b_{1i}, b_{2i}, \cdots, b_{ki}). \end{aligned}$$

This holds (1) in view of the result of Chase.

If $E({}_RR)$ (or $E(R_R)$) is flat, R is τ-absolutely pure (see [6]). On the other hand, in [11] Morita proved that if R is left Noetherian and $E({}_RR)$ is flat, then $E(R_R)$ is also flat. Hoshino and Takashima [6] generalized this result by proving that if R is τ-absolutely pure and left τ-coherent, then $E(R_R)$ is flat. Now, in this note we study a necessary and sufficient condition under which τ-absolutely pure rings have flat right injective hulls.

Theorem 1.2. *@If $E({}_RR)$ is flat (or more generally, R is τ-absolutely pure), then the following conditions are equivalent.*

(1) $E(R_R)$ it is flat.

(2) If $\boldsymbol{a} \in \oplus_{i=1}^n R$ and $\boldsymbol{v} \in l_{R^n}(\boldsymbol{a})$, then there exists a right R-submodle K of $\oplus_{i=1}^n R$ such that $\boldsymbol{a}R \subseteq K \subseteq r_{R^n}(\boldsymbol{v})$ satifying $l_{R^n}(K)$ is τ-finitely generated and $r_{R^n}(l_{R^n}(K)) \subseteq r_{R^n}(\boldsymbol{v})$.

Proof. Assume (1). Let $\boldsymbol{a}$ and $\boldsymbol{v}$ be the same as in the condition (2). By Proposition 1.2 there exist $\boldsymbol{b}_1, \boldsymbol{b}_2, \cdots, \boldsymbol{b}_k \in l_{R^n}(\boldsymbol{a})$ such that $\cap_{i=1}^k r_{R^k}(\boldsymbol{b}_i)$ is contained in $r_{R^n}(\boldsymbol{v})$. Put $K = \cap_{i=1}^k r_{R^n}(\boldsymbol{b}_i)$. Then,

$$\boldsymbol{a}R \subseteq K \subseteq r(\boldsymbol{v}), \text{ and } K = r_{R^n}(\textstyle\sum_{i=i}^n R\boldsymbol{b_i}).$$

Now, $l_{R^n}(K) = l_{R^n}(r_{R^n}(\sum_{i=i}^n (R\boldsymbol{b_i}))$ is τ-finitely generated, since R is τ-absolutey pure and hence $\sum_{i=i}^n R\boldsymbol{b_i}$ is a τ-dense submodule of $l_{R^n}(K)$. This holds (2), since $r_{R^n}(l_{R^n}K)) = K \subseteq r_{R^n}(\boldsymbol{v})$.

Coversely, assume (2). There exist $\boldsymbol{b}_1, \boldsymbol{b}_2, \cdots, \boldsymbol{b}_k$ in $l_{R^n}(K)$ such that $\sum_{i=i}^n R\boldsymbol{b}_i$ is a τ-dense submodule of $l_{R^n}(K)$. Then,

$$\mathrm{Hom}_R(l_{R^n}(K)/\textstyle\sum_{i=i}^k R\boldsymbol{b_i}, R) = 0.$$

Thus, we have (1) from Lemma 1.1, since

$$(\cap_{i=1}^k r_{R^n}(\boldsymbol{b}_i)) = r_{R^n}(l_{R^n}(K)) \subseteq r_{R^n}(\boldsymbol{v}).$$

2. Flat extension rings.

Now, in the following we shall consider extension rings of R which become flat as right (or left) R-modules.

Proposition 2.1. *Let Q be a left coherent extension ring of R. Then, the following conditions are equivalent.*

(1) Q_R is flat.

(2) For every $\boldsymbol{a} \in \oplus_{i=1}^n R$, there exist finite elements $\boldsymbol{b}_1, \boldsymbol{b}_2, \cdots, \boldsymbol{b}_k \in l_{R^n}(\boldsymbol{a})$ such that $l_{Q^n}(\boldsymbol{a}) = \sum_{i=1}^k Q\boldsymbol{b_i}$.

Especially, if Q is a maximal left quotient ring of R, R becomes left τ-coherent when the above equivalent conditions are satisfied.

Proof. It suffices to prove (1) $\Rightarrow$ (2). Assume Q_R is flat and $\boldsymbol{v} \in l_{Q^n}(\boldsymbol{a})$. There exists $(u_1, u_2, \cdots, u_k) \in \oplus_{i=1}^k Q$ and $\boldsymbol{b}_1, \boldsymbol{b}_2, \cdots, \boldsymbol{b}_k \in l_{R^n}(\boldsymbol{a})$ such that

$$\begin{aligned}\boldsymbol{v} &= (\textstyle\sum_{j=1}^k u_j b_{j1}, \sum_{j=1}^k u_j b_{j2}, \cdots, \sum_{j=1}^k u_j b_{jn})\\ &= u_1\boldsymbol{b}_1 + u_2\boldsymbol{b}_2 + \cdots + u_k\boldsymbol{b}_k,\end{aligned}$$

where we use the similar notation as in the proof of Proposition 1.2. On the other hand, since Q is left coherent, there exist $\boldsymbol{q}_1, \boldsymbol{q}_2, \cdots, \boldsymbol{q}_t$ in $\oplus_{i=1}^n Q$ such that $l_{Q^n}(\boldsymbol{a}) = \sum_{i=1}^t Q\boldsymbol{q}_i$. From the above argument for every $\boldsymbol{q}_i$ there exist finite elements $\boldsymbol{b}_{i1}, \boldsymbol{b}_{i2}, \cdots, \boldsymbol{b}_{ik_i}$ in $l_{R^n}(\boldsymbol{a})$ such that

$$\boldsymbol{q}_i \in Q\boldsymbol{b}_{i1} + Q\boldsymbol{b}_{i2} + \cdots + Q\boldsymbol{b}_{ik_i},\ i = 1, 2, \cdots, t.$$

Then, we have (1), since

$$l_{Q^n}(\boldsymbol{a}) = \textstyle\sum_{i=1}^t (Q\boldsymbol{b}_{i1} + Q\boldsymbol{b}_{i2} + \cdots + Q\boldsymbol{b}_{i_k}).$$

Furthermore, assume Q is a maximal left quotient ring of R. It is evident that $\sum_{i=1}^t (R\boldsymbol{b}_{i1} + R\boldsymbol{b}_{i2} + \cdots + R\boldsymbol{b}_{ik_i})$ is a τ-dense left R-submodule of $l_{Q^n}(\boldsymbol{a})$ and hence R is left τ-coherent, since

$$l_{Q^n}(\boldsymbol{a}) \supseteq l_{R^n}(\boldsymbol{a}) \supseteq \textstyle\sum_{i=1}^t (R\boldsymbol{b}_{i1} + R\boldsymbol{b}_{i2} + \cdots + R\boldsymbol{b}_{ik_i}).$$

Remark 2.1. In the case of Proposition 2.1 we are able to see that $\prod Q_R$ is flat by a slight calculation.

A submodule N of a τ-torsion free left R-module is said to be *ratoinally closed*, if M/N is τ-torsion free.

Proposition 2.2. *Assume R is a ring such that R_R is injective. Then, every finitely generated left R-submodule of $\oplus_{i=1}^n R$ is rationally closed.*

Proof. Assume M is generated by $\boldsymbol{a}_1, \boldsymbol{a}_2, \cdots, \boldsymbol{a}_k$. Let $\boldsymbol{b} \in l_{R^n}(r_{R^n}(M))$ and put $K = \{\boldsymbol{b} * \boldsymbol{x} \mid \boldsymbol{x} \in \oplus_{i=1}^n R\}$. Further, put

$$@@N = \{(\boldsymbol{a}_1 * \boldsymbol{x}, \boldsymbol{a}_2 * \boldsymbol{x}, \cdots, \boldsymbol{a}_k * \boldsymbol{x}) \mid \boldsymbol{x} \in \oplus_{i=1}^n R\}.$$

Then, by the same argument as in the proof of Proposition 1.2 we are able to define an R-homomorphism $\theta : N \to K$ by

$$\theta((\boldsymbol{a}_1 * \boldsymbol{x}, \boldsymbol{a}_2 * \boldsymbol{x}, \cdots, \boldsymbol{a}_k * \boldsymbol{x})) = \boldsymbol{b} * \boldsymbol{x}.$$

Since R_R is injective, θ is extended to $\hat{\theta} : \oplus_{i=1}^k R \to R$. Then, $\hat{\theta}$ is realized by the left multiplication of an element $(u_1, u_2, \cdots, u_k)$ of $\oplus_{i=1}^k R$. Then, we can conclude that $\boldsymbol{b} = u_1\boldsymbol{a}_1 + u_2\boldsymbol{a}_2 + \cdots + u_n\boldsymbol{a}_k$. This implies $l_{R^n}(r_{R^n}(M)) = M$ and hence $\oplus_{i=1}^n R/M$ is torsionless.

The following corollary provides a generalizatin of a result obtained in Theorem 2.1 of [1].

Corollary 2.1. *Let Q be a maxima left quotient of R. If Q is left coherent and every finitely generated left Q-submodule of $\oplus_{i=1}^n Q$ is rationally closed. Then, the following conditins are equivalent.*
(1) Q_R is flat.
(2) R is left τ-coherent.

Proof. In view of Proposition 2.1 it suffices to prove that (2) implies (1). Assume R is left τ-coherent and $\boldsymbol{a} \in \oplus_{i=1}^n R$. There exist $\boldsymbol{b}_1, \boldsymbol{b}_2, \cdots, \boldsymbol{b}_k$ in $\oplus_{i=1}^n R$ such that $\sum_{i=1}^k R\boldsymbol{b}_i$ is a τ-dense submodule of $l_{R^n}(\boldsymbol{a})$. Since Q is a maximal left quotient ring, it is easily checked that $l_{R^n}(\boldsymbol{a})$ is a τ-dense submodule of $l_{Q^n}(\boldsymbol{a})$ and hence so is $\sum_{i=1}^k R\boldsymbol{b}_i$. Consequently, $\sum_{i=1}^k Q\boldsymbol{b}_i$ is a τ-dense Q-submodule of $l_{Q^n}(\boldsymbol{a})$. On the other hand, $\sum_{i=1}^k Q\boldsymbol{b}_i$ is a rationally closed Q-submodule of $l_{Q^n}(\boldsymbol{a})$, so that we have $l_{Q^n}(\boldsymbol{a}) = \sum_{i=1}^k Q\boldsymbol{b}_i$. Hence by Proposition 2.1 Q_R is flat.

Remark 2.2. The implication (2) $\Rightarrow$ (1) remains true without assuming that Q is left coherent.

A ring R is called *QF-3"*, if every finitely generated submodule of $E({}_RR)$ and $E(R_R)$ are torsionlles. In this case the maximal left quotient ring Q coincides with the maximal right quotient ring by [8]. The relation between R and Q are important to study the property of R (cf. [9], [10]). Now, in the following left and right coherent (resp. left and right τ-cohrent) rings are called coherent (resp. τ-coherent).

Theorem 2.1. *Let R be a QF-3" ring with a maximal two-sided quotient ring Q. If every finitely generated left Q-submodule and right Q-submodule of $\oplus_{i=1}^n Q$ are rationally closed, then the following conditins are equivalent.*
(1) R is left and right TF.
(2) Q is coherent and ${}_RQ_R$ is flat.

Proof. From the argument of the proof of Theorem 1.1 it is easy to check that R is TF, if and only if R is QF-3" and τ-coherent. Assume (1). Since R is TF, so is Q. Therefore, Q is τ-coherent. Then, Q becomes coherent, since for evry $\boldsymbol{q} \in \oplus_{i=1}^{n} Q$, $l_{Q^n}(\boldsymbol{q})$ is τ-finitely generated and hence finitely generated. Further, since R is τ-coherent, from Corollary 2.1 ${}_RQ_R$ is flat.

Conversely, Assume (2). Then, R is τ-coherent by Corollary 2.1 and hence R is TF, since R is QF-3".

References

1. F.W. Cateforis *Flat regular quotient rings*, Trans. Amer. Math. Soc. **138** (1969), 241-250.
2. S.U. Chase *Direct product of modules*, Trans. Amer. Math. Soc. **97** (1960), 457-473. .
3. K.R. Godearl *Embedding non-singular modules in free modules*, J. Pure Appl. Math. Algebra **1** (1971), 275-279.
4. M. Hoshino, *On Lambek torsion theories*, Osaka J. Math. **29** (1992), 447-453.
5. ______, *On Lambek torsion theories III*, Osaka J. Math. **32** (1995), 521-531.
6. M. Hoshino and S. Takashima *On Lambek torsion theories II*, Osaka J. Math. **31** (1994), 729-746.
7. L. Levy, *Torsion free and divisible modules over non-integral domains*, Canad. J. Math. **15** (1963), 132-151.
8. K. Masaike, *On quotient rings and torsionless modules*, Sci.Rep. Tokyo Kyoiku Daigaku Sec. A **11** (1970) 26-30.
9. ______, *Semi-primary QF-3 quotient rings*, Comm. Algebra **11** (1983), 377-389.
10. ______, *Reflexive modules over QF-3 rings*, Canad. Math. Bull. **35** (1992), 247–251.
11. K. Morita, *Noetherian QF-3 rings and two-sided quasi-Frobenius maximal quotient rings*, Proc. Japan Acad. **46** (1970), 837-840.
12. B. Stenström, *Rings of Quotients*, Grund. Math. Wiss. Vol. **217** Springer, Berlin, 1975.

SKEW-MATRIX RINGS AND APPLICATIONS TO QF-RINGS

Yoshihisa Nagadomi

Open Technologies Corporation, Corona No.3 Bldg. 2F 6-1-21 Hon-komagome, Bunkyo-ku, Tokyo

Kiyoichi Oshiro, Masahiko Uhara

Department of Mathematics, Yamaguchi University, Yoshida, Yamaguchi

Kota Yamaura

Graduate School of Mathematics, Nagoya University, Nagoya

In this paper, we introduce a skew-matrix ring, which generalizes the usual matrix rings. This ring was first introduced by Kupisch [8] in 1975 under the name VPE-rings through his study on Nakayama rings, and latter independently by Oshiro [9] in 1987 for the study on left H-rings and its applications to Nakayama rings (artinian serial rings). In the present paper, by refering to the latter article, we explicitely introduce the definition of a skew-matrix ring.

1. Introduction

The main purpose of our paper is to provide fundamental properties of skew-matrix rings and apply them to QF-rings. Actually, as applications, we study the following: (1) The Nakayama automorphism of a skew-matrix ring over a local QF-ring. (2) A basic indecomposable QF-ring whose Nakayama permutation corresponds to any given permutation. (3) Strongly QF-rings.

Throughout this paper, rings R considered are associative rings with identity and all R-modules are unital. For an R-module M, we use M_R ($_RM$) to stress that M is a right (left) R-module. $J(M)$ and $S(M)$ denote the Jacobson radical and the socle of M, respectively. In particular, $J(R)$, or simply J, denotes the Jacobson radical of R. $T(M)$ denotes the top of M, that is, $T(M) = M/J(M)$. For a in R and a subset S of R, $(a)_L : S \to aS$ and $(a)_R : S \to Sa$ mean the left multiplication and the right multiplication by a, respectively. Let $(R)_n$ be a matrix ring. For $x \in R$, $\langle x \rangle_{ij}$ denotes the matrix whose (i,j)-component is x and other entries are 0. For a subset X, we put $\langle X\rangle = \{\langle x\rangle_{ij} \mid x \in X\}$. For a set A, we write its cardinal by $\sharp A$. For a semiperfect ring R, $Pi(R) = \{e_1, \ldots, e_n\}$ denotes a complete set of orthogonal primitive idempotents of R. We use the notation: $l_S(T)$ $(r_S(T))$ is the left (right) annihilator of T in S.

Let R be a basic QF-ring with $Pi(R) = \{e_1, \ldots, e_n\}$. Then there exists a permutation

$$\begin{pmatrix} e_1 & \cdots & e_n \\ e_{\sigma(1)} & \cdots & e_{\sigma(n)} \end{pmatrix}$$

satisfying $S(e_iR_R) \cong T(e_{\sigma(i)}R_R)$ $\forall$ i. This permutation is called the *Nakayama permutation* of $Pi(R)$ or R. R is said to be a weakly symmetric QF-ring if its Nakayama permutation is the identity permutation. A ring automorphism τ of R is called a *Nakayama automorphism* (as a ring) if $\tau(e_i) = e_{\sigma(i)}R$ $\forall$ i.

2. Definition of a Skew-Matrix Ring

Let Q be a ring and let $c \in Q$ and $\sigma \in \mathrm{End}(Q)$ such that $\sigma(c) = c$ and $\sigma(q)c = cq \quad \forall\, q \in Q$. Let R denote the set of all $n \times n$ matrices over Q:

$$R = \begin{pmatrix} Q \cdots Q \\ \cdots \\ Q \cdots Q \end{pmatrix}.$$

We define a multiplication on R with respect to (σ, c) as follows.

For $(x_{ik}), (y_{ik})$ in R, the multiplication of these matrices is given by

$$(z_{ik}) = (x_{ik})(y_{ik}),$$

where z_{ik} is defined by the following:

$$\text{if } i \le k,\ z_{ik} = \sum_{j<i} x_{ij}\sigma(y_{jk})c + \sum_{i\le j\le k} x_{ij}y_{jk} + \sum_{k<j} x_{ij}y_{jk}c,$$

$$\text{if } k < i,\ z_{ik} = \sum_{j\le k} x_{ij}\sigma(y_{jk}) + \sum_{k<j<i} x_{ij}\sigma(y_{jk})c + \sum_{i\le j} x_{ij}y_{jk}.$$

We may write this multiplication as follows:

$$\langle a\rangle_{ij}\langle b\rangle_{kl} = 0 \ \text{ if } \ j \neq k,$$

$$\langle a\rangle_{ij}\langle b\rangle_{jk} = \begin{cases} \langle a\sigma(b)\rangle_{ik} & \text{if } \ j \leq k < i, \\ \langle a\sigma(b)c\rangle_{ik} & \text{if } \ k < j < i \ \text{ or } \ j < i \leq k, \\ \langle ab\rangle_{ik} & \text{if } \ i = j, \\ \langle abc\rangle_{ik} & \text{if } \ i \leq k < j, \\ \langle ab\rangle_{ik} & \text{if } \ k < i < j \ \text{ or } \ i < j \leq k. \end{cases}$$

It is straightforward to check that this multiplication satisfies the associative law:

$$(\langle x\rangle_{ij}\langle y\rangle_{jk})\langle z\rangle_{kl} = \langle x\rangle_{ij}(\langle y\rangle_{jk}\langle z\rangle_{kl}).$$

Furthermore, R becomes a ring under this multiplication and the usual addition of matrices. We call this ring R the *skew matrix ring* over Q with respect to (σ, c, n), and denote it by

$$R = \begin{pmatrix} Q & \cdots & Q \\ & \cdots & \\ Q & \cdots & Q \end{pmatrix}_{\sigma,c,n}$$

or simply by $(Q)_{\sigma,c,n}$. Note that, for the identity map id_Q of Q and the identity $1 \in Q$, $(Q)_{id_Q,1,n}$ is the usual matrix ring over Q, and that $(Q)_{\sigma,c,1}$ is Q itself, since σ and c do not impact.

We may consider Q as a subring of R by the map:

$$q \longmapsto \begin{pmatrix} q & 0 & \cdots & 0 \\ 0 & \ddots & \ddots & \vdots \\ \vdots & \ddots & \ddots & 0 \\ 0 & \cdots & 0 & q \end{pmatrix}.$$

Put $\alpha_{ij} = \langle 1\rangle_{ij}$ for each $i, j \in \{1, \ldots, n\}$. We call $\{\alpha_{ij}\}_{i,j\in\{1,\ldots,n\}}$ the *skew matrix units* of R. For these matrix units, we obtain the following relations:

$$(*)\cdots\begin{cases}\text{If}\quad j\neq k, & \alpha_{ij}\alpha_{kl}=0.\\[2ex] \text{If}\quad i>j, & \begin{cases}\sigma(q)\alpha_{ij}=\alpha_{ij}q\ \ \forall\ \ q\in Q,\\ \alpha_{ij}\alpha_{jk}=\begin{cases}\alpha_{ik} & \text{if}\ \ i>k\geq j,\\ \alpha_{ik}c & \text{if}\ \ k\geq i\ \ \text{or}\ \ j>k.\end{cases}\end{cases}\\[2ex] \text{If}\quad i=j, & \begin{cases}q\alpha_{ij}=\alpha_{ij}q\ \ \forall\ \ q\in Q,\\ \alpha_{ij}\alpha_{jk}=\alpha_{ik}.\end{cases}\\[2ex] \text{If}\quad i<j, & \begin{cases}q\alpha_{ij}=\alpha_{ij}q\ \ \forall\ \ q\in Q,\\ \alpha_{ij}\alpha_{jk}=\begin{cases}\alpha_{ik}c & \text{if}\ \ i\leq k<j,\\ \alpha_{ik} & \text{if}\ \ k<i\ \ \text{or}\ \ j\leq k.\end{cases}\end{cases}\end{cases}$$

Consider the set R' defined by

$$R'=\begin{pmatrix}Q\alpha_{11} & \cdots & Q\alpha_{1n}\\ & \cdots & \\ Q\alpha_{n1} & \cdots & Q\alpha_{nn}\end{pmatrix}.$$

Then R' becomes a ring by the relations $(*)$ above, and there exists a canonical matrix ring isomorphism between R and R' by the map

$$\begin{pmatrix}q_{11} & \cdots & q_{1n}\\ & \cdots & \\ q_{n1} & \cdots & q_{nn}\end{pmatrix}\longleftrightarrow\begin{pmatrix}q_{11}\alpha_{11} & \cdots & q_{1n}\alpha_{1n}\\ & \cdots & \\ q_{n1}\alpha_{n1} & \cdots & q_{nn}\alpha_{n,n}\end{pmatrix}.$$

Under this isomorphism, we often identify R and R' and represent R as

$$R=\begin{pmatrix}Q\alpha_{11} & Q\alpha_{12} & \cdots & Q\alpha_{1n}\\ Q\alpha_{21} & Q\alpha_{22} & \cdots & Q\alpha_{2n}\\ & \cdots & \cdots & \\ Q\alpha_{n1} & Q\alpha_{n2} & \cdots & Q\alpha_{nn}\end{pmatrix}_{\sigma,c,n}.$$

Furthermore using the canonical ring isomorphisms $Q\alpha_{ii}\cong Q$, we often represent R as

$$R = \begin{pmatrix} Q & Q\alpha_{12} & \cdots & Q\alpha_{1n} \\ Q\alpha_{21} & Q & \ddots & \vdots \\ \vdots & \ddots & \ddots & Q\alpha_{n-1,n} \\ Q\alpha_{n1} & \cdots & Q\alpha_{n,n-1} & Q \end{pmatrix}_{\sigma,c,n}.$$

In order to make calculations easier, we put $\beta_{ji} = \alpha_{ji}$ for each $j > i$ in these representations. Moreover, instead of $\{\alpha_{ij}\}_{i,\, j\in\{1,\dots,n\}}$, we take the sets $\{\alpha_{ij}\}_{1\leq i\leq j\leq n} \cup \{\beta_{ji}\}_{1\leq i<j\leq n}$ as the skew matrix units of R. Then

$$R = \begin{pmatrix} Q\alpha_{11} & Q\alpha_{12} & \cdots & Q\alpha_{1n} \\ Q\beta_{21} & Q\alpha_{22} & \ddots & \vdots \\ \vdots & \ddots & \ddots & Q\alpha_{n-1,n} \\ Q\beta_{n1} & \cdots & Q\beta_{n,n-1} & Q\alpha_{nn} \end{pmatrix}_{\sigma,c,n} = \begin{pmatrix} Q & Q\alpha_{12} & \cdots & Q\alpha_{1n} \\ Q\beta_{21} & Q & \ddots & \vdots \\ \vdots & \ddots & \ddots & Q\alpha_{n-1,n} \\ Q\beta_{n1} & \cdots & Q\beta_{n,n-1} & Q \end{pmatrix}_{\sigma,c,n}$$

with relations:

$$\begin{cases} \alpha_{ij}q = q\alpha_{ij} & \forall\, q \in Q, \\ \beta_{ij}q = \sigma(q)\beta_{ij} & \forall\, q \in Q, \\ \alpha_{ij}\alpha_{jk} = \alpha_{ik} & \text{if } i < j < k, \\ \beta_{ij}\beta_{jk} = \beta_{ik}c & \text{if } k < j < i, \\ \alpha_{ij}\beta_{jk} = \begin{cases} \alpha_{ik}c \\ \beta_{ik} \end{cases} & \begin{array}{l} \text{if } i \leq k < j, \\ \text{if } k < i < j, \end{array} \\ \beta_{ij}\alpha_{jk} = \begin{cases} \alpha_{ik}c \\ \beta_{ik} \end{cases} & \begin{array}{l} \text{if } j < i \leq k, \\ \text{if } j < k < i. \end{array} \end{cases}$$

Remark 2.1. We look $(Q)_{\sigma,c.n}$ for some particular cases.

(1) If $n = 2$, then the multiplication is as follows:

$$\begin{pmatrix} x_1 & x_2 \\ x_3 & x_4 \end{pmatrix} \begin{pmatrix} y_1 & y_2 \\ y_3 & y_4 \end{pmatrix} = \begin{pmatrix} x_1y_1 + x_2y_3c & x_1y_2 + x_2y_4 \\ x_3\sigma(y_1) + x_4y_3 & x_3\sigma(y_2)c + x_4y_4 \end{pmatrix}.$$

(2) If Q is a local ring, then R is a semiperfect ring since $e_iRe_i \cong Q$ for any $i = 1,\dots,n$, where $e_i = \langle 1\rangle_{ii}$.

(3) If Q is a right (left) artinian ring then so is R.

(4) If Q is a right (left) noetherian ring then so is R.

The following is an important theorem.

Theorem 2.1. *The matrix mapping $\tau = (\tau_{ij}) : R \longrightarrow R$ given by*

$$\begin{pmatrix} x_{11} & x_{12} & \cdots & x_{1n} \\ x_{21} & x_{22} & \cdots & x_{2n} \\ \cdots & \cdots & \cdots & \cdots \\ x_{n1} & x_{n2} & \cdots & x_{nn} \end{pmatrix} \longmapsto \begin{pmatrix} x_{nn} & x_{n1} & \cdots & x_{n,n-1} \\ \sigma(x_{1n}) & \sigma(x_{11}) & \cdots & \sigma(x_{1,n-1}) \\ & \cdots & & \cdots \\ \sigma(x_{n-1,n}) & \sigma(x_{n-1,1}) & \cdots & \sigma(x_{n-1,n-1}) \end{pmatrix}$$

is a ring homomorphism; in particular, if $\sigma \in \mathrm{Aut}(Q)$, then $\tau \in \mathrm{Aut}(R)$.

Proof. Straightforward. □

Now, for each $i = 1, \ldots, n$, let

$$\begin{aligned} W_i &= \langle Q\rangle_{i1} + \cdots + \langle Q\rangle_{i,i-1} + \langle Qc\rangle_{ii} + \langle Q\rangle_{i,i+1} + \cdots + \langle Q\rangle_{in} \\ &= \begin{pmatrix} & & 0 & & \\ Q & \cdots & Q\ Qc\ Q & \cdots & Q \\ & & 0 & & \end{pmatrix} \ (i\text{-th row}). \end{aligned}$$

Then W_i is a submodule of e_iR_R. For $i = 2, \ldots, n$, let $\theta_i : e_iR \longrightarrow W_{i-1}$ be the map given by

$$\begin{pmatrix} & & 0 & & \\ x_1 & \cdots & x_{i-1}\ x_i & \cdots & x_n \\ & & 0 & & \end{pmatrix} (i\text{-th row}) \longmapsto \begin{pmatrix} & & 0 & & \\ x_1 & \cdots & x_{i-1}c\ x_i & \cdots & x_n \\ & & 0 & & \end{pmatrix} (i\text{-}1\text{-th row})$$

and let $\theta_1 : e_1R \longrightarrow W_n$ be the map given by

$$\begin{pmatrix} x_1 & \cdots & x_n \\ 0 & \cdots & 0 \\ & \cdots & \\ 0 & \cdots & 0 \end{pmatrix} \longmapsto \begin{pmatrix} 0 & \cdots & \cdots & 0 \\ & \cdots & \cdots & \\ 0 & \cdots & \cdots & 0 \\ \sigma(x_1) & \cdots & \sigma(x_{n-1}) & \sigma(x_n)c \end{pmatrix}.$$

Then, for each $i = 2, \ldots, n$, θ_i is realized by the left multiplication by $\langle 1\rangle_{i-1,i}$, and θ_1 is realized by the left multiplication by $\langle 1\rangle_{n1}$. Hence we have the following result.

Proposition 2.1. *Each θ_i is a right R-module homomorphism. Moreover,* *if σ is an onto homomorphism, then each θ_i is an onto homomorphism and*

$$\operatorname{Ker}\theta_1 = \begin{pmatrix} 0 \cdots & 0 & l_Q(c) \\ 0 \cdots & 0 & 0 \\ & \cdots\cdots & \\ 0 \cdots\cdots & & 0 \end{pmatrix},$$

$$\operatorname{Ker}\theta_i = \langle l_Q(c)\rangle_{i,i-1} = \begin{pmatrix} & \overset{\overset{i-1}{\vee}}{0} & \\ 0 & l_Q(c) & 0 \\ & 0 & \end{pmatrix} \ (i\text{-th row}) \text{ for } i = 2, \ldots, n.$$

Thus we obtain a fundamental result.

Theorem 2.2. *Let Q be a local QF-ring, $\sigma \in \operatorname{Aut}(Q)$ and $c \in J(Q)$ with $\sigma(c) = c$ and $\sigma(q)c = cq \ \forall\ q \in Q$. Set $R = (Q)_{\sigma,c,n}$. Then the following statements hold.*

(1) R is a basic indecomposable QF-ring with the Nakayama permutation:

$$\begin{pmatrix} e_1 & e_2 & \cdots & e_n \\ e_n & e_1 & \cdots & e_{n-1} \end{pmatrix},$$

where $e_i = \langle 1\rangle_{ii}$ for each $i = 1, \ldots, n$. Therefore R is a basic indecomposable QF-ring with a cyclic Nakayama permutation.

(2) For any idempotent e of R, eRe is represented as a skew-matrix ring over Q with respect to $(\sigma, c, k\ (\leq n))$, where $k = \#Pi(eRe)$. Thus eRe is a basic indecomposable QF-ring with a cyclic Nakayama permutation.

(3) R has a Nakayama automorphism.

Proof.

(1) Put $X = S(Q_Q)$ $(= S({}_QQ))$. Noting that $cX = Xc = 0$, we can easily see that

$$S(e_1R_R) = \begin{pmatrix} 0 \cdots 0 & X \\ 0 \cdots 0 & 0 \\ \cdots & \\ 0 \cdots 0 & 0 \end{pmatrix} = S(Re_n), \text{ and}$$

$$S(e_iR_R) = \begin{pmatrix} 0 & & & & & & \\ 0 & \cdot & & & 0 & & \\ & \cdot & \cdot & & & & \\ & & 0 & 0 & & & \\ & & & X & 0 & & \\ & & & & 0 & \cdot & \\ & 0 & & & & \cdot & \cdot \\ & & & & & 0 & 0 \end{pmatrix} = S({}_RRe_{i-1}) \quad \text{for } i = 2, \ldots, n.$$

Hence it follows that $(e_1R; Re_n)$, $(e_2R; Re_1)$, $\ldots$, $(e_nR; Re_{n-1})$ are i-pairs. Therefore R is a basic indecomposable QF-ring.

(2) For any subset $\{f_1, \ldots, f_k\} \subseteq Pi(R)$, clearly fRf is represented as a skew matrix ring over Q with respect to (σ, c, k), where $f = f_1 + \cdots + f_k$, whence we see that eRe is represented as a skew matrix ring for any idempotent e of R.

(3) This follows from Theorem 2.1. □

3. Nakayama Permutations vs Given Permutations

Using skew matrix rings, we shall provide several examples of basic indecomposable QF-rings.

Let k be a field and consider $Q = k[x]/(x^4)$ and put $c = x + (x^4)$. Then Q is a local ring and its ideals are only Q, Qc, Qc^2, Qc^3, 0. Using Q, we shall construct an interesting example of a QF-ring with a Nakayama permutation which corresponds to any given permutation.

Let R be the skew matrix ring $(Q)_{id_Q,c,w}$ with skew matrix units $\{\alpha_{ij}\} \cup \{\beta_{ji}\}$. We take $\{n(1), n(2), \ldots, n(m)\} \subset \{1, 2, \ldots, w\}$ such that

$$1 < n(1) \leq n(2) \leq \cdots \leq n(m) \quad \text{and} \quad n(1) + n(2) + \cdots + n(m) = w$$

and put

$$\begin{aligned} &(1,1) = 1, (1,2) = 2, \ldots, (1, n(1)) = n(1), \\ &(2,1) = n(1) + 1, (2,2) = n(1) + 2, \ldots, (2, n(2)) = n(1) + n(2), \ldots \\ &\ldots, (m,1) = n(1) + n(2) + \cdots + n(m-1) + 1, \ldots, (m, n(m)) = w. \end{aligned}$$

For the sake of convenience, we use kl instead of (k,l). Then

$$\{1, 2, \ldots, w\} = \{11, 12, \ldots, 1n(1), 21, \ldots, 2n(2), \ldots, m1, \ldots, mn(m)\}.$$

We make the following partition of $\{1, 2, \dots, w\}$:

$$\{1 = 11,\ 12, \dots,\ 1n(1))\} \cup \{21,\ 22, \dots,\ 2n(2)\} \cup \cdots$$
$$\cdots \cup \{(m-1)1, \dots, (m-1)n(m-1)\} \cup \{m1,\ m2,\ \dots,\ mn(m)\}.$$

For each $i,\ j \in \{1, \dots, m\}$, we construct blocks R_{ij} of R as follows:

$$R_{ii} = \begin{pmatrix} Q\alpha_{i1,i1} & Q\alpha_{i1,i1+1} & \cdots & Q\alpha_{i1,in(i)} \\ Q\beta_{i1+1,i1} & Q\alpha_{i1+1,i1+1} & \cdots & \cdots \\ \cdots & \cdots & \cdots & Q\alpha_{in(i)-1,i(n(i))} \\ Q\beta_{in(i),i1} & \cdots & Q\beta_{in(i),in(i)-1} & Q\alpha_{in(i),in(i)} \end{pmatrix},$$

$$R_{ij} = \begin{pmatrix} Q\alpha_{i1,j1} & Q\alpha_{i1,j1+1} & \cdots & Q\alpha_{i1,jn(j)} \\ Q\alpha_{i1+1,j1} & Q\alpha_{i1+1,j1+1} & \cdots & Q\alpha_{i1+1,jn(j)} \\ \cdots & \cdots & \cdots & \cdots \\ Q\alpha_{in(i),j1} & Q\alpha_{in(i),j1+1} & \cdots & Q\alpha_{in(i),jn(j)} \end{pmatrix} \quad \text{for } i < j,$$

$$R_{ji} = \begin{pmatrix} Q\beta_{j1,i1} & Q\beta_{j1,i1+1} & \cdots & Q\beta_{j1,in(i)} \\ Q\beta_{j1+1,i1} & Q\beta_{j1+1,i1+1} & \cdots & Q\beta_{j1+1,in(i)} \\ \cdots & \cdots & \cdots & \cdots \\ Q\beta_{jn(j),i1} & Q\beta_{jn(j),i1+1} & \cdots & Q\beta_{jn(j),in(i)} \end{pmatrix} \quad \text{for } i < j.$$

Then R is represented as

$$R = \begin{pmatrix} R_{11} & \cdots & R_{1m} \\ & \cdots & \\ R_{m1} & \cdots & R_{mm} \end{pmatrix}.$$

For $i < j$, we consider the following subsets $T_{ij} \subsetneq R_{ij}$ and $T_{ji} \subsetneq R_{ji}$:

$$T_{ij} = \begin{pmatrix} Qc\alpha_{i1,j1} & \cdots & \cdots & & \cdots & \cdots & Qc\alpha_{i1,jn(j)} \\ Qc^2\alpha_{i1+1,j1} & \ddots & & & & & \vdots \\ \vdots & \ddots & \ddots & & & & \vdots \\ Qc^2\alpha_{in(i),j1} & \cdots & Qc^2\alpha_{in(i),jx} & Qc\alpha_{in(i),jx+1} & & \cdots & Qc\alpha_{in(i),jn(j)} \end{pmatrix},$$

where $x = n(i) - 1$,

$$T_{ji} = \begin{pmatrix} Qc\beta_{j1,i1} & \cdots & \cdots & Qc\beta_{j1,in(i)} \\ Qc^2\beta_{j1+1,i1} & \ddots & & Qc\beta_{j1+1,in(i)} \\ \vdots & \ddots & \ddots & \vdots \\ \vdots & & Qc^2\beta_{js,in(i)-1} & Qc\beta_{js,in(i)} \\ \vdots & & \vdots & \vdots \\ Qc^2\beta_{jn(j),i1} & \cdots & Qc^2\beta_{jn(j),in(i)-1} & Qc\beta_{jn(j),in(i)} \end{pmatrix},$$

where $s = n(i) - 1$. Put

$$T = \begin{pmatrix} R_{11} & T_{12} & \cdots & \cdots & T_{1m} \\ T_{21} & \ddots & \ddots & & \vdots \\ \vdots & \ddots & \ddots & \ddots & \vdots \\ \vdots & & \ddots & \ddots & T_{m-1,m} \\ T_{m1} & \cdots & \cdots & T_{m,m-1} & R_{mm} \end{pmatrix}.$$

Then T is a basic indecomposable artinian subring of R. In T, we put $f_k = \langle 1 \rangle_{kk}$ for each $k = 1, \ldots, w$. Then $\{f_1, \ldots, f_w\}$ is a complete set of orthogonal primitive idempotents of T; note that $f_i T f_i \cong Q$.

For each $i < j$, we define $I_{ij} \subsetneq T_{ij}$ and $I_{ji} \subsetneq T_{ji}$ as follows:

$$I_{ij} = \begin{pmatrix} Qc^2\alpha_{i1,j1} & \cdots & \cdots & \cdots & \cdots & Qc^2\alpha_{i1,jn(j)} \\ Qc^3\alpha_{i1+1,j1} & \ddots & & & & \vdots \\ \vdots & \ddots & \ddots & & & \vdots \\ Qc^3\alpha_{in(i),j1} & \cdots & Qc^3\alpha_{in(i),x} & Qc^2\alpha_{in(i),x+1} & \cdots & Qc^2\alpha_{in(i),jn(j)} \end{pmatrix},$$

$$I_{ji} = \begin{pmatrix} Qc^2\beta_{j1,i1} & \cdots & \cdots & Qc^2\beta_{j1,in(i)} \\ Qc^3\beta_{j1+1,i1} & \ddots & & \vdots \\ \vdots & \ddots & \ddots & \vdots \\ \vdots & & Qc^3\beta_{s,in(i)-1} & Qc^2\beta_{s,in(i)} \\ \vdots & & \vdots & \vdots \\ Qc^3\beta_{jn(j),i1} & \cdots & Qc^3\beta_{jn(j),in(i)-1} & Qc^2\beta_{jn(j),in(i)} \end{pmatrix}.$$

Put

$$I = \begin{pmatrix} 0 & I_{12} \cdots & \cdots & I_{1m} \\ I_{21} & 0 \ddots & & \vdots \\ \vdots & \ddots \ddots & \ddots & \vdots \\ \vdots & \ddots & \ddots & I_{m-1,m} \\ I_{m1} & \cdots \cdots I_{m,m-1} & & 0 \end{pmatrix}.$$

Then I is an ideal of T. Here we put

$$\mathrm{G} = \mathrm{T/I}$$

and put $g_i = f_i + I \ \forall\, i$. We show that G is a basic indecomposable QF-ring. Actually, by the description of G and I together with the structure of R (cf. Theorem 2.2), we can easily see that the following pairs are i-pairs:

$$\begin{aligned}&(g_{11}G;\ Gg_{1n(1)}),\ (g_{11+1}G;\ Gg_{11}),\ (g_{11+2}G;\ Gg_{11+1}), \ldots,\\ &(g_{1n(1)-1}G;\ Gg_{1n(1)-2}),\ (g_{1n(1)}G;\ Gg_{11}), \ldots,\\ &(g_{m1}G;\ Gg_{mn(m)}),\ (g_{m1+1,m1}G;\ Gg_{m1}),\ (g_{m1+2}G;\ Gg_{m1+1}), \ldots,\\ &(g_{mn(m)-1}G;\ Gg_{mn(m)-2}),\ (g_{mn(m)}G;\ Gg_{mn(m)-1}).\end{aligned}$$

Hence G is a QF-ring whose Nakayama permutation is the product of the following cyclic permutations:

$$\begin{pmatrix} g_{11} & g_{11+1} & g_{11+2} & \cdots & g_{1n(1)} \\ g_{1n(1)} & g_{11} & g_{11+1} & \cdots & g_{1n(1)-1} \end{pmatrix},$$

$$\cdots \quad \cdots \quad \cdots$$

$$\begin{pmatrix} g_{m1} & g_{m1+1} & g_{m1+2} & \cdots & g_{mn(m)} \\ g_{mn(m)} & g_{m1} & g_{m1+1} & \cdots & g_{mn(m)-1} \end{pmatrix}.$$

Next, under this QF-ring G, we shall make another type of QF-rings. In order to make it, put $Q^\star = k[x]/(x^5)$ and put $c = x + (x^5)$ in $Q^\star$. Though we already used $c = x + (x^4)$ in Q, there are no confusions in below arguments. We note that $Q^\star/S(Q^\star) \cong Q$ and each $Q\alpha_{ij,kl}$ and each $Q\beta_{ij,kl}$ are $(Q^\star,Q^\star)$-bimodule.

Now let $t > 0$. In addition to above

$$T = \begin{pmatrix} T_{11} & \cdots & T_{1m} \\ & \cdots & \\ T_{m1} & \cdots & T_{m1} \end{pmatrix}$$

we make more blocks $t\times t$-matrix T_{00}, $t\times n(i)$-matrix T_{0i} and $n(j)\times t$-matrix T_{j0}: as follows: Put

$$T_{00} = \begin{pmatrix} Q^\star & Qc^2\alpha_{11} & \cdots & Qc^2\alpha_{11} \\ Qc^2\alpha_{11} & Q^\star & \cdots & Qc^2\alpha_{11} \\ \cdots & \cdots & \cdots & \cdots \\ \cdots & \cdots & \cdots & \cdots \\ Qc^2\alpha_{11} & \cdots & Qc^2\alpha_{11} & Q^\star \end{pmatrix}$$

$$= \begin{pmatrix} Q^\star & Qc^2\alpha_{01,02} & \cdots & Qc^2\alpha_{01,0t} \\ Qc^2\alpha_{02,01} & Q^\star & \cdots & Qc^2\alpha_{02,0t} \\ \cdots & \cdots & \cdots & \cdots \\ \cdots & \cdots & \cdots & \cdots \\ Qc^2\alpha_{0t,01}\cdots & \cdots & Qc^2\alpha_{0t,0(t-1)} & Q^\star \end{pmatrix} (t\times t\textit{-matrix}),$$

where $\alpha_{oi,ok} = \alpha_{i0,k0} = \alpha_{11}\ \forall\ 1 \le i, k \le t$.

In T_{00}, for $1 \le i \le t$, we can define canonical multiplications: $\langle c^2\rangle_{1i}\langle c^2\rangle_{i1} = \langle c^4\rangle_{11}$; so $\langle c^3\rangle_{1i}\langle c^2\rangle_{i1} = \langle c^5\rangle_{11} = 0, \langle c^2\rangle_{1i}\langle c^3\rangle_{i1} = \langle c^5\rangle_{11} = 0$. Therefore T_{00} canonically becomes a ring.

Put

$$T_{01} = \begin{pmatrix} Qc^2\alpha_{11,11} & Qc\alpha_{11,11+1} & Qc\alpha_{11,11+2} & \cdots & Qc\alpha_{11,1n(1)} \\ \vdots & \vdots & \vdots & & \vdots \\ \vdots & \vdots & \vdots & & \vdots \\ Qc^2\alpha_{11,11} & Qc\alpha_{11,11+1} & Qc\alpha_{11,11+2} & \cdots & Qc\alpha_{11,1n(1)} \end{pmatrix} (t\times n(1)\textit{-matrix})$$

$$= \begin{pmatrix} Qc^2 & Qc\alpha_{12} & Qc\alpha_{13} & \cdots & Qc\alpha_{1n(1)} \\ \vdots & \vdots & \vdots & & \vdots \\ \vdots & \vdots & \vdots & & \vdots \\ Qc^2 & Qc\alpha_{12} & Qc\alpha_{13} & \cdots & Qc\alpha_{1n(1)} \end{pmatrix},$$

$$T_{10} = \begin{pmatrix} Qc^2\alpha_{11,11} & \cdots & Qc^2\alpha_{11,11} \\ Qc\beta_{11+1,11} & \cdots & Qc\beta_{11+1,11} \\ Qc^2\beta_{11+2,11} & \cdots & Qc^2\beta_{11+2,11} \\ Qc^2\beta_{11+3,11} & \cdots & Qc^2\beta_{11+3,11} \\ & \cdots & \\ Qc^2\beta_{1n(1),11} & \cdots & Qc^2\beta_{1n(1),11} \end{pmatrix} (n(1) \times t\textit{-matrix})$$

$$= \begin{pmatrix} Qc^2 & \cdots & Qc^2 \\ Qc\beta_{21} & \cdots & Qc\beta_{21} \\ Qc^2\beta_{31} & \cdots & Qc^2\beta_{31} \\ Qc^2\beta_{41} & \cdots & Qc^2\beta_{41} \\ & \cdots & \\ Qc^2\beta_{n(1)1} & \cdots & Qc^2\beta_{n(1)1} \end{pmatrix}.$$

For $1 < i$, put

$$T_{0i} = \begin{pmatrix} Qc\alpha_{11,i1} & Qc\alpha_{11,i1+1} & \cdots & Qc\alpha_{11,in(i)} \\ & \cdots & \cdots & \\ Qc\alpha_{11,i1} & Qc\alpha_{11,i1+1} & \cdots & Qc\alpha_{11,in(i)} \end{pmatrix} \ (t \times n(i)\text{-}matrix).$$

For $1 < j$, put

$$T_{j0} = \begin{pmatrix} Qc\beta_{j1,11} & \cdots & Qc\beta_{j1,11} \\ Qc^2\beta_{j1+1,11} & \cdots & Qc^2\beta_{j1+1,11} \\ Qc^2\beta_{j1+2,11} & \cdots & Qc^2\beta_{j1+2,11} \\ & \cdots & \\ Qc^2\beta_{jn(j),11} & \cdots & Qc^2\beta_{jn(j),11)} \end{pmatrix} \ (n(j) \times t\text{-}matrix).$$

And we put

$$U = \begin{pmatrix} T_{00} & T_{01} & \cdots & T_{0m} \\ T_{10} & T_{11} & \cdots & T_{1m} \\ & & \cdots & \\ T_{m0} & T_{m1} & \cdots & T_{mm} \end{pmatrix}.$$

Then we can canonically define a multiplication: $(T_{0k}, T_{k0}) \to T_{00}$, and hence we can define a canonical multiplication in U and see that U becomes a (basic indecomposable artinian) ring.

Next, we make $I_{00} \subsetneq T_{00}$, $I_{ij} \subsetneq T_{ij}$ and $I_{ji} \subsetneq T_{ji}$ $(0 \leq i < j)$ as follows:

$$I_{00} = \begin{pmatrix} 0 & Qc^3 & \cdots & Qc^3 \\ Qc^3 & \ddots & \ddots & \vdots \\ \vdots & \ddots & \ddots & Qc^3 \\ Qc^3 & \cdots & Qc^3 & 0 \end{pmatrix} \ (t \times t\text{-}matrix),$$

$$I_{01} = \begin{pmatrix} Qc^3 & Qc^2\alpha_{12} & Qc^2\alpha_{13} & \cdots & Qc^2\alpha_{1n(1)} \\ \vdots & \vdots & \vdots & & \vdots \\ \vdots & \vdots & \vdots & & \vdots \\ Qc^3 & Qc^2\alpha_{12} & Qc^2\alpha_{13} & \cdots & Qc^2\alpha_{1n(1)} \end{pmatrix} (t \times n(1)\text{-}matrix),$$

$$I_{10} = \begin{pmatrix} Qc^3 & \cdots & Qc^3 \\ Qc^2\beta_{21} & \cdots & Qc^2\beta_{21} \\ Qc^3\beta_{31} & \cdots & Qc^3\beta_{31} \\ Qc^3\beta_{41} & \cdots & Qc^3\beta_{41} \\ & \cdots & \\ Qc^3\beta_{n(1)1} & \cdots & Qc^3\beta_{n(1)1} \end{pmatrix} (n(1) \times t\text{-}matrix).$$

For $1 < i$,

$$I_{0i} = \begin{pmatrix} Qc^2\alpha_{11,i1} & \cdots & Qc^2\alpha_{11,in(i)} \\ & \cdots & \\ Qc^2\alpha_{11,i1} & \cdots & Qc^2\alpha_{11,in(i)} \end{pmatrix} (t \times n(i)\text{-}matrix).$$

For $1 < j$,

$$I_{j0} = \begin{pmatrix} Qc^2\beta_{j1,11} & \cdots & Qc^2\beta_{j1,11} \\ Qc^3\beta_{j1+1,11} & \cdots & Qc^3\beta_{j1+1,11} \\ & \cdots & \\ Qc^3\beta_{jn(j),11} & \cdots & Qc^3\beta_{jn(j),11} \end{pmatrix} (n(j) \times t\text{-}matrix).$$

For $1 \leq i < j$, we use I_{ij} and I_{ji} mentioned above:

$$I_{ij} = \begin{pmatrix} Qc^2\alpha_{i1,j1} & \cdots & \cdots & \cdots & \cdots & Qc^2\alpha_{i1,jn(j)} \\ Qc^3\alpha_{i1+1,j1} & \ddots & & & & \vdots \\ \vdots & \ddots & \ddots & & & \vdots \\ Qc^3\alpha_{in(i),j1} & \cdots & Qc^3\alpha_{in(i),x} & Qc^2\alpha_{in(i),x+1} & \cdots & Qc^2\alpha_{in(i),jn(j)} \end{pmatrix},$$

$$I_{ji} = \begin{pmatrix} Qc^2\beta_{j1,i1} & \cdots & \cdots & Qc^2\beta_{j1,in(i)} \\ Qc^3\beta_{j1,ii} & \ddots & & \vdots \\ \vdots & \ddots & \ddots & \vdots \\ \vdots & & Qc^3\beta_{s,in(i)-1} & Qc^2\beta_{s,in(i)} \\ \vdots & & \vdots & \vdots \\ Qc^3\beta_{jn(j),i1} & \cdots & Qc^3\beta_{jn(j),in(i)-1} & Qc^2\beta_{jn(j),in(i)} \end{pmatrix}.$$

Put

$$I = \begin{pmatrix} I_{00} & I_{01} & I_{02} & \cdots & \cdots & I_{0m} \\ I_{10} & 0 & I_{12} & \cdots & \cdots & I_{1m} \\ I_{20} & I_{21} & 0 & I_{23} & \cdots & I_{2m} \\ \vdots & & \ddots & \ddots & & \vdots \\ \vdots & & & \ddots & \ddots & I_{m-1,m} \\ I_{m0} & I_{m1} & \cdots & \cdots & I_{m,m-1} & 0 \end{pmatrix}.$$

Then I is an ideal of U. Put $H = U/I$ and $h_i = \langle i \rangle_{ii}$ in H. Then H is a basic indecomposable artinian ring and $\{h_i\}$ is a complete set of orthogonal primitive idempotents in H. In H, we use $0i$-row ($i0$-column) to mean i-th row (i-th column) for $1 \le i \le t$, and use p-row (column) to denote $(t+p)$-th row ($(t+p)$-th column) for $p \in \{1 = 11, 11+1, \ldots, 1n(1)\} \cup \cdots \cup \{m(1), m(1)+1, \ldots mn(m) = w\}$. Put

$h_{0i} = \langle 1 \rangle_{ii}$ for $i = 1, \ldots, t$, and
$h_p = \langle 1 \rangle_{t+p,t+p}$ for $p \in \{1 = 11, 11+1, \ldots, 1n(1)\} \cup \cdots \cup \{m(1), m(1)+1, \ldots mn(m) = w\}$.

Then we can easily see that the following pairs are i-pairs:

$(h_{01}H; Hh_{01}), \ldots, (h_{0t}H; Hh_{0t})$,
$(h_{11}H; Hh_{1n(1)}), (h_{11+1}H; Hh_{11}), (h_{11+2}H; Hh_{11+1}), \ldots,$
$\ldots, (h_{1n(1)-1}H; Hh_{1n(1)-2}), (h_{1n(1)}H; Hh_{11}), \ldots,$
$(h_{m1}H; Hh_{mn(m)}), (h_{m1+1}H; Hh_{m1}), (h_{m1+2}H; Hh_{m1+1}), \ldots,$
$(h_{mn(m)-1}H; Hh_{mn(m)-2}), (h_{mn(m)}H; Hh_{mn(m)-1})$.

Hence H is a basic indecomposable QF-ring and its Nakayama permutation is the product:

$$\begin{pmatrix} h_{01} \cdots h_{0t} \\ h_{01} \cdots h_{0t} \end{pmatrix} \begin{pmatrix} h_{11} & h_{11+1} & \cdots & h_{1n(1)} \\ h_{1n(1)} & h_{11} & \cdots & h_{1n(1)-1} \end{pmatrix} \cdots\cdots \begin{pmatrix} h_{m1} & h_{m1+1} & \cdots & h_{mn(m)} \\ h_{mn(m)} & h_{m1} & \cdots & h_{mn(m)-1} \end{pmatrix}.$$

Here we note that for $e = h_{01} + \cdots + h_{0t}$, eHe is a basic indecomposable QF-ring with identity Nakayama permutation:

$$\begin{pmatrix} h_{01} \cdots h_{0t} \\ h_{01} \cdots h_{0t} \end{pmatrix}.$$

Thus we obtain the following result.

Theorem 3.1. *For a given permutation*

$$\begin{pmatrix} 1 & \cdots & n \\ \rho(1) & \cdots & \rho(n) \end{pmatrix},$$

we can construct a basic indecomposable QF-ring R with $Pi(R) = \{e_1, \ldots, e_n\}$ *with Nakayama permutation*

$$\begin{pmatrix} e_1 & \cdots & e_n \\ e_{\rho(1)} & \cdots & e_{\rho(n)} \end{pmatrix}.$$

Example 3.1.

(1) For $w = 4$ and the partition $\{1,2,3,4\} = \{1,2\} \cup \{3,4\}$,

$$G = \begin{pmatrix} Q & Q\alpha_{12} & Qc\alpha_{13} & Qc\alpha_{14} \\ Q\beta_{21} & Q & Qc^2\alpha_{23} & Qc\alpha_{24} \\ Qc\beta_{31} & Qc\beta_{32} & Q & Q\alpha_{34} \\ Qc^2\beta_{41} & Qc\beta_{42} & Q\beta_{43} & Q \end{pmatrix} \Big/ \begin{pmatrix} 0 & 0 & Qc^2\alpha_{13} & Qc^2\alpha_{14} \\ 0 & 0 & Qc^3\alpha_{23} & Qc^2\alpha_{24} \\ Qc^2\beta_{31} & Qc^2\beta_{32} & 0 & 0 \\ Qc^3\beta_{41} & Qc^2\beta_{42} & 0 & 0 \end{pmatrix}.$$

(2) For $t = 2, w = 2$,

$$H = \begin{pmatrix} Q^\star & Qc^2 & Qc^2 & Qc\alpha_{12} \\ Qc^2 & Q^\star & Qc^2 & Qc\alpha_{12} \\ Qc^2 & Qc^2 & Q & Q\alpha_{12} \\ Qc\beta_{21} & Qc\beta_{21} & Q\beta_{21} & Q \end{pmatrix} \Big/ \begin{pmatrix} 0 & Qc^3 & Qc^3 & Qc^2\alpha_{12} \\ Qc^3 & 0 & Qc^3 & Qc^2\alpha_{12} \\ Qc^3 & Qc^3 & 0 & 0 \\ Qc^2\beta_{21} & Qc^2\beta_{21} & 0 & 0 \end{pmatrix}.$$

Remark 3.1. Theorem 3.1 is shown in Fujita [3], Hoshino [6] and Koike [7], but we gave a proof using skew-matrix rings.

4. Strongly *QF*-Rings

The following fundamental theorem is due to Oshiro-Rim [11] .

Theorem 4.1. *Let R be a basic indecomposable QF-ring such that, for any idempotent e of R, eRe is a QF-ring with a cyclic Nakayama permutation. Then there exist a local QF-ring Q, an element* $c \in J(Q)$ *and* $\sigma \in \mathrm{Aut}(Q)$ *satisfying* $\sigma(c) = c$ *and* $\sigma(q)c = cq$ *for* $\forall\, q \in Q$ *such that R can be represented as the skew matrix ring:*

$$R \cong \begin{pmatrix} Q & \cdots & Q \\ & \cdots & \\ Q & \cdots & Q \end{pmatrix}_{\sigma,c,n}.$$

Following Yukimoto [12], we call a ring R a strongly QF-ring if it satisfies the property that eRe is a QF-ring for any idempotent e in R. We give following characterizations of this ring.

Theorem 4.2. *For a basic indecomposable QF-ring R with $Pi(R) = \{e_1, \ldots, e_n\}$, the following conditions are equivalent.*

(1) *eRe is a strongly QF-ring.*

(2) *R is a weakly symmetric QF-ring or R can be represented as a skew-matrix ring over $Q = e_1Re_1$:*

$$R = \begin{pmatrix} Q \cdots Q \\ \cdots \\ Q \cdots Q \end{pmatrix}_{\sigma,c,n},$$

where $\sigma \in \mathrm{Aut}(Q)$ and $c \in J(Q)$;

(3) *R is weakly symmetric or, for any idempotent e, eRe is a QF-ring with a cyclic Nakayama permutation.*

(4) *The endomorphism ring of every finitely generated projective right R-module is a QF-ring.*

(5) *The endomorphism ring of every finitely generated projective left R-module is a QF-ring.*

For a proof of this theorem, we use the following two lemmas.

Lemma A. *Let R be a basic indecomposable QF-ringwith $Pi(R) = \{e, f\}$. Assume that hRh is a QF-ring for any idempotent h of R and the Nakayama permutation of $\{e, f\}$ is*

$$\begin{pmatrix} e & f \\ f & e \end{pmatrix}.$$

We represent R as

$$R = \begin{pmatrix} Q & A \\ B & T \end{pmatrix},$$

where $Q = eRe$, $T = fRf$, $A = eRf$ and $B = fRe$. Put $X = \{a \in A \mid aB = 0\}$ and $Y = \{b \in B \mid Ab = 0\}$. Then , as is easily seen,

(1) $X = \{a \in A \mid Ba = 0\}$, $Y = \{b \in B \mid bA = 0\}$, *and*

(2) $\langle X\rangle_{21} = \begin{pmatrix} 0 & 0 \\ X & 0 \end{pmatrix}$, $\langle Y\rangle_{12} = \begin{pmatrix} 0 & Y \\ 0 & 0 \end{pmatrix}$ *are ideals of R.*

Lemma B. *Let R be a basic indecomposable QF-ring with $Pi(R) = \{e_1, e_2, e_3\}$. If the Nakayama permutation of $Pi(R)$ is not identity and eRe is a QF-ring for any idempotent e of R, then the Nakayama permutation of R is cyclic.*

Proof. Since the Nakayama permutation of R is non-identity, we may assume that

$$\begin{pmatrix} e_1 & e_2 & e_3 \\ e_2 & e_1 & e_3 \end{pmatrix}$$

is the Nakayama permutation. We represent R as

$$R = \begin{pmatrix} Q_1 & A_{12} & A_{13} \\ A_{21} & Q_2 & A_{23} \\ A_{31} & A_{32} & Q_3 \end{pmatrix}.$$

Consider the three rings:

$$R(12) = \begin{pmatrix} Q_1 & A_{12} \\ A_{21} & Q_2 \end{pmatrix}, \quad R(23) = \begin{pmatrix} Q_2 & A_{23} \\ A_{32} & Q_3 \end{pmatrix}, \quad R(13) = \begin{pmatrix} Q_1 & A_{13} \\ A_{31} & Q_3 \end{pmatrix}.$$

The socles of these rings are the following:

$$S(R(12)) = \begin{pmatrix} 0 & S(A_{12}) \\ S(A_{21}) & 0 \end{pmatrix}, \; S(R(23)) = \begin{pmatrix} S(Q_2) & 0 \\ 0 & S(Q_3) \end{pmatrix}, \; S(R(13)) = \begin{pmatrix} S(Q_1) & 0 \\ 0 & S(Q_3) \end{pmatrix},$$

respectively. Put $X = \{x \in A_{21} \mid xA_{12} = 0\}$ and $Y = \{y \in A_{12} \mid A_{21}y = 0\}$. Then, by Lemma A, $\langle X \rangle_{21}$ and $\langle Y \rangle_{12}$ in $R(12)$ are ideals of $R(12)$. Noting this fact, together with

$$S\left(\begin{pmatrix} Q_2 & A_{13} \\ 0 & 0 \end{pmatrix}\right) = \begin{pmatrix} S(Q_2) & 0 \\ 0 & 0 \end{pmatrix} \text{ and } S\left(\begin{pmatrix} 0 & 0 \\ A_{32} & Q_3 \end{pmatrix}\right) = \begin{pmatrix} 0 & 0 \\ 0 & S(Q_3) \end{pmatrix},$$

we see that

$$\langle X_{21} \rangle_{21} = \begin{pmatrix} 0 & 0 & 0 \\ X_{21} & 0 & 0 \\ 0 & 0 & 0 \end{pmatrix} \quad \text{and} \quad \langle X_{12} \rangle_{12} = \begin{pmatrix} 0 & X_{12} & 0 \\ 0 & 0 & 0 \\ 0 & 0 & 0 \end{pmatrix}$$

are left and right ideals of R. We represent $\overline{e_2R} = e_2R/\langle X_{21} \rangle_{21}$ as

$$\overline{e_2R} = \begin{pmatrix} 0 & 0 & 0 \\ A_{21}/X_{21} & Q_2 & 0 \\ 0 & 0 & 0 \end{pmatrix}.$$

Then, we see that

$$S(\overline{e_2R}) = \begin{pmatrix} 0 & 0 & 0 \\ 0 & S(Q) & 0 \\ 0 & 0 & 0 \end{pmatrix}.$$

Since e_1R_R is injective and $S(e_1R_R) \cong S(\overline{e_2R_R})$, there exists a monomorphism φ_{21} from $\overline{e_2R}$ to e_1J. Let η_2 be the canonical epimorphism from $e_2R \to \overline{e_2R}$, and put $\pi_{21} = \varphi_{21}\eta_2$. Then π_{21} is an R-homomorphism from e_2R to e_1J, and in R, the restriction map $\pi_{21}|_{\langle A_{23}\rangle_{23}}$ is monomorphic. Similarly, considering the factor module $\overline{e_1R} = e_1R/\langle X_{12}\rangle_{12}$, we obtain an R-homomorphism $\pi_{12} : e_1R \to e_2J(R)$ such that $\pi_{12}|_{\langle A_{13}\rangle_{13}}$ is monomorphic. Therefore it follows that $\pi_{21}(\langle A_{23}\rangle_{23}) = \langle A_{13}\rangle_{13}$ and $\pi_{12}(\langle A_{12}\rangle_{12}) = \langle A_{23}\rangle_{23}$. Since $\pi_{21}(\langle A_{12}\rangle_{12}) \subseteq \langle J(Q_2)\rangle_{22}$ and $\pi_{21}(\langle A_{21}\rangle_{21}) = \langle J(Q_1)\rangle_{11}$, we can obtain m such that

$$(\pi_{21}\pi_{12})^m(e_2R) = \begin{pmatrix} 0 & 0 & A_{13} \\ 0 & 0 & 0 \\ 0 & 0 & 0 \end{pmatrix},$$

which is impossible, since $S(eR_R) = \langle S(A_{12}\rangle_{12})$. Thus the Nakayama permutation of $Pi(R)$ must be cyclic. □

Proof of Theorem 4.2. (1) ⇒ (3). Suppose that the Nakayama permutation ρ of R is non-identity. We show that ρ is cyclic. Assume that ρ is not cyclic and write $\rho = \rho_1\rho_2\cdots\rho_t$, where the ρ_i are cyclic permutations. We can assume that the size k ($< n$) of ρ_1 is bigger than 1, and we may express ρ_1 as

$$\rho_1 = \begin{pmatrix} e_1 \; e_2 \; \cdots\cdots \; e_n \end{pmatrix} = \begin{pmatrix} e_1 & e_2 & e_3 & \cdots & e_n \\ e_k & e_1 & e_2 & \cdots & e_{k-1} \end{pmatrix}.$$

Put $f_1 = e_1$ and $f_2 = e_2$ and take $f_3 \in Pi(R) - \{e_1, \ldots, e_k\}$. Consider the subring of R:

$$T = \begin{pmatrix} f_1Rf_1 & f_1Rf_2 & f_1Rf_3 \\ f_2Rf_1 & f_2Rf_2 & f_2Rf_3 \\ f_3Rf_1 & f_3Rf_2 & f_3Rf_3 \end{pmatrix}.$$

By (1), this ring is a basic indecomposable QF subring of R and we can see that

$$S(T) = \begin{pmatrix} 0 & S(f_1Rf_2) & 0 \\ S(f_2Rf_1) & 0 & 0 \\ 0 & 0 & S(f_3Rf_3) \end{pmatrix}.$$

Furthermore, since R is indecomposable, we can take f_3 such as T is indecomposable. Thus T is a basic indecomposable QF-ring and its Nakayama permutation is neither identity nor cyclic. This contradicts Lemma B.

(3) $\Rightarrow$ (2). This follows from Theorem 4.1.

(2) $\Rightarrow$ (1). If R is a weakly symmetric QF-ring, then clearly so is eRe for any idempotent e of R. If R is represented as a skew-matrix ring over $Q = e_1Re_1$ with respect to $\sigma \in \mathrm{Aut}(Q)$ and $c \in J(Q)$, then so is eRe for any idempotent e of R.

(4) $\Rightarrow$ (1) is obvious. (1) $\Rightarrow$ (4). Let P be a projective right R-module. Then P can be expressed as a direct sum of indecomposable projective modules $\{P_i\}_I$. We take a subfamily $\{P_i\}_J$ of $\{P_i\}_I$ which is a representative set of $\{P_i\}_I$. Put $T = \sum_J \oplus P_i$. Then $End(P_R)$ is Morita equivalent to $End(T_R)$. We can take an idempotent $e \in R$ such that $eR_R \cong T_R$, $End(P_R)$ is Morita equivalent to eRe. Hence, $End(P_R)$ is QF.

Accordinly, we showed the equivalences of (1) - (4) . Since (1) is left and right symmetric, (5) is also equivalent to these conditions. □

Remark 4.1. The naming of a strongly QF-ring is due to Yukimoto [12]. Yukimoto showed the implication (1) $\Rightarrow$ (3). Hoshino [5] showed the equivalences of (1), (3), (4) and (5). Our proof of Theorem 4.2 is new.

5. Block Extensions of Skew-Matrix Rings

Let Q be a local ring, and let $c \in J(Q)$ and $\sigma \in \mathrm{Aut}(Q)$ satisfying $\sigma(c) = c$ and $\sigma(q)c = cq \ \forall\ q \in Q$. In the skew matrix ring $R = (Q)_{\sigma,c,n}$, we put $e_i = \langle 1 \rangle_{ii}$ for each $i = 1, \ldots, n$. We represent R as

$$R = \begin{pmatrix} Q_1 & A_{12} & \cdots & A_{1n} \\ A_{21} & \ddots & \ddots & \vdots \\ \vdots & \ddots & \ddots & A_{n-1,n} \\ A_{n1} & \cdots & A_{n,n-1} & Q_n \end{pmatrix},$$

where $Q_i = e_iRe_i$ for each $i = 1, \ldots, n$, and $A_{ij} = e_iRe_j$ for each distinct $i, j \in \{1, \ldots, n\}$.

For $k(1), \ldots, k(n) \in \mathbb{N}$, we define the block extension

$R(k(1), \ldots, k(n))$ of R as follows:

$$R(k(1), \ldots, k(n)) = \begin{pmatrix} Q(1) & A(1,2) & \cdots & A(1,n) \\ A(2,1) & \ddots & \ddots & \vdots \\ \vdots & \ddots & \ddots & A(n-1,n) \\ A(n,1) & \cdots & A(n,n-1) & Q(n) \end{pmatrix},$$

where

$$Q(i) = \begin{pmatrix} Q_i & \cdots & \cdots & Q_i \\ J(Q_i) & \ddots & & \vdots \\ \vdots & \ddots & \ddots & \vdots \\ J(Q_i) & \cdots & J(Q_i) & Q_i \end{pmatrix} \quad (\ k(i) \times k(i)\text{-matrix}),$$

$$A(i,j) = \begin{pmatrix} A_{ij} & \cdots & A_{ij} \\ & \cdots & \\ A_{ij} & \cdots & A_{ij} \end{pmatrix} \quad (\ k(i) \times k(j)\text{-matrix}).$$

We put

$$Q(i;c) = \begin{pmatrix} Q_i & \cdots & \cdots & Q_i \\ Q_i c & \ddots & & \vdots \\ \vdots & \ddots & \ddots & \vdots \\ Q_i c & \cdots & Q_i c & Q_i \end{pmatrix} \quad (\ k(i) \times k(j)\text{-matrix})$$

and

$$R(k(1), \ldots, k(n); c) = \begin{pmatrix} Q(1;c) & A(1,2) & \cdots & A(1,n) \\ A(2,1) & \ddots & \ddots & \vdots \\ \vdots & \ddots & \ddots & A(n-1,n) \\ A(n,1) & \cdots & A(n,n-1) & Q(n;c) \end{pmatrix}.$$

Then, as is easily seen, $R(k(1), \ldots, k(n); c)$ is a subring of $R(k(1), \ldots, k(n))$. We say that $R(k(1), \ldots, k(n); c)$ is a *c-block extension* of $R = (Q)_{\sigma,c,n}$.

Under this situation, we show the following theorem which plays an important role for the study on the classification on Nakayama rings (see Baba-Oshiro [2)].

Theorem 5.1. *$R(k(1), \ldots, k(n); c)$ is isomorphic to a factor ring of the skew matrix ring $T = (Q)_{\sigma,c,k(1)+\cdots+k(n)}$.*

Proof. We represent T as

$$T = \begin{pmatrix} T(1) & T(1,2) & \cdots & T(1,n) \\ T(2,1) & \ddots & \ddots & \vdots \\ \vdots & \ddots & \ddots & T(n-1,n) \\ T(n,1) & \cdots & T(n,n-1) & T(n) \end{pmatrix},$$

where

$$T(i) = \begin{pmatrix} Q \cdots Q \\ \cdots \\ Q \cdots Q \end{pmatrix} \quad (k(i) \times k(i) \text{ matrix}), \text{ and}$$

$$T(i,j) = \begin{pmatrix} Q \cdots Q \\ \cdots \\ Q \cdots Q \end{pmatrix} \quad (k(i) \times k(j) \text{ matrix}).$$

For each i, $j \in \{1, \ldots, n\}$, let $\varphi(i) : T(i) \longrightarrow Q(i;c)$ be the canonical map given by

$$\begin{pmatrix} q_{11} & q_{12} & \cdots & q_{1k} \\ q_{21} & \ddots & \ddots & \vdots \\ \vdots & \ddots & \ddots & q_{n-1,n} \\ q_{n1} & \cdots & q_{n,n-1} & q_{nn} \end{pmatrix} \longmapsto \begin{pmatrix} q_{11} & q_{12} & \cdots & q_{1k} \\ q_{21}c & \ddots & \ddots & \vdots \\ \vdots & \ddots & \ddots & q_{n-1,n} \\ q_{n1}c & \cdots & q_{n,n-1}c & q_{nn} \end{pmatrix}$$

and let $\varphi(i,j) : T(i,j) \longrightarrow A(i,j)$ be the canonical map given by

$$\begin{pmatrix} q_{i1} & \cdots & q_{ik_j} \\ & \cdots & \\ q_{k_i1} & \cdots & q_{k_ik_j} \end{pmatrix} \longmapsto \begin{pmatrix} q_{i1}\alpha_{ij} & \cdots & q_{ik_j}\alpha_{ij} \\ & \cdots & \\ q_{k_i1}\alpha_{ij} & \cdots & q_{k_ik_j}\alpha_{ij} \end{pmatrix}.$$

Then it is easy to see that

$$\Phi = \begin{pmatrix} \varphi(1) & \varphi(1,2) & \cdots & \varphi(1,n) \\ \varphi(2,1) & \ddots & \ddots & \vdots \\ \vdots & \ddots & \ddots & \varphi(n-1,n) \\ \varphi(n,1) & \cdots & \varphi(n,n-1) & \varphi(n) \end{pmatrix}$$

is a ring epimorphism from $T = (Q)_{\sigma,c,k(1)+\cdots+k(n)}$ to $R(k(1), \ldots, k(n); c)$. □

Example 5.1. (1). Let D be a division ring. Then

$$\begin{pmatrix} D & \cdots & \cdots & D \\ D & \ddots & & \vdots \\ \vdots & \ddots & \ddots & \vdots \\ D & \cdots & D & D \end{pmatrix}_{id,0,k} \Big/ \begin{pmatrix} 0 & \cdots & \cdots & 0 \\ D & \ddots & & \vdots \\ \vdots & \ddots & \ddots & \vdots \\ D & \cdots & D & 0 \end{pmatrix} \cong D(k;0) = \begin{pmatrix} D & \cdots & \cdots & D \\ 0 & \ddots & & \vdots \\ \vdots & \ddots & \ddots & \vdots \\ 0 & \cdots & 0 & D \end{pmatrix}.$$

(2). For

$$R = \begin{pmatrix} Q & Q\alpha \\ Q\beta & Q \end{pmatrix}_{\sigma,c,2},$$

where Q is a local ring and $c \in J(Q)$ and $\sigma \in \mathrm{Aut}(Q)$, we have

$$R(2,2;c) \cong \begin{pmatrix} Q & Q & Q & Q \\ Q & Q & Q & Q \\ Q & Q & Q & Q \\ Q & Q & Q & Q \end{pmatrix}_{\sigma,c,4} \Big/ \begin{pmatrix} 0 & 0 & 0 & 0 \\ l_Q(c) & 0 & 0 & 0 \\ 0 & 0 & 0 & 0 \\ 0 & 0 & l_Q(c) & 0 \end{pmatrix}.$$

References

1. Y. Baba and K. Oshiro, On a Theorem of Fuller, *J. Algebra* 154, 1 (1993), 86–94.
2. Y. Baba and K. Oshiro, Classical Artinian Rings and Related Topics, (Lecture Note).
3. F. Fujita and Y. Sakai, Frobenius full matrix algebras and Gorenstein tiled orders, *Comm. Algebra* 34 (2006), 1181-1203.
4. T. A. Hannula, The Morita context and the construction of *QF* rings, Proceedings of the Conference on Orders, Group Rings and Related Topics, Lect. Notes Math. 353, Springer-Verlag, Heidelberg/New York/Berlin (1973), 113-130.
5. M. Hoshino, Strongly quasi-Frobenius rings, *Comm. Algebra* 28 (2000), 3585–3599.
6. M. Hoshino, Frobenius extensions and tilting complexes, *preprint.*
7. K. Koike, Self-duality of quasi-Harada rings and locally distributive rings, *J. Algebra* 2006.
8. H. Kupisch, Uber ein Klasse von Ringen mit Minimalbedingung II, *Arch. Math.* 26 (1975), 23–35.
9. K. Oshiro, Structure of Nakayama rings, *Proceedings 20th Symp. Ring Theory*, Okayama (1987), 109–133.
10. K. Oshiro, Theories of Harada in Artinian rings, *International Symposium on Ring Theory, (Kyongju, 1999) , Birkhauser Boston, Boston (2001), 279-328*

11. K. Oshiro and S. H. Rim, QFrings with cyclic Nakayama permutation, *Osaka J. Math.* 34 (1997), 1–19.
12. Y. Yukimoto, On decomposition of strongly quasi-Frobenius rings, *Comm. Algebra* 28 (2000), 1111–1113.

ON μ–ESSENTIAL AND μ–M–SINGULAR MODULES

A. Çiğdem ÖZCAN
Hacettepe University
Department of Mathematics
06800 Beytepe Ankara, Turkey
E-mail: ozcan@hacettepe.edu.tr

As a generalization of essential submodules Zhou defines a μ-essential submodule provided it has a non-zero intersection with any non-zero submodule in μ for any class μ. Let M be a module. In this article we study δ–essential submodules as a dual of δ-small submodules of Zhou where $\delta = \{N \in \sigma[M] : \text{Rej}(N, \mathcal{M}) = 0\}$ and $\mathcal{M} = \{N \in \sigma[M] : N \ll \widehat{N}\}$, and also define μ-M–singular modules as modules $N \in \sigma[M]$ such that $N \cong K/L$ for some $K \in \sigma[M]$ and L is μ–essential in K. By $\mathcal{M}$–M–singular modules and δ–M–singular modules a characterization of GCO–modules, and by $\mathcal{FC}$–M–singular modules where $\mathcal{FC}$ is the class of finitely cogenerated modules, a characterization of semisimple Artinian rings are given.

Keywords: essential submodule, singular module

1. Preliminaries

Let M be a module, $N \in \sigma[M]$ and μ a class of modules in $\sigma[M]$ which is closed under isomorphisms and submodules. Following Zhou[16] we call a submodule N a *μ–essential submodule* of $K \in \sigma[M]$ if for any nonzero μ-submodule X in K, $N \cap X \neq 0$, denoted by $N \leq_{\mu e} K$. In this article after studying some properties of μ–essential submodules we consider δ–essential submodules as a dual of δ-small submodules of Zhou by denoting the class $\delta = \{N \in \sigma[M] : \text{Rej}(N, \mathcal{M}) := \overline{Z}_M(N) = 0\}$ where $\mathcal{M} = \{N \in \sigma[M] : N \ll \widehat{N}\}$ and $\overline{Z}_M(.)$ is defined by Talebi and Vanaja as a dual of the singular submodule $Z_M(.)$. If $\mathcal{F} = \{F \in \sigma[M] : \forall 0 \neq K \subseteq F,\ \overline{Z}_M(K) \neq K\}$, then it is known that $\mathcal{M} \subseteq \delta \subseteq \mathcal{F}$. We prove a result on when an $\mathcal{F}$–essential submodule is δ–essential and a δ–essential submodule is $\mathcal{M}$–essential. Also we prove that $Tr(\mathcal{S} \cap \mathcal{M}, N) = Tr(\mathcal{S} \cap \delta, N) = Tr(\mathcal{S} \cap \mathcal{F}, N)$ where $\mathcal{S}$ is the class of simple modules in $\sigma[M]$ and Tr is used for the trace.

In the last section we define μ-M–singular modules $N \in \sigma[M]$ for a module M. N is called *μ-M-singular module* if $N \cong K/L$ for some $K \in \sigma[M]$ and $L \leq_{\mu e} K$. It is proved that M is a GCO–module (i.e. every simple M–singular module is injective in $\sigma[M]$) if and only if for every $\mathcal{M}$–M–singular module N in $\sigma[M]$, $\overline{Z}_M(N) = N$ if and only if for every δ–M–singular module N in $\sigma[M]$, $\overline{Z}_M(N) = N$. When we consider the class of all finitely cogenerated modules $\mathcal{FC}$ we prove that every finitely cogenerated R–module is projective if and only if for every $\mathcal{FC}$–R–singular R–module N, $\mathrm{Rej}(N, \mathcal{FC}) = N$ if and only if R is semisimple Artinian.

Let R be a ring with identity. All modules we consider are unitary right R-modules and we denote the category of all such modules by Mod–R. Let M be an R-module. The R–injective hull of M is denoted by $\mathrm{E}(M)$, and the *M–injective hull* of N in the category of $\sigma[M]$ is denoted by $\widehat{N}$. For the definition of $\sigma[M]$ and $\widehat{N}$ see.[15]

Let μ be a class of modules. For any module N, the *trace* of μ in N is denoted by $\mathrm{Tr}(\mu, N) = \Sigma\{Imf : f \in Hom(C, N), C \in \mu\}$. Dually the *reject* of μ in N is denoted by $\mathrm{Rej}(N, \mu) = \bigcap\{kerg : g \in Hom(N, C), C \in \mu\}$.

Let N be a submodule of M ($N \leq M$). The notations $N \ll M$, $N \leq_e M$ and $N \leq_d M$ is used for a small submodule, an essential submodule and a direct summand of M, respectively. $Soc(M)$ will denote the socle of M. A module $N \in \sigma[M]$ is said to be *M–small* (or small in $\sigma[M]$) if $N \cong K \ll L$ for $K, L \in \sigma[M]$. Then an R–module $N \in \sigma[M]$ is M–small if and only if $N \ll \widehat{N}$.

Dually, a module $N \in \sigma[M]$ is called *M–singular* (or singular in $\sigma[M]$) if $N \cong L/K$ for an $L \in \sigma[M]$ and $K \leq_e L$. Every module $N \in \sigma[M]$ contains a largest M–singular submodule which is denoted by $\mathrm{Z}_M(N)$. Then $\mathrm{Z}_M(N) = \mathrm{Tr}(\mathcal{U}, N)$ where $\mathcal{U}$ denotes the class of all M–singular modules (see[15]).

Simple modules in $\sigma[M]$ split into four disjoint classes by combining the exclusive choices [M–injective or M–small] and [M–projective or M–singular]. Also note that if a module N in $\sigma[M]$ is M–singular and projective in $\sigma[M]$, then it is zero.

Let $N \subseteq K \in \sigma[M]$. N is called *δ–M–small* in K if, whenever $N + X = K$ with K/X is M–singular, we have $X = K$ (see[7]). Zhou[17] studies δ–R–small submodules in Mod–R. By [17, Lemma 1.2], in the definition of δ–R–small submodule, K/X can be taken Goldie torsion, i.e. K/X can be a member of the torsion class of the Goldie torsion theory in Mod–R.

In this paper μ **will be a class in $\sigma[M]$ which is closed under isomorphisms and submodules**, unless otherwise stated. Any member of μ we shall call a *μ–module*. In this article we denote the following classes:

$\mathcal{S} = \{N \in \sigma[M] : N \text{ is simple}\}$,
$\mathcal{M} = \{N \in \sigma[M] : N \text{ is } M\text{-small}\}$
$\delta = \{N \in \sigma[M] : \overline{Z}_M(N) = 0\}$,
$\mathcal{F} = \{F \in \sigma[M] : \forall 0 \neq K \subseteq F,\ \overline{Z}_M(K) \neq K\}$
$\mu M - Sing = \{N \in \sigma[M] : N \text{ is } \mu\text{-}M\text{-singular}\}$
$\mathcal{FC} = \{N \in \sigma[M] : N \text{ is finitely cogenerated}\}$

Definition 1.1. Let $N \in \sigma[M]$. Following Zhou[16] N is called a *μ–essential submodule* of $K \in \sigma[M]$ if for any nonzero μ–module X in K, $N \cap X \neq 0$. It is denoted by $N \leq_{\mu e} K$.

Clearly every essential submodule is μ–essential. But the converse is not true in general.

Example 1.1. Let μ be the class of simple modules and zero modules in Mod–R. Then a submodule N of a module M is μ–essential if and only if N contained the socle of M but this is not enough to make N essential. For example in the $\mathbb{Z}$–module $\mathbb{Z} \oplus \mathbb{Z}_p$, where p is a prime, $0 \oplus \mathbb{Z}_p$ is μ–essential but not essential.

Example 1.2. Consider the class of M–small modules $\mathcal{M}$ in $\sigma[M]$. Let N be an injective module in $\sigma[M]$ with $0 \neq Rad(N) \nleq_e N$ (for example let $N = U \oplus V$ where U is injective with essential radical and V is injective simple module.) Let X be a non-zero M–small submodule of N. Then $X \ll N$ so $X = X \cap Rad(N) \neq 0$. Thus $Rad(N)$ is $\mathcal{M}$–essential but not essential in N.

The following lemma is clear from definitions.

Lemma 1.1. *Let $K \in \sigma[M]$. If every nonzero submodule of K contains a nonzero μ–module, then for any submodule N of K, $N \leq_e K$ if and only if $N \leq_{\mu e} K$.*

Corollary 1.1. *Let $N \leq K \in \sigma[M]$. If $N \leq_{\mu e} K$ and K is a μ–module, then $N \leq_e K$.*

Now we list the properties of μ–essential submodules. We omit the proofs because they are similar to those for essential submodules (see, for example[2]).

Lemma 1.2. *Let M be a module.*

a) Let $N \leq L \leq K \in \sigma[M]$. Then $N \leq_{\mu e} K$ if and only if $N \leq_{\mu e} L \leq_{\mu e} K$.

b) If $K_1 \leq_{\mu e} L_1$, $K_2 \leq_{\mu e} L_2$, then $K_1 \cap K_2 \leq_{\mu e} L_1 \cap L_2$ for $L_1, L_2 \in \sigma[M]$.

c) Let $N, L \in \sigma[M]$. If $f : N \to L$ is a homomorphism and $K \leq_{\mu e} L$, then $f^{-1}(K) \leq_{\mu e} N$.

d) If $N/L \leq_{\mu e} K/L$, then $N \leq_{\mu e} K$.

e) Let $N \in \sigma[M]$, $\{K_i\}$ an independent family of submodules of N and if $K_i \leq_{\mu e} L_i \leq N$ for all $i \in I$, then $\oplus_{i\in I} K_i \leq_{\mu e} \oplus_{i \in I} L_i$.

Example 1.3. In Lemma 1.2(e), $\{L_i\}$ need not be an independent family. For example, let μ be the class of simple modules and zero modules and put $K_1 = 0 \oplus \mathbb{Z}_p \leq \mathbb{Z} \oplus \mathbb{Z}_p = L_1$ and $K_2 = L_2 = \mathbb{Z} \oplus \overline{0} \leq L_1$. Then $K_1 \leq_{\mu e} L_1$, $K_2 \leq_{\mu e} L_2$ and $K_1 \cap K_2 = 0$ but $L_1 \cap L_2 \neq 0$.

2. δ–essential Submodules Where $\delta = \{N \in \sigma[M] \mid \overline{Z}_M(N) = 0\}$

Talebi and Vanaja[14] define $\overline{Z}_M(N)$ as a dual of $Z_M(N)$ as follows:

$$\overline{Z}_M(N) = \mathrm{Rej}(N, \mathcal{M}) = \cap\{\ker g \mid g \in \mathrm{Hom}(N, L), L \in \mathcal{M}\}$$

where $N \in \sigma[M]$. They call N an *M–cosingular (non–M–cosingular) module* if $\overline{Z}_M(N) = 0$ ($\overline{Z}_M(N) = N$). If N is M–small, then N is M–cosingular. The class of all M–cosingular modules is closed under submodules, direct sums and direct products [14, Corollary 2.2]. Note that $\overline{Z}^2_M(N) = \overline{Z}_M(\overline{Z}_M(N))$. Talebi and Vanaja study the torsion theory cogenerated by M–small modules, $\tau = (\mathcal{T}, \mathcal{F})$ where

$$\mathcal{T} = \{T \in \sigma[M] \mid \overline{Z}_M(T) = T\},$$
$$\mathcal{F} = \{F \in \sigma[M] \mid \forall 0 \neq K \leq F, \overline{Z}_M(K) \neq K\}.$$

This torsion theory is also studied by Özcan and Harmancı.[9] This is a dual of the Goldie torsion theory and not necessarily hereditary. Also $\mathcal{M} \subseteq \delta \subseteq \mathcal{F}$. Now we investigate the relationship between $\mathcal{M}$–essential, δ–essential and $\mathcal{F}$–essential submodules by inspired [17, Lemma 1.2]. First we note that the following two theorems which are characterize the torsion free class $\mathcal{F}$.

Theorem 2.1. *[9, Theorem 15] Let M be a module and assume that M has a projective cover in $\sigma[M]$. If $\overline{Z}_M(M) = M$, then $\mathcal{M} = \delta = \mathcal{F}$.*

Let N and L be submodules of a module M. N is called a *supplement* of L (in M) if $N+L=M$ and $N\cap L\ll N$. M is called *amply supplemented* if, for all submodules N and L of M with $N+L=M$, N contains a supplement of L in M.

Theorem 2.2. *[14, Theorem 3.6] Let M be a module such that every injective module in $\sigma[M]$ is amply supplemented. Then $\mathcal{F}$ is closed under factor modules and $\mathcal{F}=\{N\in\sigma[M]\mid \overline{Z}^2_M(N)=0\}$.*

For shortness we denote

(A) M has a projective cover and $\overline{Z}_M(M)=M$.

(B) Every injective module in $\sigma[M]$ is amply supplemented.

Proposition 2.1. *Consider the following conditions for $K\le N\in\sigma[M]$.*

a) $K\le_{\mathcal{F}e} N$.

b) $K\le_{\delta}$ -MN.

c) $K\le_{\mathcal{M}e} N$.

Then (a) $\Rightarrow$ (b) $\Rightarrow$ (c). If M has (B), then (b) $\Rightarrow$ (a). If M has (A), then (c) $\Rightarrow$ (a).

Proof. $(a)\Rightarrow(b)\Rightarrow(c)$ They are clear.
$(b)\Rightarrow(a)$ Let $X\le N$ with $X\cap K=0$ and $X\in\mathcal{F}$. Then $\overline{Z}^2_M(X)=\overline{Z}_M(\overline{Z}_M(X))=0$ by Theorem 2.2. Since $\overline{Z}_M(X)\cap K=0$, $\overline{Z}_M(X)=0$ by (b). Again by (b), $X=0$.
$(c)\Rightarrow(a)$ It is clear by Theorem 2.1. □

Let M be a module. Define

$$Soc_M(N)=Tr(\mathcal{S}\cap\mathcal{M},N)$$

for any module $N\in\sigma[M]$. Then $Soc_M(N)\le Soc(N)$. Clearly if $Soc_M(N)\le_e N$, then $Soc_M(N)=Soc(N)$. The following lemma shows that in the definition of $Soc_M(N)$ we can take $\mathcal{F}$–modules or δ–modules instead of $\mathcal{M}$–modules. That is

$$Soc_M(N)=Tr(\mathcal{S}\cap\delta,N)=Tr(\mathcal{S}\cap\mathcal{F},N).$$

Lemma 2.1. *Let M be a module. Any simple $\mathcal{F}$–module in $\sigma[M]$ is M–small.*

Proof. Let X be a simple $\mathcal{F}$–module in $\sigma[M]$. If $\overline{Z}_M(X)=X$, then $X\in\mathcal{T}\cap\mathcal{F}=0$, a contradiction.Then $\overline{Z}_M(X)=0$. If X is M–injective, then

$\overline{Z}_M(X) = X$. For, let L be a submodule of X such that X/L is M–small. If $L = 0$, then X is M–small, a contradiction. Hence $L = X$, that is $\overline{Z}_M(X) = X$. This contradiction implies that X is M–small. □

The following proposition can be seen by [16, Proposition 3], but we give the proof for completeness.

Proposition 2.2. *Let $N \in \sigma[M]$. $Soc_M(N)$ is the intersection of all its $\mathcal{F}$–essential submodules of N.*

Proof. Let S be a simple M–small submodule of N and K be an $\mathcal{F}$–essential submodule of N, then $S \cap K \neq 0$. Therefore $S \leq K$. It follows that the intersection of all $\mathcal{F}$–essential submodules contains all simple M–small submodules and hence it contains their sum. Thus $Soc_M(N)$ is contained in the intersection of all $\mathcal{F}$–essential submodules of N.

If $N = Soc_M(N)$, then the proof is completed. Suppose that $N \neq Soc_M(N)$. Let $n \in N - Soc_M(N)$. Then there exists a submodule K maximal with respect to $K \supseteq Soc_M(N)$ and $n \notin K$. If we can show that $K \leq_{\mathcal{F}e} N$, then n lies outside an $\mathcal{F}$–essential submodule, and so $Soc_M(N)$ is the intersection of all $\mathcal{F}$–essential submodules of N.

Suppose that $L \cap K = 0$ for some nonzero submodule L of N with $L \in \mathcal{F}$. Consider the natural epimorphism $\pi : N \to N/K$. Then $L \cong \pi(L) \leq N/K$. Since K is maximal with respect to $K \supseteq Soc_M(N)$ and $n \notin K$, N/K has a minimal submodule contained in every nonzero submodule. Also since $L \in \mathcal{F}$, then $L \cap Soc_M(N) \neq 0$ by Lemma 2.1. But $L \cap Soc_M(N) \leq L \cap K = 0$, a contradiction. □

Hence intersections of $\mathcal{M}$–essential, δ–essential and $\mathcal{F}$–essential submodules are equal.

Tr$(\mathcal{M}, N)$ is investigated in[7] and denoted by $Z^*_M(N)$. Then it can be seen that

$$Soc(Z^*_M(N)) = Soc_M(N).$$

There are some examples of modules M such that $Soc_M(N) \neq 0$, $Soc_M(N) \neq Soc(N)$ and $Soc_M(N) = Soc(N)$.

Example 2.1. 1) If M is a cosemisimple module (i.e. every simple module is M–injective) and $N \in \sigma[M]$, then $Soc_M(N) = 0$, because $Soc(N) = Soc_M(N) \oplus T$ where T is a direct sum of simple M–injective submodules of N.

2) If R is a small ring (for example a commutative integral domain) then every finitely generated R–module is small.[12] This implies that $Soc_R(N) = Soc(N)$ for every R–module N.

3) Let R be the ring $\begin{bmatrix} F & F \\ 0 & F \end{bmatrix}$ where F is a field. Then $Z^*(R_R) = Soc(R_R)$ by [6, Example 11]. This imples that $Soc(Z^*(R_R)) = Soc_R(R_R) = Soc(R_R)$. Since $Soc(R_R) \leq_e R_R$, every δ–essential right ideal is essential.

4)[4] Let Q be a local quasi-Frobenius ring and $J = J(Q)$ (the Jacobson radical of Q), $S = Soc(Q_Q) = Soc({}_QQ)$. Then $W = \begin{bmatrix} Q & Q/S \\ J & Q/S \end{bmatrix}$ is a well–defined ring by the usual matrix addition, equality and the following multiplication

$$\begin{bmatrix} u & v+S \\ j & w+S \end{bmatrix} \begin{bmatrix} x & y+S \\ k & z+S \end{bmatrix} = \begin{bmatrix} ux+vk & uy+vz+S \\ jx+wk & jy+wz+S \end{bmatrix}$$

where $u, v, w, x, y, z \in Q$ and $j, k \in J$. W is a right and left Artinian ring. $J(W) = \begin{bmatrix} J & Q/S \\ J & J/S \end{bmatrix}$ and $Soc({}_WW) = \begin{bmatrix} S & Soc(Q/S) \\ 0 & 0 \end{bmatrix}$, $Soc(W_W) = \begin{bmatrix} S & 0 \\ S & 0 \end{bmatrix}$. By [12, Theorem 3] or [8, Proposition 2.8], it can be shown that $Z^*(W_W) = l_W(Soc({}_WW)) = \begin{bmatrix} J & Q/S \\ J & Q/S \end{bmatrix}$ where $l_W(.)$ is the left annihilator over W. Since $S \leq J$, then $Soc(W_W) \leq Z^*(W_W)$. This implies that $Soc_W(W_W) = Soc(W_W)$.

5)[5] Let $R = F[x;\sigma]$ be the twisted polynomial rings where F is a field of characteristic $p > 0$ and $\sigma : F \to F$ is the endomorphism given by $\sigma(a) = a^p (a \in F)$. The ring R consists of all polynomials $a_0 + xa_1 + x^2a_2 + \ldots + x^na_n$ where n is a non–negative integer, $a_i \in F$ $(0 \leq i \leq n)$, multiplication is given by the relation $ax = x\sigma(a) (a \in F)$ Note that R is a principal right ideal domain [5, p.597]. Let A denote the ideal xR of R. Clearly A is a maximal right ideal of R and the R–module R/A is not injective because $R/A \neq (R/A)x$ (see [13, Proposition 2.6]). In [5, Proposition 9], it is given an example of a field F such that the R–module R/sR is injective for all $s \in R - xR$. Thus some simple R–modules are injective and some are not. In particular, for the principal right ideal domain R, $Z^*(M_1) = M_1$ and $Z^*(M_2) = 0$ for some simple R–modules M_1 and M_2. In this case, $Z^*(M_1 \oplus M_2) = M_1 \oplus 0 \neq 0, M_1 \oplus M_2$ (see [8, p.4918]). Hence

$Soc_R(M_1 \oplus M_2) = Soc(Z^*(M_1 \oplus M_2)) = M_1 \oplus 0 \neq Soc(M_1 \oplus M_2)$.

3. μ–M–Singular Modules

Definition 3.1. Let M be a module and $N \in \sigma[M]$. N is called *μ–M–singular* if $N \cong K/L$ for some $K \in \sigma[M]$ and $L \leq_{\mu e} K$. In case $M = R$, we use μ–singular.

The class of μ–M–singular modules is closed under submodules, homomorphic images, direct sums and isomorphisms.

Hence every module $N \in \sigma[M]$ contains a largest μ–M–singular submodule which we denote by $Z_{\mu M}(N) = \mathrm{Tr}(\mu M\text{-}Sing, N)$ where $\mu M\text{-}Sing$ is the class of all μ–M–singular modules. Then $Z_M(N) \leq Z_{\mu M}(N)$. If $N \in \sigma[M]$ is μ–M–singular (i.e. $Z_{\mu M}(N) = N$) and a μ–module, then N is M–singular. For, let $N \in \mu$ and $N \cong K/L$ where $K \in \sigma[M]$, $L \leq_{\mu e} K$. We claim that $L \leq_e K$. Let $0 \neq X \leq K$ and assume that $L \cap X = 0$. Then $X \cong (L \oplus X)/L \leq K/L$ and so $X \in \mu$. Since $L \leq_{\mu e} K$ we have a contradiction. This proves that N is M-singular.

If $Z_{\mu M}(N) = 0$, then N is called *non–μ–singular* in $\sigma[M]$ or *non–μ–M–singular*.

Proposition 3.1. *Let N be a μ–M–singular module and $f \in Hom_R(M, N)$.*

(1) If M is quasi–projective and $f(M)$ is finitely generated, then $ker f \leq_{\mu e} M$.

(2) If M is projective in $\sigma[M]$, then $ker f \leq_{\mu e} M$.

Proof. (1) We may assume $f(M) \cong L/K$ where $L \in \sigma[M]$ is finitely generated and $K \leq_{\mu e} L$. Since $L \in \sigma[M]$ and L is finitely generated, then M is L–projective. Hence there exists a homomorphism $g : M \to L$ such that $\pi g = f$ where π is the natural epimorphism $L \to L/K$. Then $ker f = g^{-1}(K) \leq_{\mu e} M$ by Lemma 1.2.
(2) By the proof of (1). □

Proposition 3.2. *Let P be a projective R–module and $X \leq P$. Then P/X is μ–singular if and only if $X \leq_{\mu e} P$.*

Proof. If $I \leq R_R$ and R/I is μ–singular, then $I \leq_{\mu e} R$ by Proposition 3.1. Now let P/X be μ–singular and assume $X \not\leq_{\mu e} P$. Let F be a free module such that $F = P \oplus P'$, $P' \leq F$. Then $F/(X \oplus P') \cong P/X$ is μ–singular and $X \oplus P' \not\leq_{\mu e} F$. So we may assume without loss of generality P is free, i.e. $P = \oplus R_\lambda$, each R_λ is a copy of R. Take R_λ. Then $R_\lambda/(R_\lambda \cap X) \cong (R_\lambda + X)/X \hookrightarrow P/X$ is μ–singular. So $R_\lambda \cap X \leq_{\mu e} R_\lambda$. This implies that $\oplus R_\lambda \cap X \leq_{\mu e} \oplus R_\lambda = P$, i.e. $X \leq_{\mu e} P$. □

From the properties of μ–singular modules and the above propositions the following can be seen easily.

Proposition 3.3. *For an R–module N the following are equivalent.*
a) N is μ–singular (in Mod–R).
b) $N \cong F/K$ with F a projective (free) R–module and $K \leq_{\mu e} F$.
c) For every $n \in N$, the right annihilator $r(n)$ is μ–essential in R.

Recall that a submodule N of a module M is said to be *closed* in M if N has no proper essential extension in M, denote $N \leq_c M$.

Lemma 3.1. *Let M be a module and $N \in \sigma[M]$. If $Z_{\mu M}(N) = 0$ and $K \leq_c N$, then $Z_{\mu M}(N/K)$=0.*

Proof. Clear by definitions. □

From now on we consider the condition that for every M–singular module N, Rej$(N, \mu) = N$ and give a characterization of GCO-modules and semisimple Artinian rings by considering the classes $\mathcal{M}, \delta$ and $\mathcal{FC}$.

Theorem 3.1. *Let M be a module. Consider the following conditions.*
a) Every μ–module is projective in $\sigma[M]$.
b) For every M–singular module N, Rej$(N, \mu) = N$.
c) For every μ–M–singular module N, Rej$(N, \mu) = N$.
d) For every simple M–singular module N, Rej$(N, \mu) = N$.
Then (a) ⇒ (b) ⇔ (c) ⇒ (d) . If μ is closed under factor modules, then (a)-(d) are equivalent.

Proof. (a)⇒ (b) Let N be an M–singular module. Let $g : N \to L$ where $L \in \mu$. Then $N/kerg \in \mu$. By (a), $N/kerg$ is projective in $\sigma[M]$. Since N is M–singular, we have that $N = kerg$. Hence Rej$(N, \mu) = N$.
(b)⇒ (c) Let N be a μ–M–singular module and $g : N \to L$ a homomorphism where $L \in \mu$. Then $N/kerg \in \mu$. This implies that Rej$(N/kerg, \mu) = 0$. Since $N/kerg$ is μ–M–singular and a μ–module, it is M–singular. Then by (b), $N = kerg$. Hence Rej$(N, \mu) = N$.
(c)⇒ (b) and (b)⇒ (d) are clear.
(d)⇒ (a) Assume that μ is closed under factor modules. Let $N \in \sigma[M]$

be a μ–module. We claim that N is semisimple. Let $x \in N$ and K be a maximal submodule of xR. Then xR/K is a simple μ–module. By (d) it cannot be M–singular. Hence xR/K is projective in $\sigma[M]$. This implies that K is a direct summand of xR. Hence N is semisimple. Because of the above process, any simple submodule of N is projective in $\sigma[M]$. It follows that N is projective in $\sigma[M]$. □

If we consider the class $\mathcal{M}$ of all M–small modules we have a characterization of GCO–modules: A module M is called a *GCO–module* if every simple M–singular module is injective in $\sigma[M]$. (see[1]).

Corollary 3.1. *Let M be a module. Then the following are equivalent.*

a) Every M–small module is projective in $\sigma[M]$.

b) Every M–singular module is non–M–cosingular.

c) Every $\mathcal{M}$–M–singular module is non–M–cosingular.

d) M is a GCO–module.

e) Every δ–M–singular module is non–M–cosingular.

Proof. (d) $\Leftrightarrow$ (a) is by[7] and (b) $\Leftrightarrow$ (d) is by.[10]

Simple modules are either M–injective or M–small. Hence (a)-(d) are equivalent by Theorem 3.1. (e) $\Rightarrow$ (*b*) is clear. Since $\mathcal{M} \subseteq \delta$, every δ–M–singular module is $\mathcal{M}$–M–singular. Hence (c) $\Rightarrow$ (*e*) is clear. □

For the class δ of all M–cosingular modules, we immediately have the following corollary. The equivalencies of (a), (b) and (d) are given in.[10]

Corollary 3.2. *Let M be a module. Consider the following conditions.*

a) Every M–cosingular module is projective in $\sigma[M]$.

b) For every M–singular module N, Rej$(N, \delta) = N$.

c) For every δ–M–singular module N, Rej$(N, \delta) = N$.

d) M is a GCO–module.

Then (a) $\Rightarrow$ (b) $\Leftrightarrow$ (c) $\Rightarrow$ (d). If δ is closed under factor modules (see Theorem 2.1), then (a)-(d) are all equivalent.

Talebi and Vanaja[14] are also studied the modules M such that every M–cosingular module is projective in $\sigma[M]$.

A module M is called *finitely cogenerated* if Soc(M) is finitely generated and essential submodule of M. Let $\mathcal{FC}$ be the class of all finitely cogenerated R–modules. Note that $\mathcal{FC}$ is closed under submodules.

Corollary 3.3. *The following are equivalent for a ring R.*

a) Every finitely cogenerated R–module is projective.
b) For every singular R–module N, Rej(N, $\mathcal{FC}$*) = N.*
c) For every $\mathcal{FC}$*–singular R–module N, Rej(N,* $\mathcal{FC}$*) = N.*
d) R is semisimple Artinian.

Proof. (a) $\Rightarrow$ (b) $\Leftrightarrow$ (c) By Theorem 3.1.
(d) $\Rightarrow$ (a) is clear.
(b) $\Rightarrow$ (d) Let E be an essential right ideal of R. Suppose that a is an element of R but a does not belong to E. Let F be a right ideal of R maximal with respect to the properties that E is contained in F and a does not belong to F. Then $(aR + F)/F$ is simple singular. By (b), we have a contradiction. Hence R is semisimple Artinian. □

A ring R is a quasi-Frobenius ring (briefly QF–ring) if and only if every right R–module is a direct sum of an injective module and a singular module.[11] In this result we may take μ–singular modules instead of singular as the following result shows.

Theorem 3.2. *The following are equivalent for a ring R.*
a) R is a QF–ring.
b) Every right R–module is a direct sum of an injective module and a μ*–singular module.*

Proof. (a)$\Rightarrow$ (b) It is clear.
(b)$\Rightarrow$ (a) Let M be a projective R–module. Then M is a direct sum of an injective module and a μ–singular module. Since projective μ–singular modules are zero, M is injective. Then R is a QF–ring (see for example[1]). □

Acknowledgment

The author is supported by the project of Hacettepe University of number 05 G 602 001. The referee deserves thanks for careful reading and useful comments.

References

1. N.V. Dung, D.V. Huynh, P.F. Smith, R. Wisbauer, *Extending Modules*, Pitman RN Mathematics 313, Longman, Harlow, 1994.
2. K.R. Goodearl, *Ring Theory*, Marcel-Dekker, 1976.
3. S.H. Mohamed and B.J. Müller, *Continuous and discrete modules*, London Math.Soc. LN.147, Cambridge University Press, NewYork Sydney, 1990.

4. K. Oshiro, Lifting modules, extending modules and their applications to QF–rings, *Hokkaido Math. J.* 13 (1984) 310–338.
5. B. Osofsky, On twisted polynomial rings, *J. Algebra* 18 (1971) 597–607.
6. A.Ç. Özcan, Some characterizations of V-modules and rings, *Vietnam J.Math.*, 26(3) (1998) 253-258.
7. A.Ç. Özcan, On GCO-modules and M-small modules, *Comm. Fac. Sci. Univ. Ank. Series A1*, 51(2) (2002), 25-36.
8. A.Ç. Özcan and P.F. Smith, The Z* functor for rings whose primitive images are artinian, *Comm. Algebra*, 30(10) (2002), 4915–4930.
9. A.Ç. Özcan and A. Harmancı, The torsion theory generated by M–small modules, *Algebra Coll.*, 10(1), (2003), 41–52.
10. A.Ç. Özcan, The torsion theory cogenerated by δ-M-small modules and GCO-modules, *Comm. Algebra* 35, 623-633 (2007).
11. M. Rayar, On small and cosmall modules, *Acta Math. Acad. Sci. Hungar.*, 39(4) (1982) 389–392.
12. M. Rayar, A note on small rings, *Acta Math. Hung.*, 49(3–4) (1987), 381–383.
13. D.W. Sharpe and P. Vamos, *Injective Modules*, Cambridge University Press: Cambridge, 1972.
14. Y.Talebi, N. Vanaja, The torsion theory cogenerated by M-small modules, *Comm. Algebra*, 30(3) (2002) 1449–1460.
15. R. Wisbauer, *Foundations of Module and Ring Theory*, Gordon and Breach, Reading, 1991.
16. Y. Zhou, Relative socle, relative radical, and chain conditions , *Math. Japonica*, 38(3) (1993) 525–529.
17. Y. Zhou, Generalizations of perfect, semiperfect, and semiregular rings, *Algebra Coll.*, 7(3) (2000) 305-318.